MARINE BIOLOGY

MARINE FINFISH AQUACULTURE AND THE ENVIRONMENT

Marine Biology

Additional books in this series can be found on Nova's website
under the Series tab.

Additional e-books in this series can be found on Nova's website
under the e-book tab.

MARINE BIOLOGY

MARINE FINFISH AQUACULTURE AND THE ENVIRONMENT

STEWART WHITE
EDITOR

nova publishers
New York

For permission to use material from this book please contact us:
Telephone 631-231-7269; Fax 631-231-8175
Web Site: http://www.novapublishers.com

Library of Congress Cataloging-in-Publication Data

ISBN: 978-1-63463-537-0

Published by Nova Science Publishers, Inc. † New York

CONTENTS

Preface		**vii**
Chapter 1	Marine Cage Culture and the Environment: Twenty-first Century Science Informing a Sustainable Industry *Carol Seals Price and James A. Morris*	**1**
Chapter 2	Guidelines for Ecological Risk Assessment of Marine Fish Aquaculture *Colin E. Nash, Peter R. Burbridge and John K. Volkman*	**157**
Index		**227**

PREFACE

Chapter 1 – This report provides a comprehensive review of some predominant environmental risks that marine fish cage aquaculture poses in the marine environment and designs and practices now in use to address these environmental risks in the U.S. and elsewhere. The report includes a critical review of the large body of scientific work published since 2000 on the environmental impacts of marine finfish aquaculture around the world.

Chapter 2 – Risk assessment is frequently defined as a tool for making decisions under conditions of uncertainty. Therefore it is far from being an exact science and yet it thrives on exact information. At this period in the emergence of the field of marine fish aquaculture, the scientific and technical knowledge of aquaculture's effects on the environment are a growing compendium of information, but at times the information appears to be ambiguous, conflicting, and incomplete. This is because it is being gleaned continuously from a diverse range of marine ecosystems where there are different fish and shellfish species being produced under a variety of systems and practices which, in some cases, are themselves still being developed.

But the time when knowledge is apparently deficient is arguably the right time for seriously addressing risks, and the constraint of incomplete scientific and technical knowledge has to be offset by the practical knowledge and professional experience of individuals in order to make balanced judgments. In the near future it will be possible to model the interactive ecological effects of marine aquaculture activities, and decisions will be made based on risks quantified numerically for different ecosystems. However, for the moment, informed decisions will have to be made by risk managers with the help of responsible individuals guided mostly by their past experiences and research.

The purpose of this document is to provide a basic set of guidelines for risk managers and other decision makers to use all information available to assess the different ecological risks of marine fish aquaculture in a variety of marine ecosystems. Ten areas of substantive risk in the interaction between marine fish aquaculture are perceived by the public and public administrators to be of most concern. In no order of priority they are: increased organic loading, increased inorganic loading, residual heavy metals, transmission of disease organisms, residual therapeutants, biological interaction of escapes with wild populations, physical interaction with marine wildlife, physical impact on marine habitat, using wild juveniles for grow-out, and harvesting industrial fisheries for aqua-feeds.

In this technical memorandum each of these 10 areas of risk is assessed for its degree of potential adversity, together with its mitigation, in an identical step-by-step process. This common analytical framework, which is first described and illustrated in detail, was developed by the World Health Organization many years ago, and its generic nature and international acceptance make its use for the risk assessment of marine fish aquaculture an ideal application.

Appendices A through J are templates that outline the approach for conducting a risk assessment for all 10 perceived issues. With the help of a flowchart, each template identifies the biological end points or entities and their attributes, both locally and far field, which might be affected for that respective area of risk. It also identifies appropriate methodologies that can be used for measuring or monitoring the effects of exposure to each specific risk.

The chances of any risk occurring can differ greatly in accordance with the natural characteristics of the local ecosystem and its geographic location. Therefore, each template contains a biological overview of its respective risk and briefly discusses factors that may enhance or mitigate the risk's occurrence. In addition, for the benefit of risk managers and risk assessors in all parts of the world, the 10 risks are framed in a matrix to suggest different orders of relevance for their application in different climatic zones. Each template contains a list of documents where further pertinent information can be found.

Finally, this technical memorandum contains a glossary of the risk assessment terms and marine terms used throughout, a list of the international participants of the workshop, and a list of references specific to marine aquaculture and risk assessment, but broader in nature.

In: Marine Finfish Aquaculture and the Environment
Editor: Stewart White
ISBN: 978-1-63463-537-0

Chapter 1

MARINE CAGE CULTURE AND THE ENVIRONMENT: TWENTY-FIRST CENTURY SCIENCE INFORMING A SUSTAINABLE INDUSTRY*

Carol Seals Price and James A. Morris

EXECUTIVE SUMMARY

Technological innovation has made it possible to grow marine finfish in the coastal and open ocean. Along with this opportunity comes environmental risk. As a federal agency charged with stewardship of the nation's marine resources, the National Oceanic and Atmospheric Administration (NOAA) requires tools to evaluate the benefits and risks that aquaculture poses in the marine environment, to implement policies and regulations which safeguard our marine and coastal ecosystems, and to inform production designs and operational procedures compatible with marine stewardship.

There is an opportunity to apply the best available science and globally proven best management practices to regulate and guide a sustainable United States (U.S.) marine finfish farming aquaculture industry. There are strong economic incentives to develop this industry, and doing so in an environmentally responsible way is possible if stakeholders, the public and regulatory agencies have a clear understanding of the relative risks to the environment and the feasible solutions to minimize, manage or eliminate those risks. This report spans many of the environmental challenges that marine finfish aquaculture faces. We believe that it will serve as a useful tool to those interested in and responsible for the industry and safeguarding the health, productivity and resilience of our marine ecosystems.

This report aims to provide a comprehensive review of some predominant environmental risks that marine fish cage culture aquaculture, as it is currently conducted, poses in the marine environment and designs and practices now in use to address these environmental risks in the U.S. and elsewhere. Today's finfish aquaculture industry has learned, adapted and improved to lessen or eliminate impacts to the marine habitats in which it operates. What

* This is an edited, reformatted and augmented version of NOAA Technical Memorandum NOS NCCOS 164, issued by the National Oceanic and Atmospheric Administration, December 2013.

progress has been made? What has been learned? How have practices changed and what are the results in terms of water quality, benthic, and other environmental effects? To answer these questions we conducted a critical review of the large body of scientific work published since 2000 on the environmental impacts of marine finfish aquaculture around the world. Our report includes results, findings and recommendations from over 420 papers, primarily from peer-reviewed professional journals. This report provides a broad overview of the twenty-first century marine finfish aquaculture industry, with a targeted focus on potential impacts to water quality, sediment chemistry, benthic communities, marine life and sensitive habitats. Other environmental issues including fish health, genetic issues, and feed formulation were beyond the scope of this report and are being addressed in other initiatives and reports. Also absent is detailed information about complex computer simulations that are used to model discharge, assimilation and accumulation of nutrient waste from farms. These tools are instrumental for siting and managing farms, and a comparative analysis of these models is underway by NOAA.

We anticipate this report will be useful to both industry and coastal and ocean managers. Farm owners and operators can learn about the current state of knowledge regarding environmental effects of cage culture and apply it to guide siting and other farm management practices. Coastal managers and community planners can use this information to make decisions about the environmentally responsible economic opportunities that aquaculture offers. Federal, state and local regulatory agencies can apply the analysis as they develop and implement permitting and monitoring processes for coastal marine finfish farming and the emerging U.S. offshore aquaculture industry. The scientific research community can use this report to guide future studies which will improve our knowledge of how fish cages function within the marine ecosystem, help improve farm efficiency and further decrease environmental effects. This report also provides a scientific basis for national and international outreach and education.

We hope this analysis will prove useful for integrating current scientific knowledge about the real, potential and perceived effects of marine finfish cage culture and support continued development of an industry which is both economically and environmentally sustainable. Below is a synopsis of observations and trends observed for various environmental effects discussed in this report.

Water Quality

The primary potential effects to water quality associated with marine cage culture include dissolved nitrogen and phophorus, turbidity, lipids and dissolved oxygen fluxes. Usually there are no measurable effects 30 meters beyond the cages when farms are sited in well-flushed waters. Nutrient spikes and declines in dissolved oxygen sometimes are seen following feeding events, but there are few reports of long-term risk to water quality from marine aquaculture. The trend of many studies over the last 20 years indicates that improvements in feed formulation and feeding efficiency are the major reasons for decreased nutrient loading and acceptable water quality in and near farms, and explains why significant enrichment to the water column at offshore farms is generally not detected. Impaired water quality may be observed around farms in nearshore or intertidal habitats where flushing is minimal and at farms using feeds that include unprocessed raw fish rather than formulated feeds. Protection of water quality will be best achieved by siting farms in well-flushed waters.

Benthic Effects

There is a great deal of scientific information about the biogeochemical processes in sediments near fish farms and how those processes may be driven under nutrient enrichment. Excess feed and fish waste are discharged from the farms and, if they accumulate, may alter the chemical processes of decomposition and nutrient assimilation. Well-managed farms may exhibit little perturbation and, where chemical changes are measured, impacts are typically confined to within 100 meters of the cages. Benthic chemical recovery is often rapid following harvest. In contrast, heavily impacted sites may have anaerobic conditions persisting in the sediment and extending hundreds of meters beyond the farm perimeter.

Organic matter can accumulate on the bottom and push the benthos to an anaerobic and ultimately azoic state. In heavily farmed or depositional areas, remediation of highly enriched sediments may take several years.

Impacts can be avoided by placing farms in deep, well flushed areas over erosional seafloor. This results in a net movement of organic matter away from the farm, dispersing nutrients over a broader area for decomposition and assimilation. While this approach protects the immediate farm perimeter, care must be taken to monitor areas downstream of the farms to detect far-field effects, especially in habitats sensitive to nutrient enrichment and sedimentation. The accumulation of some organic matter below farms is to be expected, especially toward the end of a grow-out period when farm biomass is at its peak. Visual observations of benthic conditions below farms are a valuable tool throughout a crop cycle for assessing whether operations are within the capacity of the ecosystem. Farms located in deep water with continuous or episodic benthic scouring of organic waste will be less likely to exhibit sediment degradation. As with water quality, benthic geochemical impacts are most pronounced at enclosed, nearshore or coastal farm sites with insufficient depth and flow to disperse organic wastes.

In some areas, nutrients from farm discharge may be absorbed as food for wild marine organisms. In other locations, deposition beneath cages will have minimal, ephemeral or acceptable effects. Still other farms may need fallowing periods ranging from a few months to one or two years to recover the benthos. Site specific characteristics such as hydrodynamics, trophic status of the water column benthic shear, sediment composition, water depth and nutrient loading will interact to determine which of these will be the scenario for any given farm site.

Effective sediment monitoring protocols should include key site-specific indicator parameters like sulfide, redox potential and total organic carbon to allow for early detection of impacts. Within an adaptive management framework, a good monitoring program can be used to adjust farm management to avert serious and persistent impacts to the benthos. Monitoring and research to quantify downstream, far-field and long-term effects of fish farms beyond the immediate cage perimeter will continue to be important. The use of stable isotopes as tracers of farm waste output at larger spatial and temporal scales is a promising tool to help in this area. Continued efforts comparing different monitoring technologies and protocols to provide reliable, accurate and cost effective methods of assessing enrichment and biogeochemical impacts will be beneficial to both the industry and regulatory entities. Image analysis and acoustic methods are being successfully tested in the field, thus offering cheap and quick alternatives to traditional geochemical analysis.

Marine Life

The broader ecological role of aquaculture operations within the marine environment must be considered since fish farms in the open ocean must co-exist with a host of wild organisms including phytoplankton, benthic fauna, wild fish, marine mammals and corals. If farm nutrients accumulate and persist in the water column or sediment, marine organisms can be impacted. At appropriately-sited and well-managed farms, natural processes can be sufficient to assimilate nutrients. In nutrient limited marine environments these inputs may even fertilize marine food webs, boosting overall productivity.

At some farm sites, a phytoplankton response to nutrient loading was reported, but generally this is a low risk and causal linkages to algal blooms are not evident. Because a change in primary productivity linked to fish farm effluents would have to be detected against the background of natural variability, it is difficult to discern effects unless they are of great magnitude and duration. At larger scales, the occurrence of many anthropogenically derived nutrients in coastal marine waters, also make it difficult to attribute increased primary productivity directly to aquaculture. Hydrology of farms located near shore or in semi-enclosed water bodies which may be poor farm sites must be carefully examined to prevent eutrophication and increased primary productivity in coastal areas and habitats. A knowledge gap continues to be how dissolved nutrients are dispersed and assimilated over large marine areas, and how ecosystem productivity may be affected under increasing production from multiple farms.

Changes in the benthic community are evident when sediments become enriched with organic farm waste nutrients. At well flushed sites in deep water and with efficient feed management, ecological impacts tend to be minimal and confined to the area just beneath the cages. Under light organic enrichment an increase in benthic species abundance and biodiversity may be observed and can be a net benefit to the community. At moderately impacted farms, effects may extend to 100 meters beyond the farm edge. In enriched sediments, the benthic species composition and diversity shift toward tolerant generalists like capitellid polychaetes. The far-field effects of aquaculture to the ecological functionality of food webs and secondary production have not been studied, are difficult to ascertain and should be an area of future monitoring and research efforts.

The excess food and waste released from fish cages may be food for wild fish, especially benthic feeders. Cages may also provide shelter and foraging habitat for wild fish. These characteristics may be beneficial to the local and regional environment. Wild fish and other marine life often aggregate around fish cages and this may be considered a beneficial impact to marine life at some locations. As fish are attracted to farms, the potential for negative and positive interactions with human fishers may increase and farm management or regulatory steps should be considered to minimize conflicts. Likewise, marine fish and mammalian predators may also be attracted to farms. Little research has documented the extent to which marine predators target wild fish around farms, but this would be useful for understanding ecological interactions between farming and marine life.

At modern fish farms, impacts to predatory sharks and marine mammals are being minimized with improved net technologies and removal of dead fish from cages to prevent predation on cultured fish. Siting away from known aggregation sites and installing rigid predator exclusion nets are effective at preventing negative impacts to cultured fish, farm structures and marine predators. Acoustic deterrent devices are not consistently useful against sea lions and seals and may have deleterious impacts to non-target marine mammals. In the U.S., nonlethal interventions to prevent marine mammal predation are preferred. At marine fish farms, entanglement in the farm structures may pose a slight threat to sea turtles, dolphins, whales and seabirds. Keeping lines taut and the water free of debris are effective at minimizing or eliminating conflict with marine mammals and turtles.

The potential effect of marine cage culture to corals, seagrass and mangrove forests are of concern to resource managers and scientists. These ecosystems may be sensitive to nutrient enrichment and sedimentation, making them potentially vulnerable to farm effluent. If farms are located upstream of sensitive habitats, careful monitoring should be implemented for early detection of any perturbation.

Chemicals

The use of antibiotics, therapeutants and antifoulants at marine fish farms has declined greatly (up to 95%) in the last 20 years, resulting in decreased potential for secondary harmful effects of these chemicals on the marine environment. Vaccination, improvements in fish husbandry and best management practices are proven alternatives for achieving and maintaining fish health. Antifoulant chemicals are being replaced largely with onshore de-fouling or mechanical methods for controlling biofouling. Heavy metals from feed and antifoulants are known to accumulate beneath cages, but are often in low concentrations and sequestered in the sediment.

Management Tools

Beyond good site selection, fallowing and integrated multi-trophic aquaculture (IMTA) are two management tools that can be used to further reduce the potential environmental effects of marine fish farms. Fallowing is the practice of relocating or not re-stocking marine fish cages to allow the sediment below to undergo natural recovery, both geochemically and ecologically, from the impacts of nutrient loading. Under ideal conditions, farms should not require a fallowing period for the purpose of sediment recovery. Currently, this practice is widely and successfully implemented around the world as a method for preventing long-lasting damage to the benthic environment.

IMTA is the practice of culturing species from multiple trophic levels in systems that allow for the assimilation of fish waste particulates and dissolved nutrients into additional valuable crops, thereby reducing environmental discharge and expanding the economic base of a farming operation. The species most commonly selected for IMTA with marine fish are seaweeds, oysters and mussels, but lobsters, sea urchins, sea cucumbers and other invertebrates are also being used. Though largely experimental, the culture of additional marketable seafood products may provide dual benefits of economic profitability and reduction in nutrient enrichment.

The correlation of latitude, geographic area and trophic status of the receiving waters with the degree of biological and geochemical response to farm discharge is a critical area for further investigation. Comparative meta-analyses of environmental impacts are needed. The question of environmental impacts of any farm should be considered within a holistic context taking into account the array of oceanographic, hydrological and ecological characteristics of the site and the structural, technological and production aspects of the farm. One pattern that does emerge is that decreasing environmental risk from aquaculture appears to be driven by prudent siting of operations outside of shallow, enclosed, coastal and nearshore waters lacking dispersive current regimes, coupled with modern feed, aquatic health and farm management. This observation is important as it suggests that farming with minimal or acceptable environmental effects is possible in many ecosystems as long as proper safeguards are in place to minimize nutrient and chemical discharge and to manage its immediate and cumulative impacts. These safeguards may be in the form of regulatory oversight or industry-developed best management practices. Ideally, a combination of the two approaches would be most beneficial.

This report provides a broad perspective on a range of potential environmental impacts and their relative intensity, which should be coupled with detailed, site-specific information to make good management decisions about a proposed or operational farm site during its lifetime. As farming expands, cumulative impacts may become more apparent, and thus, robust monitoring protocols are necessary and should be proactively designed to discern both near and far-field environmental impacts.

INTRODUCTION

America's marine finfish aquaculture industry is poised to expand in the near future. Demand for seafood is on the rise and cannot be met by wild catch fisheries (Halwart et al. 2007). Domestic aquaculture production of seafood is certain to play an increasing role in fulfilling the need for reliable marine protein sources (National Oceanic and Atmospheric Administration 2007). The United States (U.S.) is committed to the growth of a modern ocean aquaculture industry that is both profitable and environmentally responsible (Gulf of Mexico Fishery Management Council 2009, National Oceanic and Atmospheric Administration 2011). To accomplish this, the most current knowledge will inform the public, support sustainable industrial practices, guide regulatory processes and strategically direct research. This technical memorandum comprises in depth summaries of the latest information and scientific research on the water quality and benthic effects of marine finfish cage culture on coastal and ocean environments. The report is a useful tool for a regulatory, industry, and research stakeholders making decisions about permitting, siting, and operating marine fish farms and to guide monitoring and further research.

Discharge from marine finfish farms and associated issues of siting such operations are among the most important environmental questions facing this industry. The National Oceanic and Atmospheric Administration (NOAA) has a variety of regulatory and marine management mandates that affect permitting of finfish farms in U.S. state and federal waters.

For example, NOAA is required to consult with the U.S. Army Corps of Engineers (COE) on its permit applications to address the requirements of the Magnuson-Stevens Fishery Conservation and Management Reauthorization Act, the Endangered Species Act and the Marine Mammal Protection Act. Siting and water quality models, water column and benthic monitoring data and best management practices are among the tools that NOAA and other agencies use in evaluating marine finfish farm permit applications and in monitoring operational farms.

The U.S. has everything required to develop a significant marine finfish aquaculture industry in coastal and open ocean waters including excellent locations, scientific expertise, state-of-the-art technology, innovative equipment and feed manufacturers and willing investors. Globally, aquaculture produces about half of the seafood people eat, but only 5% of U.S. seafood comes from domestic aquaculture. In the U.S., the aquaculture industry has not developed due to an uncertain permit processes at the state and federal levels, concerns about environmental effects and conflicting coastal uses. These factors contribute to trade imbalance, export of innovative technology and loss of potential jobs.

Over the last few decades, many reports, peer- reviewed journal articles, books and papers have been written about the environmental effects of marine finfish aquaculture. To keep our effort manageable, we narrowed our scope to focus on work published in English since 2000. Over 420 reports and papers, mostly from peer-reviewed professional journals, are cited in this report. Several key reports generated by government agencies, academic or research institutions, and private organizations are also cited. These are often good sources of summary information and data review, and provide relevant overviews of environmental issues. Only reports which included scientific citations were included. Not included in our analysis are newspaper or magazine articles or opinion pieces lacking scientific citations and review. Our sources include a vast body of work reflecting a global realization of the importance of understanding the interplay between this industry and the marine environment. While not an exhaustive compilation, this effort does provide a comprehensive look at the state of knowledge in the marine finfish aquaculture industry with regards to the environmental effects of cage culture. The collected literature originates from research conducted in countries around the world, covers a range of cultured fish species, includes many new and practical farm management approaches and addresses ecological processes at many scales.

This technical memorandum contains five main sections each covering specific environmental effects. We have organized the research in the first three sections by geographical location based upon the predominant climate regions of the world (Peel et al. 2007). Specifically, we have used the following latitudinal demarcations to arrange the works into tropical (from the equator to the Tropics of Cancer and Capricorn at 23.5° north and south), subtropical (from the Tropics of Cancer and Capricorn to 35°north and south and including the entire Mediterranean Sea) and temperate (poleward from 35° to 66.5 ° north and south). We do not include papers from aquaculture production in the polar region.

Additionally, each section is organized in one of two ways. Shorter sections with less literature are organized by relative impact level. That is, studies that show no impacts are generally covered first, followed by a summary of work that did find impacts organized by climate zones. Lengthier sections are arranged by climate zones first and then sorted geographically, usually from north to south and east to west. This organizational structure allows us to provide a geographical context for the research, and also tends to group often extensive content by species and to some degree culture techniques, as these are relatively similar within large areas. For example, most of the papers published in the northern Europe temperate zone address salmon culture in large off-shore net pen operations, while those in the subtropical Mediterranean report on sea bass and sea bream, often reared closer to shore. Though we did not conduct any quantitative analysis, this geographical organization also allows us to observe broad locational trends in terms of environmental impacts and factors affecting the level of those impacts. This grouping also makes it possible to consider and compare different management approaches implemented in aquaculture industries around the world. The reader should bear in mind that regional management approaches may influence how impacts are reported and interpreted. Within geographical area, work published from the same country is presented together, generally in chronological order. Finally, available summary reports or papers covering broadly applicable topics are included.

The **Water Quality** section includes sections on nitrogen, phosphorus, dissolved oxygen, turbidity and lipids. Next the **Benthic Effects** section focuses on abiotic impacts with sections addressing nitrogen phosphorus and carbon enrichment and sediment biogeochemical changes. A section on **Marine Life** reviews biodiversity effects on primary production, the benthic community, pelagic species and in sensitive habitats. The **Chemicals** section covers antibiotics, therapeutants, antifoulants and heavy metals. The final section, **Management Tools**, provides information about fallowing and integrated multi-trophic aquaculture.

This report was initiated by scientists at the NOAA Center for Coastal Fisheries and Habitat Research, National Centers for Coastal Ocean Science, National Ocean Service in 2011 in collaboration with the NOAA Fisheries Office of Aquaculture. The draft report was revised in 2012-2013 following review by NOAA scientists and other experts. Also, the draft paper served as a key reference paper for a NOAA sponsored workshop in March 2012. Participants at the workshop provided peer review which was additionally incorporated into the final report.

SECTION 1. WATER QUALITY

The environmental effects of marine finfish cage aquaculture to water quality are of primary concern to anyone interested in developing an ecologically responsible industry, and many review articles have addressed this topic (Wu 1995, Goldburg et al. 2001, Pearson and Black 2001, Hargrave 2003, Goldburg and Naylor 2005, Braaten 2007, Pittenger et al. 2007, Grigorakis and Rigos 2011). Regional efforts, many sponsored by governmental entities, are also underway to address questions about effects to water quality at varying spatial scales (Nash 2001, Wildish et al. 2004, Nash et al. 2005, Huntington et al. 2006, Costa-Pierce et al. 2007, Halwart et al. 2007, International Union for Conservation of Nature 2007, Olsen et al. 2008). This section summarizes results from research and monitoring projects at marine fish farms around the world investigating predominant water quality effects including dissolved nitrogen and phosphorus loading, dissolved oxygen depletion, turbidity and lipids.

While results and impacts vary among farm sites and species, there is agreement that the last twenty years has seen a significant improvement in the management of marine cage operations resulting in improved water quality. For example, in a recent review of more than 20 research papers, Holmer (2010) found none that detected significant enrichment to the water column at offshore farms. Improvements in feed formulation and feeding efficiency are repeatedly cited as major reasons for decreased nutrient loading and decreased impacts to water quality in and near farms. Siting farms in well- flushed, non-depositional waters with depth at least twice that of the net pen is recommended to ensure good water quality (Beveridge 2004, Belle and Nash 2008). Continued research to understand the complex array of forces driving nutrient dispersion in and around fish farms, including bathymetry, current flow, tidal fluctuation, and Earth's rotation (Karney and Venayagamoorthy *in press*), will provide additional tools to develop sustainable farming practices. Few comparative analyses have investigated the correlation between farm site (e.g., depth, latitude, current profile) and management characteristics (e.g., species cultured, volume of cages, biomass, feeding rate), and observed water quality impacts (Sarà 2007), so additional work in this area would provide useful guidance. The dispersal and assimilation of farm nutrients via the sea surface microlayer is another area warranting study. It is important to consider the trophic status and background nutrient flux of the receiving water, as well as other sources of nutrient loading, when assessing the relative contribution of marine fish farm discharge to the environment.

Nitrogen

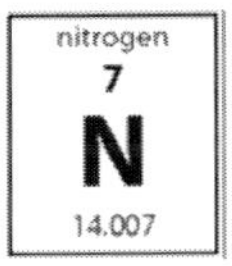

The trend of increasing nitrogen levels in coastal waters due to anthropogenic sources is a concern worldwide, especially because it may contribute to algal blooms and eutrophication (Cloern 2001, Galloway et al. 2004, Anderson et al. 2008, Holmer et al. 2008, Tett 2008). Marine cage aquaculture operations are a recognized source of nitrogenous discharge released in the form of uneaten food, feces and metabolic wastes including ammonia and urea (Cole 2002, Nash et al. 2005, Huntington et al. 2006, International Union for Conservation of Nature 2007, Pittenger et al. 2007).

Research and modeling have determined the amounts of nitrogen released from marine fish cages and the potential water quality and environmental effects of dissolved nitrogen (Hargrave 2003). Recently, Norði et al. (2011) calculated that about 63% of nitrogen fed at a rainbow trout *Oncorhynchus mykiss* farm in the Faroe Islands was lost as dissolved nitrogen. Olsen et al. (2008) constructed a mass balance estimate of nitrogen flow from a hypothetical Norwegian salmon farm producing 1000 metric tons of fish per year. Their estimated annual loading of 44 kg of nitrogen per metric ton was comparable to rates measured in Scottish farms. Islam (2005) provides a summary of nitrogen budgets in marine cage aquaculture. He reports that 68–86% of the nitrogen input as feed is eventually released to the environment, and notes that the percent of nitrogen lost varies due to the type of feed used, the feed conversion ratio of the cultured organism and feeding efficiency. Strain and Hargrave (2005) used mass balance to calculate that the total dissolved nitrogen released from farms in an inlet in southwestern New Brunswick was 33 kg of waste nitrogen per metric ton of fish produced. Total annual discharge depended upon production levels, but was determined to be a significant contributor to nutrient loading. A nitrogen budget for marine cage culture of mutton snapper *Lutjanus analis* and cobia *Rachycentron canadum* estimated that 79% of the nitrogen fed to the fish was released into the water (Alston et al. 2005). Wu (1995) and Pearson and Black (2001) cited nitrogen loss in European salmon *Salmo salar* farms between 52-95%, but noted an improvement due to advances in feeding efficiency. Nitrogen discharge to the water column continues to be cited as a major potential impact of marine cage culture (Cloern 2001, Hargrave 2003, Nash et al. 2005, Pittenger et al. 2007, Holmer et al. 2008, Olsen et al. 2008) and its potential to affect water quality have been studied around the world in various marine habitats.

In some cases, no nitrogen related effects to the water column are detected (Table 1).

Temperate

Tlusty et al. (2005) monitored water quality at Newfoundland salmon farms and found no increased nutrification of the water column despite collecting over 25,000 water samples and relatively low flushing rates of 5-20 days. Similar results are reported for salmon farms in Chile (Soto and Norambuena 2004), where no significant elevations in dissolved nitrogen were observed across nine farm and control sites. Likewise, Nordvarg and Johansson (2002) concluded that a fish farm in the Baltic's Åland archipelago had no measurable effect on dissolved nitrogen levels.

In some cases, elevated levels of dissolved nitrogen were detected, but no environmental effects were found. In Maine, Blue Hill Bay was assessed to determine the feasibility of adding more fish cages (Sowles 2005). While nitrogen levels in the bay were found to be elevated due to aquaculture, it was concluded that the bay had the capacity to assimilate additional nitrogen loading from fish farming. Modeling to assess the feasibility of establishing cage farms in the Straits of Juan de Fuca predicted increased nitrogen in the farm plume, but did not predict enrichment or phytoplankton blooms (Rensel et al. 2007). Nash et al. (2005) concluded that extensive monitoring of net pens in Europe, Canada and the U.S. indicated only modest increases in dissolved nitrogen around fish cages, with farmers siting their operations in well flushed areas to avoid seasonal nutrient enrichment. Norði et al. (2011) measured dissolved ammonium levels up to 4.3 times higher at a trout farm in the Faroe Islands than at reference stations, but they remained within concentrations typical at efficiently flushed farm sites and no stimulation of primary production was evident.

Sub-tropical

Off the Italian coast, Doglioli et al. (2004) modeled the regional dispersion patterns of nitrogen from an eight-cage (200 tons/year) sea bream and sea bass farm. They concluded that dissolved nitrogen concentration remained low due to flushing by strong currents. Predicted results agreed well with field sampling, validating their model. In an Adriatic sea bass *Sparus auratus* and sea bream *Dicentrarchus labrax* farm, Matijevic et al. (2009) reported slightly elevated dissolved nitrogen at farm versus reference sites, but only in surface waters. Schembri et al. (2002) found no consistent or significant changes to water quality, including dissolved nitrogen levels, at a tuna fattening operation in Malta. Neofitou and Klaoudatos (2008) measured elevated nitrogen at sea bass and sea bream cages in the Aegean Sea.

Levels decreased quickly downstream of the cages (300 m), however, and never exceeded the permitted concentrations. Also in the Aegean Sea, Pitta et al. (2005) compared water column effects in farmed and reference sites and found significantly elevated nitrogen levels in bottom water (10 m) during September. However, they concluded that the levels of nitrogen were within the range of normal values reported in the Aegean and may have resulted from resuspension of benthic sediments below the thermocline. Surface and mid-water levels of nitrogen were not significantly different. In another study at northwestern Aegean sea bream and sea bass farms, increased nitrogen levels were found nearest cage sites, but dissipated within 30 m of the farm (Mantzavrakos et al. 2005).

Table 1. Levels of dissolved nitrogen discharge reported and modeled at fish cage sites. At sites with minimal impact levels, the authors reported measurable increases in dissolved nitrogen, but these were statistically insignificant or were not thought to have significant environmental implications

Impact Level	Reference	Location	Species Cultured
None Detected	Tlusty et al. 2005	Canada	Atlantic salmon
	Soto and Norambuena 2004	Chile	Atlantic salmon
	Nordvarg and Johansson 2002	Baltic Sea	Atlantic salmon
	Helsley 2007	Hawaii	Pacific threadfin
	Alston et al. 2005	Puerto Rico	Mutton snapper & Cobia
	Benetti et al. 2005	The Bahamas	Cobia
	Schembri et al. 2002	Malta	Bluefin tuna
Minimal	Norði et al. (2011	Faroe Islands	Atlantic salmon
	Rensel et al. 2007	Puget Sound	Atlantic salmon
	Sowles 2005	USA	Atlantic salmon
	Nash et al. 2005	Pacific Northwest	Atlantic salmon
	Matijevic 2009	Adriatic Sea	Sea bass & Sea Bream
	Neofitou and Klaoudatos 2008	Aegean Sea	Sea bass & Sea Bream
	Mantzavrakos et al. 2005	Aegean Sea	Sea bass & Sea Bream
	Pitta et al. 2005	Aegean Sea	Sea bass & Sea Bream
	Doglioli et al. 2004	Italy	Sea bass & Sea Bream
Significant	Aguado-Giminez et al. 2006	Mediterranean	Bluefin tuna
	Dalsgaard and Krause-Jensen 2006	Mediterranean	Sea bass & Sea Bream
	Hung et al. 2008	Taiwan	Unknown
	McKinnon et al. 2008	Australia	Barramundi
	Strain and Hargrave 2005	Canada	Atlantic salmon

Tropical

A mutton snapper and cobia farm in Puerto Rico monitored water quality, and found no difference in dissolved nitrogen between farm and control samples (Alston et al. 2005). Likewise, no increase in dissolved nitrogen was measured at a submersible cage stocked with cobia in the Bahamas (Benetti et al. 2005) or Pacific threadfin *Polydactylus sexfilis* cages in Hawaii (Helsley 2007).

Some research indicates there can be environmental effects from increased dissolved nitrogen.

Sub-tropical

Aguado-Giminez et al. (2006) calculated the estimated dissolved nitrogen outputs of a bluefin tuna *Thunnus thynnus* fattening operation in the Mediterranean could be 2-5.6 times greater than a comparable sea bream farm, with the potential for significant environmental impacts during peak production periods. Dalsgaard and Krause-Jensen (2006) conducted bioassays using macroalgae and phytoplankton at four sea bream and sea bass farms in the Mediterranean. For both algae and plankton, growth was highest at cage sites and elevated within 150 m of the cages. Tissue nitrogen content of the algae was highest in samples growing closest to the cages, indicating a clear transfer of nitrogen from the farm to the algae.

Tropical

In Tapong Bay, Taiwan (Hung et al. 2008), removal of mariculture structures after decades of farming from the semi-enclosed lagoon resulted in a significant decrease in dissolved nitrogen in the water column contributing to an overall improvement in ecosystem quality. Monitoring at a barramundi *Lates calcarifer* farm in Queensland, Australia measured elevated dissolved nitrogen concentrations of 251 μg/L, exceeding the trigger value of 160 μg/L set in the Queensland Water Quality Guidelines (McKinnon et al. 2008). High nitrogen levels resulted during the wet season and nitrogen flux was locally mitigated by mangrove trees. By comparison, reference sites tended to have lower mean dissolved nitrogen levels with less variability.

Nash et al. (2005) suggested that dissolved nutrient release may affect attached macroalgae, but that eutrophication due to aquaculture will be a distant, or far-field, rather than a local affect due to dispersal of dissolved nutrients by currents. Pittenger et al. (2007) reviewed research and monitoring at marine fish cage sites concluding that discharges from farms, including nitrogen, represent a significant influx of nutrients to the marine environment. They concluded that dilution is not a sufficient strategy to address the issue and recommend that proper siting, adherence to best management practices, improved feed formulation and integrated multi-trophic aquaculture (IMTA) — farming species from different trophic levels together to promote the uptake of waste nutrients — will be key to minimizing impacts to water quality.

Elevated dissolved nitrogen in the water column around fish farms is typically a localized effect (within a hundred meters), often with seasonal variation. Siting farm operations in deep waters with sufficient flushing rates will minimize water quality impacts. Advances in feed formulation and feeding practices have reduced nitrogen loading to the environment (Stickney 2002, Braaten 2007, Pittenger et al. 2007, Belle and Nash 2008, Olsen et al. 2008). Better understanding of temporal, species-specific differences in waste discharge due to variations in growth and metabolic rates will also provide a more accurate understanding of nitrogen effluent and how to manage environmental risk, especially when multiple species are cultured in close proximity (Piedecausa et al. (2010). Questions remain about the cumulative impacts of discharge from multiple, proximal farms, potentially leading to increased primary production and eutrophication at regional and far-field scales. Research focusing on such trophic implications is discussed in the **Marine Life** chapter of this report.

Phosphorus

phosphorus
15
P
30.974

Although nitrogen is generally the limiting nutrient in many ocean waters, the trend of increasing phosphorus levels in coastal waters due to anthropogenic sources is also of concern because primary production in some marine systems suchas tropical oceans is phosphorus limited (Cloern2001, Nordvarg and Hakanson 2002). Increasedphosphorus may contribute to algal blooms andeutrophication. Marine cage aquaculture operationsare a recognized source of phosphorus released inuneaten food, feces, and metabolic wastes in theform of phosphate (Cole 2002, Nash et al. 2005,Huntington et al. 2006, International Union forConservation of Nature 2007, Pittenger et al. 2007,Holmer et al. 2008, Tett 2008).

Earlier environmental impact summaries of marine cage culture included little about phosphorus. Wu (1995) reported that up to 82% of the phosphorus in fish feed was lost to the environment while Pearson and Black (2001) found that 34-41% of phosphorus in feed was released in dissolved form. More recently, Islam (2005) reviewed phosphorus budgets of marine cage aquaculture and reported an average of 71.4% of the phosphorus in feed being released to the environment, with the percentage varying with the species cultured, the type of feed used, the feed conversion ratio and feeding efficiency. Strain and Hargrave (2005) used mass balance calculations to estimate the total fish farm derived nutrient output of an inlet in southwestern New Brunswick. Total dissolved phosphorus released from farms was calculated at 4.9 kg of waste phosphorus per ton of fish produced. Total discharge depended upon production levels, but the researchers concluded it was a significant contributor to nutrient loading. The release of phosphorus into ocean waters and the potential for deleterious nutrification effects continues to be monitored, researched and modeled in marine ecosystems around the world. Similar to the case of nitrogen in the previous section, these efforts have reported a range of results (Table 2).

Some research has found negligible changes in phosphorus levels around marine fish farms.

Temperate

Two years of monitoring nutrients at salmon farms in Newfoundland found no changes in water quality (Tlusty et al. 2005) due to farm discharge. A simulation of the Strait of Juan de Fuca (Rensel et al. 2007) predicted no adverse effects to water quality due to fish farming, especially as sunlight is thought to be the limiting factor for primary production in that body of water. Dissolved phosphorus from salmon farms in the Pacific Northwest was not identified as a concern in Nash (2001), primarily because the system is nitrogen limited. Similarly, an ecological carrying capacity assessment including evaluation of the phosphorus profiles in Blue Hill, Maine concluded that the bay could assimilate nutrients from additional net pens (Sowles 2005). Soto and Norambuena (2004) evaluated 29 salmon farm sites in

Chile and found no effect of farming on dissolved phosphorus concentrations compared to reference locations.

Subtropical

Schembri et al. (2002) reported that no consistentor significant changes to dissolved phosphorus levelswere found at a tuna fattening operation in Malta.Likewise, sampling at Mediterranean fish farms(Pitta et al. 2005) found no increase in dissolvedphosphorus compared to reference areas.

Table 2. Levels of dissolved phosphorus discharge reported and modeled at fish cage sites. At sites with minimal impact levels, the authors reported measurable increases in dissolved phosphorus, but these were statistically insignificant or were not thought to have significant environmental implications

Impact Level	Reference	Location	Species Cultured
None Detected	Tlusty et al. 2005	Canada	Atlantic salmon
	Soto and Norambuena 2004	Chile	Atlantic salmon
	Nash 2001	Pacific Northwest	Atlantic salmon
	Alston et al. 2005	Puerto Rico	Mutton snapper & Cobia
	Benetti et al. 2005	The Bahamas	Cobia
	Helsley 2007	Hawaii	Pacific threadfin
	Pitta et al. 2005	Aegean Sea	Sea bass & Sea Bream
	Schembri et al. 2002	Malta	Bluefin tuna
Minimal	Neofitou and Klaoudatos 2008	Aegean Sea	Sea bass & Sea Bream
	Matijevic 2009	Adriatic Sea	Sea bass & Sea Bream
	Mantzavrakos et al. 2005	Aegean Sea	Sea bass & Sea Bream
	Doglioli et al. 2004	Italy	Sea bass & Sea Bream
	Sowles 2005	USA	Atlantic salmon
	McKinnon et al. 2008	Australia	Barramundi
Significant	Strain and Hargrave 2005	Canada	Atlantic salmon
	Nordvarg and Johansson 2002	Baltic Sea	Atlantic salmon
	Piedecausa et al. (2010	Mediterranean	Bluefin tuna
	Aguado-Giminez et al. 2006	Mediterranean	Bluefin tuna
	Dalsgaard and Krause-Jensen 2006	Mediterranean	Sea bass & Sea Bream
	Hung et al. 2008	Taiwan	Unknown

Neofitou and Klaoudatos (2008) did find differences in dissolved phosphorus at fish farms in the Aegean Sea, but effects were reduced within 300 m of the cages and levels did not exceed those that could lead to eutrophication (0.01 mg/L). At Greek sea bream and sea bass farms, sampling stations closest to farms had increased dissolved phosphorus, but levels were decreased at 30 m from the cages (Mantzavrakos et al. 2005). Doglioli et al. (2004) modeled the dispersion of nutrients from aquaculture cages in the Italian Mediterranean predicting that prevailing currents provided flushing sufficient to avoid nutrient accumulation. Comparison with field data verified that phosphorus levels in the water due to fish farming remained very low. Sampling at sea bream and sea bass farms in the Adriatic found that dissolved phosphorus levels were only slightly elevated at farm versus reference stations, and only in the upper water column (Matijevic et al. 2009).

Tropical

Ocean fish farms in Hawaii (Helsley 2007), The Bahamas (Benetti et al. 2005) and Puerto Rico (Alston et al. 2005) reported no significant increases in dissolved phosphorus near cages. Monitoring at a barramundi farm in Queensland, Australia (McKinnon et al. 2008) found seasonally elevated dissolved phosphorus levels, but these did not exceed governmental water quality trigger value (20 μg/L).

In contrast, some studies document increased phosphorus levels due to marine fish farming.

Temperate

Nordvarg and Johansson (2002) measured phosphorus at farm sites in the Åland archipelago in the Baltic Sea. They found that farms in semi- enclosed bays had elevated levels as did some areas during high fish production, concluding that fish farming may have significant impacts on coastal areas. Complementing this work, Nordvarg and Hakanson (2002) developed a mass balance model for phosphorus for siting farms in this area and other phosphorus-limited coastal waters.

Subtropical

Dalsgaard and Krause-Jensen (2006) used macroalgal and phytoplankton assays to monitor nutrient release at fish farms in the Mediterranean. Growth was higher in samples taken closest to the fish cages, suggesting that nutrients are locally available for primary production. The phosphorus output from a tuna fattening operation was calculated to be about 3-5 times higher than from sea bream or sea bass farms because of differences in digestibility and feed formulation (Aguado-Gimenez et al. 2006). Model simulations comparing phosphorus outputs from sea bream, sea bass and Atlantic bluefin tuna (Piedecausa et al. (2010) indicated significant differences in nutrient waste production among species — with the tuna being the highest — which must be taken into account when managing the marine environment for multiple aquaculture facilities.

Tropical

The cessation of aquaculture activities in a semi-enclosed bay in Taiwan resulted in significantly decreased dissolved phosphorus levels, contributing to improved overall water quality (Hung et al. 2008).

In summary, increased dissolved phosphorus is generally not considered to be a primary concern for marine cage aquaculture (Nash et al. 2005, Costa-Pierce et al. 2007), mostly because primary production in most marine waters is nitrogen, not phosphorus, limited. With proper siting, effluents are flushed away from cage sites, diluted within a few hundred meters and dispersed for natural assimilation. Improvements in feeding efficiency and feed formulation will lessen the amounts of phosphorus released (Stickney 2002, Braaten 2007, Pittenger et al. 2007, Belle and Nash 2008). As with nitrogen, flushing of nutrients away from the immediate cage perimeter tends to minimize measurable impacts to local water quality, but there is need for monitoring and evaluation studies investigating possible cumulative impacts to downstream areas, especially in regions where multiple operational farms may be sited.

Dissolved Oxygen

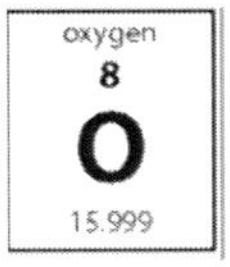

Sufficient dissolved oxygen in the water column is essential to aquaculture operations and has been extensively studied and monitored in all types of culture operations. Oxygen concentrations in the water column near farm operations are lowered primarily through fish respiration, but also due to microbial metabolism. Wu (1995) reported only localized or insignificant affects to dissolved oxygen. Yet, concern remains that marine cage culture may significantly decrease dissolved oxygen concentrations capable of causing local short term impacts (International Union for Conservation of Nature 2007, Pittenger et al. 2007, Tett 2008).

Several recent studies reported no significant effects of marine cage culture on dissolved oxygen.

Temperate

In their assessment of Pacific Northwest salmon farms Brooks and Mahnken (2003) found little risk to the environment from dissolved oxygen depletion. In Scotland, modeling was used to predict the likely effects of fish farming to biological oxygen demand in 135 loch basins (Gillibrand et al. 2006) to assess the risk of oxygen depletion in these deep water environments. The results suggested that farming was unlikely to contribute significantly to hypoxic events in the majority of lochs.

Subtropical

In the Aegean Sea no effects were detected at tuna and sea bass farms (Basaran et al. 2007, Yabanli and Egemen 2009, Aksu et al. (2010).

Tropical

Vargas-Machuca et al. (2008) found no impacts to dissolved oxygen at snapper farms off Mexico's Pacific coast. Sampling at submerged cobia cages in Puerto Rico and The Bahamas found no significant effects on dissolved oxygen levels, with concentrations consistently above 5 mg/L (Alston et al. 2005, Benetti et al. (2010). Raw water quality monitoring data from the Kona Blue Water Farms Almaco jack *Seriola rivoliana* open ocean facility suggest no dissolved oxygen impacts from cage culture were detected (www.kona-blue.com/ emonitoring.php), but no statistical tests were provided.

Some studies document decreased dissolved oxygen concentrations near fish farms.

Temperate

Hargrave's (2005) edited book focuses on environmental impacts of marine cage culture on the northeastern coast of North America and three of those papers addressing dissolved oxygen are discussed here. First, Page et al. (2005) provide a summary of monitoring efforts in New Brunswick. Decreases in oxygen are reported in and near salmon cages with the greatest declines occurring within fish cages, primarily due to lack of tidal flushing. Page et al. (2005) also present an oxygen depletion index, which models the potential for caged fish to deplete oxygen concentrations at multiple scales under varying flushing regimes. This is a tool that farmers and regulators can use to plan siting and fish stocking density. Next, Strain and Hargrave (2005) modeled nutrient fluxes and ecosystem processes around salmon farms in the Bay of Fundy. Their model estimated that fish farms could decrease oxygen concentration by up to 1.4 mg/L at individual farms, but concluded that ecosystem effects were likely minimal. Finally, Sowles (2005) investigated water quality parameters as part of an assessment of current and potential aquaculture impacts in Blue Hill Bay, Maine. Dissolved oxygen levels in the bay varied seasonally and geographically, but were found to be above the threshold value of 6 mg/L. Nash et al. (2005) found that long-term monitoring in the northeast Pacific showed maximum dissolved oxygen reductions of 2 mg/L under high density cage culture. Overall, however, dissolved oxygen reductions were generally less than 0.5 mg/L.

Dissolved oxygen at a rainbow trout farm in the Faroe Islands decreased 11- 26% from July to September compared to a reference station, although generally the water was supersaturated (Norði et al. (2011). The decreased oxygen levels were due to fish respiration and current velocity was sufficient to avoid severe oxygen depletion.

Subtropical

A study in Turkey's Güllük Bay (Demirak et al. 2006), monitored dissolved oxygen at seven sea bass cages and three control sites. Dissolved oxygen at the cage sites was significantly lower than control sites, but remained above the 4 mg/L criteria.

Tropical

Monitoring in Queensland, Australia also found decreased dissolved oxygen at barramundi cage sites often tied to seasonal tidal fluctuation, but the extent and severity were similar to unfarmed sites (McKinnon et al. 2008).

In conclusion, a meta-analysis of 30 peer-reviewed articles (Sarà 2007) found that dissolved oxygen was generally not affected by aquaculture operations. In general, low dissolved oxygen is not a serious problem in offshore fish farms (Braaten 2007) and changes in dissolved oxygen due to open ocean cage culture are not detected or are negligible. Seasonal, tidal and diurnal fluxes often cause more changes in dissolved oxygen than do fish farms. Thus, proper siting of farms in areas with sufficient flushing rates is recommended. Oxygen bubblers (Srithongouthai et al. 2006, Endo et al. 2008), mechanical aeration (Goldburg and Triplett 1997) and lowering sea cages below the ocean surface (Dempster et al. 2009) are management tools that can be implemented on farms to minimize or eliminate dissolved oxygen depletion.

Turbidity

Particulates or dust from feed and fish waste are two primary sources of turbidity associated with cage culture (Pergent et al. (1999, Ruiz et al. 2001, Hargrave 2003, International Union for Conservation of Nature 2007). Scraping of with cage culture (Pergent et al. (1999, Ruiz et al. 2001, Hargrave 2003, International Union for Conservation of Nature 2007). Scraping of biofouling may also result in temporary decrease in water clarity (Hargrave 2003, Alston et al. 2005). In general, high flushing rates will minimize increases in turbidity at cage sites. However, when flushingrates are low due to tidal or seasonal shifts in water currents (Tanaka and Kodama 2007, McKinnon et al. 2008) or due to siting in areas with decreased flow, feed and waste suspended in the water column may increase turbidity. Increased turbidity may result in lower light penetration affecting phytoplankton production (Harrison et al. 2005) and may affect photosynthesis of benthic aquatic vegetation like seagrasses (Cole 2002).

Prolonged changes in turbidity associated with freshwater aquaculture facilities have been documented (Sarà 2007), but less data is available for marine cage culture. Because of high flushing rates in open ocean conditions, turbidity is likely to be more of a concern at nearshore sites than open ocean sites. This is especially true in coastal waters in the vicinity of critical habitats such as corals and seagrass beds which could be light limited.

Temperate

A recent study conducted in Maine (Sowles 2005)found that light penetration levels (an indicatorof turbidity) at both cage and control sites metwater quality targets. Conversely, Harrison et al.(2005) report that in southwestern New Brunswick,Canada, Secchi depth readings at cages weresignificantly lower than at control sites.

Subtropical

Secchi readings at three control sites in Turkey were higher than at sea bass and sea bream sites (Aksu and Kocatas 2007), but this brief report did not include statistical values.

Tropical

Similarly, data figures of turbidity monitoring reported from the Kona Blue Water Farms open ocean facility in Hawaii suggest that no impacts from cage culture were detected even at high production levels (www.kona-blue.com/ emonitoring.php), but no statistical tests were conducted. At the Snapperfarm site, turbidity values at the cobia cage were normal for ocean waters around Puerto Rico and did not differ from the control site (Alston et al. 2005). An environmental assessment of a barramundi farm in Queensland, Australia (McKinnon et al. 2008) reported turbidity differences between cage and control sites, but these were transient and the authors attributed them to seasonal and tidal differences in flushing rather than aquaculture operations. Generally, turbidity impacts at marine cage sites are not included among high priority concerns in the reviewed literature. Proper siting to ensure flushing and improvements in feed composition and feeding efficiency are generally the two management guidelines recommended to minimize aquaculture effects on turbidity. Using fish feed with low fines (i.e., feed dust particulates) and automated feeders that do not overly erode feeds will also help to minimize turbidity effects.

Lipids

Lipids are an essential component of fish feeds and are a primary source of organic waste discharged from fish farms (Hargrave 2003, Nash et al. 2005, Huntington et al. 2006, Trushenski et al. 2006, Pittenger et al. 2007, Rust et al. (2010). Fish feeds vary significantly in composition, withlipids comprising 4–40% (Tucker and Hargreaves 2008) of commercial diets. As the industry hasexpanded, so has the amount of fish oil released infeed products (Pittenger et al. 2007). Advances infeed formulation has resulted in some vegetable oilreplacement of fish oil in feed (Nash et al. 2005, Rust et al. 2010).

Little research or monitoring data is available directly addressing lipid levels in the water in or near marine cages. Cole (2002) reported that in New Zealand lipids are often seen floating at the surface after feeding. Klaoudatos et al. (2000) reported that the mean lipid output from a 200 ton cage fish farm in Greece was 0.357 kg/day. This is the only report discovered which includes a value for lipid release at cage sites. A study (Bodennec et al. 2002) found that three ichthyotoxic unialgal species grown in media supplemented with fish feed showed altered lipid composition which could increase their toxicity. Overall, lipid output as a surface or water column pollutant has received little attention and does not appear to be a significant environmental concern.

SECTION 2. BENTHIC EFFECTS

This section examines the most significant particulate nutrients and organic matter released from marine cage operations — nitrogen, phosphorus and carbon — and the resulting sedimentation, interstitial oxygen depletion and chemical alterations of impacted sediments. Fish food and feces are flushed from marine cages and descend through the water column. With proper siting and good water circulation, accumulation of waste from cage operations may be minimal. However, if the solid wastes are not flushed from the site, they may accumulate at rates beyond the assimilative capacity of the sediments below and around farms. In enriched sediments, the increased respiration from microbial decomposition leads to decreased oxygen in the sediments as well as changes in sediment chemistry.

Earlier reviews identify these sediment impacts as leading concerns (Wu 1995, Goldburg and Triplett 1997, Pearson and Black 2001). More recent assessments of potential impacts of marine aquaculture also identify benthic organic enrichment and sediment chemistry change as primary environmental impacts of marine cage aquaculture in the U.S. (Goldburg et al. 2001, Nash 2001, Stickney 2002, Nash et al. 2005, Phillips 2005, Pittenger et al. 2007, Johnson et al. 2008), Canada (Hargrave 2003, Wildish et al. 2004, Johannessen et al. 2007), Europe (Black et al. 2002, International Council for the Exploration of the Seas 2002, Huntington et al. 2006, Braaten 2007, Devlin et al. 2007, The Mediterranean Science Commission 2007, Tett 2008), the Mediterranean (International Union for Conservation of Nature 2007), New Zealand (Cole 2002), South America (Buschmann et al. 2009) and globally (Beveridge 2004, Halwart et al. 2007, Holmer et al. 2008a, Tucker and Hargreaves 2008, Holmer 2010). In this section, the effects of marine cage culture to benthic biogeochemistry are summarized. The biotic implications of organic enrichment are addressed in the following section, **Marine Life**.

Nutrient Enrichment

Excess feed and feces are the predominant sources of nutrient outflow from marine farms (Beveridge 2004, Belle and Nash 2008, Holmer et al. 2008b). Biological debris from on-site biofouling removal may also contribute to nutrient loading but this input tends to be episodic and relatively small compared to the loading from feed and feces. The total nutrient loading at a farm site depends upon a variety of factors including the size of fish being fed, the number of cages in operation, stocking density, the type of feed being used and farm management measures implemented to maximize feeding efficiency (Holmer et al. 2005). As fish mature to harvest size, more feed is required. Larger farms with more cages will discharge higher net nutrient loads. Water bodies with multiple farms may be additively affected.

Modern extruded fish feeds result in less waste than compacted pelleted feeds, and both are more efficient than using raw and unprocessed fish as feed (Rust et al. 2010). Farms which carefully monitor each feeding, either directly or with cameras, can also decrease the nutrient loading into the surrounding marine environment. Siting plays an essential role in determining the amount of nutrient accumulation in the sediments below cages. Farms located in areas with good flushing and net erosive flow conditions (versus net depositional conditions) will show fewer or no benthic effects near the farm site. In such sites, far-field or regional accumulation of nutrients is possible, especially if many farms are present.

Nitrogen and Phosphorus

Eutrophication in coastal waters may be driven by nutrient enrichment directly to the water column and by nutrients sequestered in and then resuspended from the sediments. Reviews of research on the release and accumulation of nitrogen and phosphorus from fish farms in sediments report that, when averaged globally, 20– 463 kg of nitrogen and 5–80 kg of phosphorus are released per metric ton of fish produced (Wu 1995, Islam 2005, Pittenger et al. 2007). This wide range of loading estimates reflects the variety of species cultured, feed sources and farm practices. These reviews indicate that as much as 95% of the nitrogen and phosphorus input originally as food can be released into the environment. About half of that may be in the form of solids waste which ultimately can end up in onsite or nearby sediments. Today, nitrogen and phosphorus release and accumulation are monitored routinely at farm sites around the world and research is underway to refine estimates of loading and the potential environmental impacts. Here we review recent studies which report the measured discharge of these nutrients from fish farms in North and South America, Europe, the Mediterranean and Asia.

Temperate Regions

North America

Research has been conducted at Canadian fish farms to understand nitrogen and phosphorus accumulation in sediments below and near salmon farms. As in the U.S., it is generally thought that improved management practices have resulted in decreased nutrient loading. A fish growth model and mass balance calculations were used to estimate the nitrogen flux from salmon farms in New Brunswick at the level of individual farms and at larger scales including multiple farms (Strain and Hargrave 2005). They estimate that 9 kg of nitrogen and 2.3 kg of phosphorus per ton of fish production accumulate in the sediment over a three year grow out cycle, with the highest levels occurring in the first and third years. These values are believed to be widely applicable to salmon farms in other parts of Canada and are in close agreement with similar estimates from a comparable study in Scotland. The authors also wished to assess the wider area impacts of nutrient discharge since, even in net depositional areas, most of the waste may be transported away from farms. They calculated the total amount of discharge from fish farms within each of four hydrologically separate Coastal Management Regions. Nitrogen and phosphorus discharge estimates varied greatly between the regions because of differences in approved production limits for each region. Yearly waste also varied as fish grew from smolts to harvest size. Predicted inlet-scale changes in nutrient loading which take into account inlet volume and turnover time were also presented. Nitrogen flux attributable to salmon farms in farmed bays was calculated to be between 1.7-330% higher than natural processes, with variability within and between the Management Regions. This study reflects that aquaculture may contribute significantly to alterations in nutrient loading and cycling, especially in heavily farmed areas. Because large impacts close to farms may be smaller when averaged over larger geographical regions, the authors highlight the importance of considering multiple scales when assessing potential environmental impacts of nutrient enrichment since site-specific factors such as water depth and flushing rates may significantly influence the magnitude of near-field versus far-field effects (Hargrave 2003, Wildish et al. 2004).

In the past, benthic enrichment beneath marine fish cages in the U.S. was a concern at early sites in Washington and Maine. However, relocation of farm sites and industry improvements have resulted in less enrichment mostly confined to the areas just below the cages (Goldburg and Triplett 1997, Nash 2001, Stickney 2002, Nash et al. 2005, Phillips 2005). Modern diets have improved feed retention, and best management practices have contributed to reducing unconsumed waste feed at U.S. salmon farms to less than 5%, thus decreasing nutrient accumulation beneath cages. An assessment of nitrogen inputs to Blue Hill Bay, Maine estimated that marine aquaculture discharged 42-49 metric tons of nitrogen to the system annually (Sowles 2005). This represented less than 10% of the nitrogen loading to the bay and an ecological carrying capacity assessment indicated the area could support additional net pens. A field and modeling study of hydrographic conditions in the Strait of Juan de Fuca suggest that marine cage culture would result in minimal benthic accumulation of nutrients because flushing rates are high enough to disperse farm wastes (Rensel et al. 2007).

South America

Significant nitrogen and phosphorus accumulation was found in 29 fish farms in Chile (Soto and Norambuena 2004). Phosphorus levels were nearly six times higher at farms (115 mmol/kg) than at control sites (21 mmol/kg). Average nitrogen levels (124 mmol/kg) in farm sediments was about 4 times that at control sites (32 mmol/kg), but high variability among farm sites suggested that other geographical sources of nitrogen were also a factor. Buschmann et al. (2009) later used a mass balance approach to estimate that 15% of nitrogen and 65% of phosphorus feed inputs to Chilean salmon farms settles in the sediments below cages.

Northern Europe

Early feeding practices at many European fish farms have been replaced with more efficient feeds and feed management, resulting in greatly decreased nutrient enrichment to benthic sediments (Huntington et al. 2006). Summary tables compiled by Huntington et al. (2006) show salmonid and sea bass farms release between 31-62% of nitrogen and 11-34% of phosphorus in feed as soluble waste, and that an estimated 22 g of nitrogen and 9.5 g of phosphorus as particulate waste are produced per 1000 g fish harvested. Olsen et al. (2008) estimate that 38.2% of the nitrogen and 30.7% of the phosphorus in feed are assimilated by cultured fish, with the remainder being lost as waste to the environment. They also modeled nutrient discharge from a hypothetical fish cage producing 1000 metric tons of salmon, calculating benthic loading of 14 tons of particulate nitrogen and 5.2 tons of particulate phosphorus. Braaten (2007) reported that 19% of nitrogen and 50% of phosphorus in salmon feed are ultimately deposited in the sediment below cages. Sediment and sediment trap sampling over two years at a Norwegian salmon farm found that particulate nitrogen and phosphorus concentrations were higher under cages and out to 550 m during the production cycle, compared to samples collected prior to fish stocking and distant (up to 3000 m away) reference sites (Kutti et al. 2007a). Sediment phosphorus levels were mostly above 1500µg/g dry weigh sediment near the farm, but less than 1000 µg/g dry weigh sediment further out. Particulate nitrogen accumulation in the sediments, however, was not evident. Recently, Norði et al. (2011) sampled nitrogen budgets at a rainbow trout farm in the Faroe Islands finding that sediment nitrogen doubled (from 0.2 up to 0.41 mmol/ gram dry weight sediment) during farming, mostly from fecal waste. A Scottish review of the environmental impacts of marine aquaculture concluded that while a few farm sites are nutrient enriched, most are not (Black et al. 2002). This was partially explained by improved feed efficiency, as less than 5% of feed is currently lost directly as waste, compared to earlier (prior to 1990) estimates of 20% (Black et al. 2008).

Australasia

Cole (2002) summarized earlier studies in New Zealand which documented nutrient enrichment from salmon farms, but he notes that many fish farms in this area are in deep water.

Sub-Tropical Regions

North America

In Baja California, sampling at tuna sea cages found increased nitrogen levels in sediments ranging from 0.02-0.12% (Diaz-Castaneda and Valenzuela-Solano 2009). The highest levels were found in a more enclosed bay.

The Mediterranean

The last ten years have seen a great deal of research focusing on nutrient enrichment in the Mediterranean resulting from fish farming. The release of large amounts of nitrogen and phosphorus is of concern due to the potential for eutrophication and large-scale ecological impacts (Huntington et al. 2006, Cardia and Lovatelli 2007, International Union for Conservation of Nature 2007, Grigorakis and Rigos 2011).

In Spain, Aguado-Giminez and Garcia-Garcia (2004) found increased phosphorus beneath sea bream/sea bass cages, but similar increase in nitrogen was not detected presumably due to high current speeds and low stocking density. A follow up study using two sampling methods found significant differences in sediment nitrogen and phosphorus directly below and adjacent to cages, but the effects were not evident within 200-500 m of the cages (Aguado-Gimenez et al. 2007). Aguado-Gimenez et al. (2006) used bioenergetics modeling to estimate that Spanish tuna fattening farms release 26-49 mg nitrogen and 28-75 mg phosphorus per kg of fish per day. For particulate nitrogen these levels are comparable to output from sea bream and sea bass farms, but for phosphorus this is more than double. Further modeling comparisons of these three species and their nutrient output (Piedecausa et al. (2010) predict that tuna fattening releases the highest amounts of particulate nitrogen (up to 700 g nitrogen per day) and phosphorus (1000 g phosphorus per day), but this release is concentrated in a few months of the year. In contrast, sea bream and sea bass farms operate year round with continuous waste discharge. Dominguez et al. (2001) monitored sediments beneath sea bream cages in the Canary Islands and found little or no accumulation of nitrogen or phosphorus after the first year of farm operation.

In Italy, sediments in seagrass beds 20 or 100 m from fish farms were compared to beds at reference stations (Cancemi et al. 2000). Sediment pore water nitrogen (up to 10 times) and phosphorus (up to three times) concentrations were higher at farms compared to a reference sites. Nutrient accumulation was highest directly under the cages and decreased with distance from the farm. Similarly, total sediment phosphorus was very high beneath the cages, but decreased as a function of distance from the farm. Another Italian study found high levels of nutrient enrichment in sampling stations beneath cages with a clear pattern of nutrient decline out to 200 m (Porrello et al. 2005). Vizzini and Mazzola (2006) measured high levels of nitrogen isotopes in sediments taken off the coast of Sicily near fish farming operations, but could not separate farm effects from other anthropogenic sources of enrichment.

In the Adriatic Sea, Matijevic et al. (2008) found sediment phosphorus levels at tuna farms were up to five times higher than sites without farms. Similarly, Matijevic et al. (2009) and Kovac et al. (2001) reported increased phosphorus and nitrogen in sediments below sea bass and sea bream farms.

Using nitrogen isotope analysis of invertebrate tissue to differentiate between nitrogen enrichment caused by fish farm waste and sewage in the Adriatic, Dolenec et al. (2007)

demonstrated that nitrogen from the farm was being assimilated into higher trophic levels in this ecosystem.

A study in Greece found increased phosphorus in sediments closest to farm stations, especially in the summer months (Mantzavrakos et al. 2005). Apostolaki et al. (2007) studied sediments in seagrass meadows in Greek, Spanish and Italian waters near sea bass and sea bream farms. While phosphorus levels declined with distance from farms, organic nitrogen levels were similar between stations.

Karakassis et al. (2005) used a simple model to estimate the amount of nitrogen and phosphorus released into the entire Mediterranean from fish farming in comparison to other human activities. They estimated that less than 5% of the annual waste discharge comes from fish farming, and conclude that this industry poses less of a long-term threat to the Mediterranean than other anthropogenic sources. Because sea bream and sea bass discharges an estimated 1.5 times more nitrogen and phosphorus into the environment than Atlantic salmon, the environmental effects of farming these two species in the Mediterranean concerns some researchers (Grigorakis and Rigos 2011).

Asia

Nitrogen flux in sediments at sea bream and yellowtail tuna *Thunnus albacores* net pen sites in a Japanese bay were 2.5 times higher than at reference sites outside of the farm, caused primarily because vertical mixing was minimal during the highest nutrient loading (Tsutsumi et al. 2006). Similarly, sampling at a Japanese sea bream farm found a nearly fourfold increase in sediment nitrogen flux at cage sites compared to background levels (Yokoyama et al. 2009).

Tropical Region

Caribbean

Monitoring at cobia and mutton snapper cages in Puerto Rico concluded that 13% of nitrogen input as feed was released as solid waste, but there was no significant difference in sediment nitrogen levels compared to a control site (Alston et al. 2005).

Photo courtesy of USGS.

Asia

In a nutrient enrichment study at a Chinese multi-species fish farming site, total nitrogen and phosphorus in sediments were 129% and 1316% higher than at reference stations (Gao et al. 2005). Sediments sampled in and around milkfish *Chanos chanos* pens in the Philippines were also significantly enriched with both nitrogen and phosphorus, but because these pens enclosed the seafloor they physically restricted water flow (Holmer et al. 2002).

Australasia

Monitoring at a barramundi farm in Queensland, Australia found no net accumulation of nitrogen under cage sites and the percent of nitrogen in sediments was within normal ranges for similar environments (McKinnon et al. 2008).

The deposition and accumulation of nitrogen and phosphorus in benthic sediments below marine fish cages is widely studied and highly variable (Table 3). Modern feed formulations and feeding practices have reduced the output of these nutrients to the environment. Siting of farms in deep areas with sufficient flushing and considering the cumulative effects of multiple farms are among the best tools for minimizing long-term harmful impacts. At most modern farm facilities, nitrogen and phosphorus enrichment in benthic sediments is minimal and restricted to the areas beneath and within 500 m of the cages.

Carbon

Organic carbon is released from marine cages in the form of uneaten food and in fish feces. Reviews of the environmental effects of marine aquaculture often identify carbon deposition and accumulation as a major concern, because its microbial degradation can lead to oxygen depletion and other chemical changes in the sediment as discussed in the next section. Wu (1995) summarized that 80-84% of the carbon in feed is released to the environment, with around 23% accumulating in the sediments beneath the cages. These values are similar to the range of 29-78% or 4.1–78 g carbon/m^2/day reported by Pearson and Black (2001). In contrast, Nash (2001) estimated that only about 8.8% of the carbon in feed is discharged from salmon pens to settle in sediments because of improvements in feed formulation and feeding efficiency. The accumulation of organic carbon in sediments below marine fish cages has been identified as a concern in the U.S. (Goldburg et al. 2001, Nash 2001, Stickney 2002, Nash et al. 2005, Pittenger et al. 2007, Johnson et al. 2008), Canada (Hargrave 2003, Wildish et al. 2004, Holmer et al. 2005), South America (Costa-Pierce et al. 2007), Europe (International Council for the Exploration of the Seas 2002, Huntington et al. 2006, Black et al. 2008, Holmer et al. 2008b), the Mediterranean (International Union for Conservation of Nature 2007, The Mediterranean Science Commission 2007, Grigorakis and Rigos 2011) and globally (Halwart et al. 2007, Holmer 2010).

Table 3. Nitrogen and phosphorus loading rates at marine fish farms. Values are presented in the units in the original reference

Sediment Nitrogen Loading	Reference	Location	Species Cultured
9 kg/ton fish produced	Strain and Hargrave 2005	New Brunswick	Salmon
1.7 to 330% natural flux	Strain and Hargrave 2005	New Brunswick	Salmon
Double levels at control sites	Soto and Norambuena 2004	Chile	Salmon
13% lost from feed	Alston et al. 2005	Puerto Rico	Cobia
129% higher than control site	Gao et al. 2005	China	Sea bream
26-49 mg/kg fish/day	Aguado-Gimenez et al. 2006	Spain	Tuna
133 kg/ton fish produced	Islam 2005	Various	Various
2-48% lost from feed	Huntington et al. 2006	Europe	Various
Sediment Phosphorus Loading	**Reference**	**Location**	**Species Cultured**
2.3 kg/ton fish produced	Strain and Hargrave 2005	New Brunswick	Salmon
6-9 times level at control sites	Soto and Norambuena 2004	Chile	Salmon
1316% higher than control site	Gao et al. 2005	China	Sea bream
28-75 mg/kg fish/day	Aguado-Gimenez et al. 2006	Spain	Tuna
5 times levels at control sites	Matijevic et al. 2008	Adriatic Sea	Tuna
36-74% lost from feed	Huntington et al. 2006	Europe	Various
25 kg/ton fish produced	Islam 2005	Various	Various

Temperate Regions

North America

Much research has been conducted in Canada to understand carbon release, accumulation and impacts at marine fish farms. In the Bay of Fundy, sampling in two intensive salmon aquaculture areas found increased organic carbon (about 8%) in sediments below cages compared to inlets with no farming (Pohle et al. 2001). All sites showed decreasing carbon sediment levels over a four year sampling period implying improved environmental conditions on a large scale. Similarly, Sutherland et al. (2001) deployed sediment traps at salmon net pens in British Columbia and found higher carbon flux (≈180 mg/m^2/hour) and concentrations (637 µg/mg) adjacent to pens compared to the control site (≈75 mg/m^2/hour and 442 µg/mg) 500 m away. Schendel et al. (2004) also studied carbon accumulation in sediments along horizontal and vertical transects away from a salmon farm. Surface sediment carbon concentration below the salmon farm was 4%, but quickly dropped to about 2% at 30 m and out to 300 m. Sediment carbon concentration decreased to 2.9% at 5-10 cm below the surface. Another study in New Brunswick also found increased carbon in sediments below salmon cages (Chou et al. 2004). In this study, the background organic carbon concentration was 1.6%, while farm sites ranged from 1.8-9.1% with the highest carbon concentration found in anoxic sediments. In an experiment to validate the use of acoustic detection technology to assess benthic enrichment, Wildish et al. (2004) measured sediment carbon at salmon farms and reference sites. Background carbon levels ranged from 2 -2.8%, while sediments beneath farms ranged from 13.7-25.7%, indicating highly enriched sediments. The backscatter profile from the cage site also indicated enrichment, but further validation and refinement of the methodologies is required. Using mass balance calculations, Strain and

Hargrave (2005) estimated that 76 kg of waste carbon are released per ton of salmon produced during a full grow out cycle. Their results suggested that the sediment accumulated underneath farms represents a small fraction of the total discharged waste. Further modeling at a larger scale was used to predict the total amount of waste that could be generated by many farms in an inlet with multiple farms. Carbon flux was estimated to reach 160% that of natural levels in areas of intensive fish culture. Thus, the long-term, regional organic matter accumulation may be significant, especially in areas with intensive fish farming. This study highlights the need to consider impacts of aquaculture at multiple scales.

In the U.S., (Goldburg and Triplett 1997) cited studies conducted at fish farms in Maine where increased carbon deposition to the sediments was not measurable within 20 m of the cages. In contrast, they reported that increased deposition at a Puget Sound farm could be seen in the benthos out to about 150 m. Studies and data reviewed by Nash (2001, 2003) indicated that generally carbon was elevated in sediments around Pacific Northwest salmon farms to about 30 m beyond the cages, although effects could extend to over 200 m depending upon the degree of flushing. Therefore, the environmental impacts of settling and accumulation of bio-deposits, including organic carbon, were assessed as high risk compared to other environmental concerns posed by salmon farming. The sedimentation rate at salmon farms ranged from 15-100 g total volatile solids/m^2/day, about half of which would be deposited as carbon.

The question of the scale of carbon sedimentation impact and the ability to predict the capacity of larger systems to assimilate waste nutrients is addressed in an extensive review paper by Holmer et al. (2005). The high levels of sediment carbon accumulation documented at some fish farms are well above natural levels found in most coastal sediments. Carbon to nitrogen ratios may be useful in differentiating carbon derived from fish farms from other sources of loading (e.g., municipal waste). Improvements in feed formulation and feeding efficiency contribute to the management of sedimentation rate and extent of waste carbon. Wildish (2004) compiled near-field carbon sedimentation rates measured at salmon farms around the world. These ranged from 1-181 g/m^2/ day, and other reviewed studies using mass balance approaches also fell within this range. Differences in values depended upon sampling methodology, farming practices and physical site characteristics. Far field effects, including carbon burial rates at larger scales, were reviewed by Hargrave (2003). Hydrodynamic processes at the sediment-water interface (i.e., benthic shear) can erode and resuspend carbon rich sediments and laterally transport carbon to locations remote from farm sites. This will tend to spread organic waste over larger areas, yet it also provides a mechanism for aerobic assimilation of waste nutrients within the marine ecosystem. Detailed information about site specific hydrology, bathymetry and local nutrient dynamics are needed to provide insight into long- term processes over large areas.

South America

An assessment of benthic conditions at salmon farms in two southern Chilean fjords reported organic carbon mineralization rates of 2.16-4.53 g carbon/m^2/day, calculated based upon measured oxygen flux (Mulsow et al. 2006). A large-scale study in southern Chile (Soto and Norambuena 2004) sampled two to five salmon farms in nine farming areas and found significant sediment carbon accumulation (413 mmol/kg) compared to control (192 mmol/kg) sites. However, other anthropogenic sources of carbon were thought to contribute to some of the carbon loading and variability among locations.

Photo courtesy of NOAA.

Northern Europe

Norði et al. (2011) reported a significant increase in sediment organic carbon from 1.9 mmol/g dry weight sediment prior to farming, up to 5.55 mmol/g dry weight sediment during production at a trout farm in the Faroe Islands. Effects were most prevalent within 30 m of the cages. During a two year study at a large Norwegian salmon farm Kutti et al. (2007a) found that organic carbon flux measured in sediment traps was nine times higher at the farm (365 g/m^2) than at a reference site 3 km away. Sedimentation rates were also higher in the second production year. However, no increase in particulate organic carbon was measured in the sediments and there was no evidence of accumulated organic waste, most likely due to high bottom current speeds. Although most of the particulate waste settled in traps within 250 m of the farm, carbon isotope analysis indicated that some waste could be found in traps up to 900 m away, possibly due to sediment resuspension. Mass balance equations, using the current feed and food conversion ratio data, for a hypothetical Norwegian salmon farm producing 1000 metric tons per year estimated benthic carbon loading of 2300 g/m^2/year (Olsen et al. 2008). After accounting for assimilation and respiration, carbon waste was calculated to be 24% of feed input. A study in Norway used four sampling techniques to quantify sediment enrichment at 80 sites near salmon cages (Carroll et al. 2003), including diver surveys, imagery, sediment chemistry and faunal analysis. The sediment survey found degraded conditions (based upon total organic carbon levels) under 32% of the cages. The study results showed that elevated sediment carbon levels measured at farms had dissipated at a distance of 50-100 m, even at sites with degraded sediments beneath cages. Further analysis indicated that fallowing and current speed correlated positively with environmental quality. Shakouri (2003) collected sediment samples below a salmon cage in an Icelandic fjord, finding only slight carbon enrichment (1.7% sediment dry weight) directly beneath the cage. At 95 m (1.2%) and 600 m (1.3%), no further effect was detected.

In Scotland, the use of a high energy diet formulation resulted in a 12% decrease in carbon levels in salmon feces (Chen et al. 2003). The Scottish Environmental Protection Agency (SEPA) has set 9% sediment carbon as the allowable criteria level (Black et al. 2008). Active salmon farms in Scotland reported organic matter concentrations ranging from 3.1-22.9% beneath cages with a decreasing trend out to 50 m with reference stations ranging from 0.9-13.5% (Black et al. 2012).

Australasia

Morrisey et al. (2000) sampled sediment traps beneath salmon farms in New Zealand finding carbon deposition rates from 463-967 mmol C/ m^2/day compared to about 50 mmol C/m^2/day at control sites. This data was used in conjunction with water velocity data to model potential impact to sediments. At the lowest currents actually measured, this amount of carbon flux was predicted to cause impact with significant accumulation of carbon to the sea floor.

Sub-Tropical Regions

North America

A study at a tuna sea cage farm in Baja California, Mexico found that organic carbon sediment concentrations ranged from 0.2% to 2.53% with the highest levels in the area of the bay with the tuna pens (Diaz-Castaneda and Valenzuela-Solano 2009). Because the samples were taken 250 m from the cages, carbon deposition levels directly beneath the cages are unknown.

The Mediterranean

Many studies have been conducted in the last ten years in Italy to assess carbon enrichment at marine fish cages. La Rosa et al. (2001) measured an elevation of benthic carbon at sea bass cage sites (4949 μg/g) compared to control sites (3013 μg/g). Following removal of the cages, carbon levels dropped to concentrations comparable to control levels within a few months. Vezzulli et al. (2002) found higher levels of carbon (6327 μg/g) in benthic sediments at a sea bream farm that had been operational for 15 years compared to a control site (4596 μg/g) 200 m away. Studies such as these are important for assessing the potential cumulative effects of fish farming over longer periods. Carbon enrichment (up to 6275 μg/g) was reported at an Italian sea bream and sea bass farm, with sediment impact apparent within six weeks of fish stocking and was proportional to feeding rates (La Rosa et al. 2004). Sampling at a sea bream and drum *Argyrosomus regius* farm off Italy also found evidence of carbon loading (up to 6.32%) in sediments directly beneath the cages, but these effects declined steeply within 50 m of the cages (Porrello et al. 2005). Below a bluefin tuna fattening farm off Sicily, no significant differences were found in sediment carbon levels compared to a nearby control site (Vezzulli et al. 2008). The water depth at this open sea site was 46 m and average current was 6 cm/s, and these were considered to be the primary factors that minimized nutrient impacts despite high cultured fish biomass and feed inputs.

Modeling has been used in Italy to predict carbon loading to coastal sediments from fish farming. Doglioli et al. (2004) developed a three dimensional model, LAMP3D, to determine that a sea bream and sea bass farm in the Ligurian Sea would add about 0.085 g C/m^2/day — a level expected to have little negative environmental effect. The hypothetical site was 1.5 km offshore in 40 m of water, with strong wind and water currents. Similarly, the potential environmental impacts of a sea bass and sea bream farm expansion were modeled using the MERAMOD® model (Brambilla et al. 2007). Maximum carbon flux was estimated at 1350 g/m^2/ year, with up to 150 g/m^2/year being added by the addition of four more cages. The impact area was projected to increase from 5.6 ha to 7.3 ha.

Work has also been done in the eastern Mediterranean to assess carbon loading, with mixed results. Belias et al. (2003) found organic carbon levels of 1.8, 1.9 and 7.2% at three

Greek fish farms, compared to nearby reference levels around 1%. The farm with the highest enrichment level also had the highest production level of the three farms sampled. Similarly, Aksu et al. (2010) found variable carbon enrichment (0.03–10.65%) when sampling sediments at eight farms in the Aegean Sea. Again, the highest mean sediment carbon levels were found at the farms with high production and high carbon inputs to the environment. In the northern Adriatic Sea, Kovac et al. (2001) also found higher levels of carbon in sediments below sea bass cages. Apostolaki et al. (2007) tested sediments at three coastal Mediterranean fish farms. Percent total organic carbon tended to be higher beneath cages compared to controls, but there was variability between the farms and no site reached carbon levels higher than 2.6%. Sampling at a sea bass and sea bream farm in the middle Adriatic was unable to detect any consistent differences in carbon sediment concentration between sites below the cages and a control site 1 km away (Matijevic et al. 2009). Similarly, Aksu et al. (2010) conducted a four year study of the nutrient loading impacts of a tuna fattening farm off Turkey and were unable to detect any carbon accumulation in the sediment. Water depth and strong currents were credited with the low measures of nutrient accumulation.

Asia

A study in western Japan estimated that carbon flux to the seafloor at a net pen site averaged 2.11 g C/ m^2/day, which was 2.5 times higher than the natural flux outside of the fish farm (Tsutsumi et al. 2006). A feeding experiment in Japan calculated carbon flux at cage sites ranging from 0.13-3.1g C/m^2/day, compared to background levels ranging from 0.16-0.8g C/m^2/day (Yokoyama et al. 2009). The total organic carbon in the sediment at cage sites was 56–79 mg C/g dry sediment compared to 12–19 mg C/g at control sites. Reduction of overfeeding was successful in decreasing organic loading while maintaining fish production levels. Another Japanese study (Pawar et al. 2002) found that fish farm sediment quality was directly related to organic carbon input from farms. Here carbon input from fish feed ranged from 2-57 kg C/m^2/year.

Tropical Region

Caribbean

Monitoring at a cobia farm in Puerto Rico found that total carbon content of sediment at cage and control sites ranged from 4-6% and were not significantly different, but the organic carbon fraction was not differentiated (Alston et al. 2005). Sediment sampling for nine months at the cage sites found carbon levels to be generally very consistent, with a spike in June that coincided with peak feeding rates and biomass. Continued monitoring was recommended to track sediment carbon levels especially if the farm expanded its operational capacity.

Asia

At fish farms in Hong Kong, sediment sampling found that carbon levels were 83% higher than at reference sites 600 m away, with samples taken at 100 m away being similar to reference sites (Gao et al. 2005). Following removal of mariculture operations in a small enclosed lagoon in Taiwan, the sediment C/N ratio increased from 7.3 to 8.1, but this was not statistically significant (Hung et al. 2008). Prior to cessation of farming, the enriched sediments were 4-9% organic carbon. Alongi et al. (2003) conducted a study of carbon flow from fish cage aquaculture in Malaysian mangrove estuaries. Although greater carbon levels were found in areas with fish cages, most of the variability was driven by tides and water depth. Resuspension of sediments by boat traffic was also proposed as a factor in benthic clearing. They estimated that the fish cages represented 2% of total carbon input, as these waters are heavily impacted by other sources of pollution. In the Philippines, Holmer et al. (2002) found that sediment carbon levels increased under milkfish pens throughout the growout production cycle, ranging from 1.8-0.95%. Sediment carbon was four times higher inside cages than outside, with concentrations decreasing with both horizontal distance (100 m) from the cage and vertical depth into the sediment layer.

Australasia

Monitoring at a barramundi farm in Queensland found that carbon was not accumulating beneath the cages (McKinnon et al. 2008). Here, the yearly farm carbon input represented only 9% of total carbon loading, with mangrove litter accounting for 85%.

Giles (2008) reports that carbon levels in sediments below fish farms range from 0.2–26.1% in 17 studies (including studies cited in our report) from around the world, reflecting the great variability of potential enrichment (Table 4). Background carbon levels in most coastal sediments is <5% (Holmer et al. 2005), so the accumulation and effect of organic carbon from fish farms continue to be an environmental concern for marine aquaculture. Improvements in feed formulation in the last decades have decreased nutrient loading (Soto and Norambuena 2004, Buschmann et al. 2009), as has siting in areas with sufficient flushing. Aquaculture likely represents only a small fraction of the anthropogenic carbon and other nutrients building up in coastal waters (Goldburg and Triplett 1997, Cloern 2001, Anderson et al. 2002, Pittenger et al. 2007), and continued research, technological advancements and improved farm management will lead to further improvements.

Technologies to capture solid waste on screens below fish cages are being developed but are not currently a viable alternative for waste management (Buryniuk et al. 2006, Heinig et al. 2006). Bioremediation techniques using bacterial augmentation of sediments below fish cages to mobilize carbon are also being investigated. An experiment at a commercial sea bass farm yielded significant reduction in carbon levels in sites treated with a mixture of microorganisms and oxygen releasing compounds (Vezzulli et al. 2004), providing promising results for further development of this approach to managing nutrient enrichment at the farm scale. At a sea bream farm in an enclosed Japanese bay, an experiment to assess bioremediation of benthic sediment by stocking artificially cultured *Capitella* polychaetes early in the production cycle was successful (Tsutsumi 2007). Organic carbon concentration in the sediments beneath the cage decreased from nearly 30 mg/g dry sediment to less than 10 mg/g dry sediment within a few months after stocking the polychaetes. This approach offers another potential avenue for ameliorating environmental impacts of carbon accumulation beneath fish cages, but likely has limited applicability in very deep or high current open ocean sites.

Sediment Biogeochemistry

The output of fish feed and feces from marine fish cages, their accumulation on the seabed and effects on natural biogeochemical processes are among the environmental concerns for the marine aquaculture industry. The potential impacts to benthic sediment and chemistry have been identified as an important issue for the U.S. (Goldburg et al. 2001, Stickney 2002, Clement and Janowicz 2003, Nash et al. 2005, Pittenger et al. 2007, Johnson et al. 2008), Canada (Hargrave 2003, Wildish et al. 2004), South America (Buschmann et al. 2009), Europe (Black et al. 2002, Gillibrand et al. 2002, International Council for the Exploration of the Seas 2002, Huntington et al. 2006, Braaten 2007, Olsen et al. 2008, Holmer 2010), the Mediterranean (International Union for Conservation of Nature 2007, The Mediterranean Science Commission 2007, Borg et al. (2011, Grigorakis and Rigos 2011) and globally (Halwart et al. 2007, Holmer et al. 2008b, Hall et al. (2011).

In response to these concerns a great deal of research has documented and quantified the changes to the benthic chemistry that are directly attributable to marine finfish cage culture. In this section, we first provide an overview of the natural biogeochemical processes that occur in most seafloor sediments involving the flux and reactivity of numerous chemicals. Next we review and summarize the current research on the biogeochemical impacts of fish farms to underlying marine sediments. We also present information about which sedimentation parameters or measurements are relatively easy and cheap to collect while also being reliably indicative of significant negative impacts. Different sampling methodologies are compared. The next section, **Marine Life**, addresses the ecological impacts that nutrient enrichment has on seafloor and pelagic biodiversity.

Table 4. Reported organic carbon loading rates at marine fish farms. Values are presented in the units in the original reference

Sediment Carbon Loading	Reference	Location	Species Cultured
15-100 g TVS/ m^2/day	Nash 2001	Pacific Northwest	Salmon
13.7 – 25.7% organic carbon	Wildish et al. 2004	Canada	Salmon
8% organic carbon	Pohle et al. 2001	Bay of Fundy	Salmon
1.8 – 9.1% organic carbon	Chou et al. 2004	New Brunswick	Salmon
180 mg/ m^2/day	Sutherland et al. 2001	British Columbia	Salmon
2.16 – 4.53 g/ m^2/day	Mulsow et al. 2006	Chile	Salmon
365 g/ m^2	Kutti et al. 2007	Norway	Salmon
3.1 – 22.9% organic carbon	Black et al. (2012	Scotland	Salmon
463 – 967 mmol/ m^2/day	Morrisey et al. 2000	New Zealand	Salmon
6.32% organic carbon	Porrello et al. 2005	Italy	Sea bream & drum
1.8 – 7.2% organic carbon	Belias et al. 2003	Grece	Sea bream & bass
2.6% organic carbon	Apostolaki et al. 2007	Mediterranean	Sea bream & bass
2.11 g/ m^2/day	Tsutsumi et al. 2006	Japan	Sea bream
4-6% total carbon	Alston et al. 2005	Puerto Rico	Cobia
5.55 mmol/g dry weight	Norði et al. (2011	Faroe Islands	Trout
1.8 – 10.95% organic carbon	Holmer et al. 2002	Philippines	Milkfish
0.2-26.1% organic carbon	Giles 2008	17 studies worldwide	Various
4 – 9% organic carbon	Hung et al. 2008	Taiwan	Not reported

Marine Sediment Biogeochemistry

The following overview of marine sediment biogeochemistry is derived from reports by Schulz (2000), Wildish et al. (2004), Holmer et al. (2005), Hargrave et al. (2008), and Valdemarsen et al. (2010) which provide more technical detail. Most marine benthic sediment processes are driven by the decomposition of organic matter and are mediated by a variety of bacteria. The organic waste beneath fish farms is primarily derived from particulate organic carbon from feed and fecal waste. Most of the carbon is labile (soluble and available for decomposition), although some portion may be refractory (bound up and relatively inert). Oxygen transport into benthic sediments occurs by direct diffusion from bottom water as well as biological processes such as bioturbation from benthic feeding activity and burrowing by benthic dwellers. The oxygen is used in a series of oxidation- reduction, or redox, reactions in the sediment. Redox reactions involve the exchange of electrons between products and reactants, resulting in either reduced or oxidized products. The direction of the redox reaction is driven by oxygen availability. As oxygen is consumed for microbial respiration and ultimately depleted vertically into the sediment depths, anaerobic bacteria prevail, causing the redox reactions to be driven toward the reduced state. This series of redox reactions results in carbon dioxide production, nitrification of ammonia, and reduction of manganese, iron and sulfur. In organically enriched sediments, sulfate reduction may account for all the carbon oxidation, resulting in toxic sulfide compounds with negative impacts to benthic organisms. At the greatest sediment depths, the redox reactions convert organic carbon into methane. In unimpacted sediments with overlying water currents, the aerobic layer may penetrate several centimeters. A well-defined intermediate zone with reduced oxygen extends down before reaching the deeper anoxic layers which are in a chemically reduced state. Table 5, adapted from Holmer et al. (2005), summarizes the predominant bacteria-mediated redox reactions occurring in benthic sediment profiles with organic carbon as the original energetic input.

Redox potential (Eh) is measured in millivolts (mV) using electrodes inserted directly into sediments. Oxidized sediments (dissolved oxygen >2 mg/L) tend to have redox values above 100 mV. As oxygen is depleted and hypoxic conditions become established (dissolved oxygen is 0-2 mg/L), Eh ranges between 100 and -150 mV. Under anoxic conditions (dissolved oxygen is 0 mg/L), Eh values are less than -150 mV. Redox reactions are correlated with pH. In normoxic sediments, pH tends to be around 8. Deoxygenated, reducing sediments may be around pH of 7. Under anoxic and fully reduced conditions, pH may be as low as 5.

Beneath poorly flushed or heavily stocked fish farms, the sedimentation rate may be increased and organic matter may accumulate followed quickly by increased microbial respiration. As microbial oxygen demand is increased, sediments may reach hypoxic (<2 mg/L of oxygen) or anoxic states. Oxygen uptake (+), carbon dioxide release (+), redox (-), pH (-), and sulfide (+) levels all change predictably as the oxidation/ reduction reactions are driven toward the reduced state. Figure 1, adapted from Hargrave et al. (2008; Figure 5), summarizes the many interconnected biological and geochemical processes that can be measured in benthic sediments to assess the level of enrichment. This figure also summarizes ecological responses, which are addressed in the next section.

Table 5. Biogeochemical processes in marine sediments, Adapted from Holmer et al. (2005)

Sediment Depth	Reaction	Process	Product	Reaction	Process	Product
	CH_2O + O_2	Oxygen Respiration	H_2O + CO_2			
	CH_2O + NO_3^-	Denitrification	N_2+ CO_2	NH_4^++ O_2	Nitrification	NO_3^-
	CH_2O + Mn_4^+	Manganese Reduction	Mn_2++ O_2	Mn^{2+} + O_2	Manganese Oxidation	Mn_4^+
	CH_2O + Fe_3^+	Iron Reduction	Fe^{2+}+ CO_2	Fe^{2+}+ O_2	Iron Oxidation	Fe_3^+
	CH_2O + SO_4^{2-}	Sulfate Reduction	H_2S + CO_2	H_2S + O_2	Sulfide Oxidation	SO_4^{2-}
	CH_2O + CH_2O	Methane Production	CH_4+ CO_2	CH + O_2	Methane Oxidation	CO_2

In heavily impacted sediments, even the surface may be completely anoxic and covered in heavily enriched organic matter which persists because there is minimal aerobically mediated decomposition. Bubbling of methane gas may be visible. Although the methane itself is relatively non-toxic, the bubbles also can transport hydrogen sulfide which is produced under anaerobic conditions and is toxic to fish and other marine life, into the water column above.

The hydrography and sediment type determine the biogeochemical processes beneath marine cages. For example, depositional sites are at greater risk of becoming enriched compared with net erosional areas. Depositional sites are generally characterized by higher silt/clay and overall organic content, compared to areas with net erosion where sediments usually contain higher levels of sand and low organic matter. The concentrations of other minerals, notably iron, may also play an important role in sediment biogeochemical processes beneath fish farms and these dynamics are being investigated to aid in identification of sites with sediment compositions that may buffer organic enrichment effects (Valdemarsen et al. 2009, 2010). In general, redox and sulfide concentration are thought to be the most consistently reliable indicators of the enrichment level of sediments and are often selected as a sediment parameters for monitoring purposes. Both are relatively easy to measure in the field, even in deep water.

Sedimentation and Geochemical Effects

Temperate Regions

North America

A great deal of research has been conducted in the last decade in Canada to address the biogeochemical impacts of marine farms to benthic sediments. Reviews of far-field (Hargrave 2003) and near-field (Wildish et al. 2004, Holmer et al. 2005) enrichment effects provide excellent summaries of sediment processes and impact data collected mostly in the 1970-90s.

The best biochemical predictors of benthic enrichment at Canadian salmon farms include increased total sulfide, oxygen uptake, carbon dioxide release and decrease in redox potential. Improvements in feed formulation and feeding management have generally decreased organic loading at Canadian farms. The spatial and temporal severity of impact may be driven by hydrographic and nutrient dynamics at varying scales.

Efforts in Canada have expanded to focus on development of innovative imaging technology and refining the assessment and modeling of far- field impacts. For example, a study by Tlusty et al. (2000) compared the transport and solubility loss of fecal versus food waste from steelhead trout *Oncorhynchus mykiss* pens. They conclude that the impacts below the cages will be greater for food waste because organic matter from feed particles is dispersed less during sinking. The authors suggest that the observed bimodality for sinking rates and organic matter input potential for feed versus feces should be factored into mathematical models that seek to quantify organic input to benthic sediments below farms. This was supported by DEPOMOD model validation work which predicted that waste feed would settle out proportionally higher near cages (within ~50 m), whereas fecal waste would account for organic matter sedimentation more distant from farm sites (Chamberlain and Stucchi 2007). Wildish et al. (2001) compared chemical and biological approaches for monitoring organic enrichment near salmon farms in the Bay of Fundy. Chemical analysis found lower redox potential and higher sulfide concentrations beneath farm sites than at reference sites. The biological parameters also reflected an impacted sediment community. Thus, both sampling methods supported the conclusion that benthic sediments at the farm site were enriched and in a reducing state. However, the chemical analysis proved to be a much cheaper approach costing about 10 minutes of field time per sample taken compared to 22 minutes of field time plus 184.5 minutes of lab time per biological sample.

In British Columbia, vertical sediment flux at a fish farm was highest at the bottom edge of net pens (0.77 $g/m^2/h$) compared with control sites (0.48 $g/\ m^2/h$; Sutherland et al. 2001). Additionally, the stable carbon isotope (δ13C) signature of the feed pellets, suspended particles and sediment trap materials was determined. A distinct signal from feed was observed in water column samples during feeding events and in the sediment traps nearest the net pens. These results are useful in the development of protocols for the use of stable isotopes as tracers of aquaculture waste products across larger scales.

Another comparative study by Wildish et al. (2003) found that sediment profile imaging (SPI) was able to detect organic enrichment. Sediments sampled from a New Brunswick farm site had average redox potentials of -148 mV and sulfide levels of 32,000 μM, compared to reference sites with redox potentials averaging 158 mV and sulfide levels of 1300 μM. The redox discontinuity layer (RPD), where anoxic conditions replace aerobic processes, was 0 cm at the farm sites compared to 9.9 cm at the reference sites, indicating anoxic conditions at the sediment surface. A digital camera was used to photograph the sediment cores and image analysis software was used to classify the geochemical state of each sample. The results of the two methodologies were in close agreement with each other as well as with biological classification conducted simultaneously. While image analysis may not be applicable in all sediment types, it does provide a less expensive alternative for assessing sediment impacts below farms located over soft sediments. The use of SPI to assess benthic enrichment has been verified by researchers in the Mediterranean as well (Holmer et al. 2002, Holmer et al. 2005) and acoustic assessment of organic enrichment has been conducted at salmon farms in the Bay of Fundy (Wildish et al. 2004). Backscatter images from multibeam and sidescan acoustic surveys were able to detect the footprints of sediment enrichment below active cages. Ground truthing from two years of surveys confirmed that sediment below the fish farms was enriched. Average redox potential was -148 mV (in 2001) and -111 mV (in 2002) compared to 158 and 100 mV at reference sites. Sulfide levels at the farm sites were 30,000 and 2500 μM compared to 1300 and 350 μM at reference sites. Abandoned cage sites were identifiable on the images by lower backscatter.

Monitoring of benthic chemical remediation at intensive salmon farms in British Columbia demonstrated that high levels of sulfides (>6000 μmol) and decreased redox potential were ubiquitously measurable at the perimeter of the net pens (Brooks et al. 2003). However, sediment recovery occurred in a short period (six months or less) following harvest. These results suggest that, at adequately flushed sites, the effects of even high production farms may be relatively ephemeral with appropriate fallowing. In contrast, another remediation study (Brooks et al. 2004) conducted at a nearby salmon farm, used regression analysis to estimate a remediation time of 65 months for that site. Average redox potential was -124 mV, sulfide level was 1349 μM, the total volatile solids was 13%, and RPD was initially only 1.2 cm. Collectively these indicate an enriched anaerobic benthos with effects generally confined to within 100 m of the pen sites. Repeated sampling indicated steady, but slow recovery. Yet, chemical remediation of this site took much longer compared with other salmon farms in the area. The explanation for this is uncertain, but the authors suggest that siting over depositional bottoms with high silt and clay content or other sediment geochemical processes could be contributing factors. Similarly, the fraction of small particle size sediments (<63 μm fines including silt and clay) was correlated with redox potential and anaerobic sediment condition as part of an effort by Chou et al. (2004) to refine regression models used by the New Brunswick environmental monitoring program to assess sediment impacts from marine aquaculture sites.

Anderson et al. (2005) found correlations of lower redox potential and higher sulfides with organic enrichment under salmon farms in Newfoundland. However, these relationships were not consistently evident at reference sites and other shellfish aquaculture locations, possibly due to effects of low temperature on organic matter decomposition and sulfate reduction. Therefore, the authors concluded that in Newfoundland site-specific comparisons must account for wide variability in temperature dependent sediment characteristics if singular measurements of redox potential or sulfide levels are to be used as indicators of benthic enrichment.

Other researchers found sediment sulfide concentrations to be good indicators of enrichment (Chamberlain and Stucchi 2007). In this study at fish farms in British Columbia, sulfide concentrations were consistently high at farm sites sampled over an entire production cycle. Sulfide levels rose from <200 μM near the edge of the farm at the outset of sampling, to a final level of 7870 μM beneath the cage site. In general, the highest impacts were measurable within 125 m of the farm.

The potential far-field effect of salmon aquaculture has been investigated in Canada. A study in the Letang Region of New Brunswick analyzed core and sediment samples from 1990–2002 from three interconnected bays with various degrees of bottom stress (Milligan and Law 2005). They found that the floc limit, an indicator of fine-grained particulate matter deposition, increased over the 10–20 year period when salmon farming expanded in the area. The researchers found that over large areas, local hydrodynamics may prevent accumulation of sediment directly beneath fish cages, yet may direct fine organic matter particulates to the surrounding environment.

In the U.S., much concern about benthic sediment impacts of salmon farms has been raised. Nash (2001, 2003) and Nash et al. (2005) identify the effects of bio-deposits on sediment chemistry, including redox and free sulfides, as one of the highest environmental risk issues for salmon cage culture. Sites with inadequate circulation and poor farm management practices are more likely to have depositional rates that exceed the ability of the sediment to break down organic matter. Nash reported sedimentation rates at salmon farms ranging from 15.1-100 g of total volatile solids/ m^2/day, a measure of organic loading. At a multi-species fish farm in the Gulf of Maine, sediment sampling found no differences in redox potential measurements at the cage sites compared with intermediate and remote control sites (Langan 2007). Increased organic matter accumulation and benthic carbon flux were measured directly beneath cages at a salmon farm in Maine, but these effects were not measurable 10 m beyond the pen (Findlay et al. (1995). In this case, storm activity was thought to be an important contributor to re-suspension of sediments.

In the temperate waters of Baja California, redox potential of sediments sampled 250 m from a bluefin tuna ranching farm ranged between -110 to -302 mV (Diaz-Castaneda and Valenzuela-Solano 2009). This was the nearest to the pen that samples could be collected, so impacts at the cage sites are unknown. Feeding practices, which include the use of raw fish as the primary diet at the farm are thought to input large quantities of organic matter, and despite favorable hydrodynamics, sediment impacts appeared to be significant.

South America

Mulsow et al. (2006) used SPI analysis and micro- electrode technologies to assess the impact sediment geochemistry impacts at salmon farms in two fjords in southern Chile. Generally, both approaches led to similar conclusions regarding impact levels. At farms in the smaller, enclosed, depositional fjord site, the layer of uneaten food was 0.74 cm, the RPD depth was 0-4 cm, redox reached 0 between 4-7 mm below the sediment surface and hydrogen sulfide increased only a few millimeters below the surface. In comparison, at cage sites in the larger, better flushed fjord, the food layer was minimal even directly at the cages (.05-.07 cm), the RPD ranged from 1.7-7.87 cm, redox values decreased slowly with sediment depth and hydrogen sulfide levels never exceeded normal values. A large-scale comparison of 29 active salmon farm sites in Chile found few differences in sediment surface chemistry between farm and control sites (Soto and Norambuena 2004), including oxygen concentration, pH, and redox potential. However, change in redox potential between the water-sediment interface was found to be significantly different, with farm sites showing a -109.8 mV reduction in redox potential compared to a +2.6 mV change at control sites. Additionally, percent organic matter was higher at farm (4.41%) than control (2.09%) sites. The authors concluded that their findings suggested that salmon farming did impact sediment chemistry, but that generally enrichment from farms was a local phenomenon determined largely by site conditions, was confined to the cage shadow and had no broader ecosystem impact.

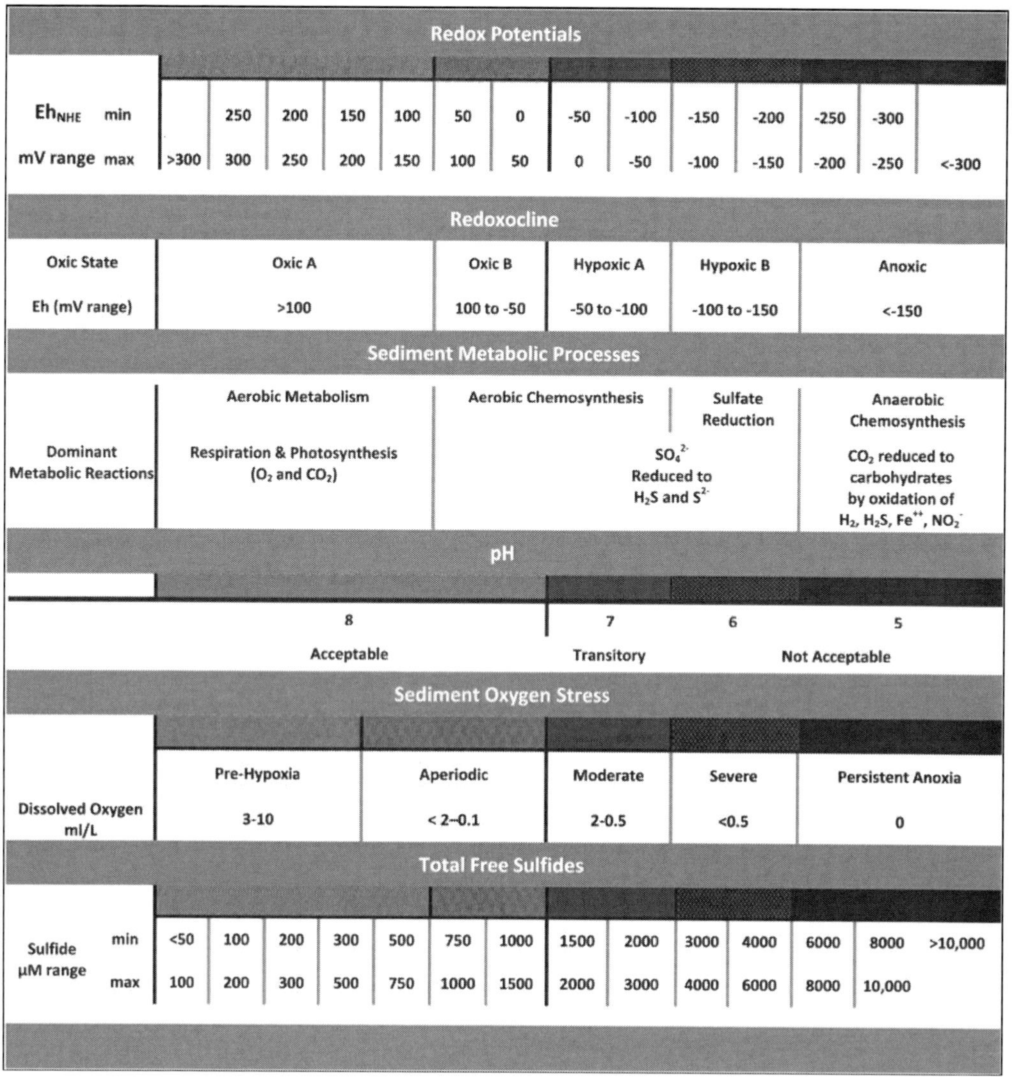

Figure 1. Relationships between biogeochemical processes in marine sediments. Adapted from Hargrave et al. (2008).

Northern Europe

Work in Europe is underway to improve monitoring and management of benthic sediments below fish farms. Sampling at a Norwegian salmon farm showed redox values down to -126 mV beneath the cages and out to 227 m (Kutti et al. 2007b). Redox values near the cages tended to decrease during the production cycle, while redox was always positive at reference stations. A comparative study was conducted in Norway (Carroll et al. 2003) to determine the ability of four different monitoring methods to detect benthic impacts at five farm sites. At all farms, sediment chemistry measurements including decreased redox potential (-125 to -184 mV) and pH (6.13-7.16) indicated significant enrichment below and within 20 m of the fish cages. At all sites, partial or full recovery was generally evident at 50 m, compared to samples from reference sites. The other sampling methodologies were faunal surveys, sediment profile image analysis and diver surveys including photographic data.

Results of all four methods agreed that the direct impact zone was adjacent to the cages, but some differences existed in defining more subtle effects. Another Norwegian study describes the integration of multiple parameters, including sediment chemistry, to monitor farm sediment conditions and promote long-term viability of fish farming sites (Hansen et al. 2001). Site condition was evaluated based upon a scoring system taking into account sedimentation rates, chemical analysis (redox potential, pH, and gas production) and biodiversity. Scores from all parameters were included in the final determination of benthic condition. In terms of geochemistry, sediment condition was rated as impacted once redox potential approached zero and pH went below 7.7. This integrated approach to an environmental monitoring system is part of the regulatory process, and continues to be evaluated with the goal of decreasing sampling cost and effort while maintain data quality and accuracy.

Using 15 years of Norwegian sampling data, Schaanning and Hansen (2005) determined that graphic plots of electrode measurements of redox potential versus pH or sulfide provide a reliable visual tool for the assessment of enrichment condition. This methodology is also cost effective and can be implemented in many types of benthic substrates. Their investigation found highly variable sediment conditions beneath farms. Mean redox potential below cages was -161 mV compared to +82 mV at reference sites, and pH was 7.22 compared to 7.85. Based upon plotted data the authors suggested that sediment redox potential above -100 mV in combination with pH above 7.1 represented acceptable sediment conditions. Methane gas ebullition was consistently present when sediment pH dropped below 7.1. At most farm sites, impacts were limited to the immediate vicinity (within 100 m) of the farms. Siting in high current speeds, deep water and maintaining space between cages were identified as factors to avoid long-term enrichment. Similarly, a study in Finland found that farms sited in a shallow bay with less water exchange were prone to organic sedimentation and hypoxia (Kraufvelin et al. 2001). In a Danish fjord, sediment sampling below estuarine rainbow trout *Oncorhynchus mykiss* farms found highly increased oxygen flux (300 mmol/m^2/ day) due to organic matter loading (Christensen et al. 2000). Effects were limited to 100 m from the cage sites. Mineralization of food and waste was also documented, indicating that the organic matter was being broken down. This input of nitrogen and phosphorus to the water column from sediment efflux was noted as an important source of nutrients to this system, particularly during summer months when land-derived nutrient loading for primary production is decreased.

In Scotland, models are being refined to predict key biogeochemical processes and sediment recovery times at farm sites (Black et al. (2012). DEPOMOD is the model currently used to track organic waste dispersion from fish farms. Modeling output and monitoring data from five operational farms were used to first test a module for estimating sediment recovery times as a function of organic matter decomposition rates based upon organic matter accumulation, microbial oxygen demand and water column oxygen supply ratios (using a Findlay-Watling approach). Although this method accurately modeled impacts during the operational phase, it did not adequately predict recovery time. Next the Respiratory Quotient (RQ) box model was implemented which, in addition to oxygen demand, accounted for background levels and resuspension of organic matter, and sulfide production and oxygen demand from reductive processes. This model better predicted recovery rates of sediments at the five farms, and modeled recovery rates correlated well with biological indices.

Australasia

A study in New Zealand recently analyzed 12 years of monitoring data at five salmon farms in low and high flow regimes to evaluate the reliability of several indicators as predictors of benthic enrichment (Keeley et al. (2012). Both decreasing redox and increasing sulfide correlated well with progressive enrichment and anoxia, especially under low flow conditions. Also in New Zealand, sediments from active farms, abandoned sites and control areas were sampled as part of a larger project validating a benthic impact model (Morrisey et al. 2000). Sedimentation rates at the active farm site were up to 63 $g/m^2/day$, compared to <13.5 $g/m^2/$ day at control sites. Benthic flux of oxygen, nitrogen and sulfur indicated that sediments were in an enriched state. Sediment profiles at the disused sites suggested extended recovery time (3.3-12 years), because low current velocities limited recovery to *in situ* decomposition with little resuspension or hydrodynamic dispersal. Similarly, sedimentation rates at a South Australian tuna ranch ranged between 8-92 g dry weight/m^2/day at farm sites, compared to control sites with lower sedimentation rates between 7-28 g dry weight/m^2/day (Lauer et al. 2009). Ammonium and phosphate fluxes were also significantly increased reflecting mineralization rates consistent with benthic nutrient enrichment. Because these benthic impacts lessened after harvest, the results support continuation of government fallowing requirements.

Wildish et al. (2003) validated the use of sediment profile imaging (SPI) at fish farms in Tasmania. Here redox potential was decreased from 311 mV at reference locations to -7 mV below cages and sulfide was increased from 0.01 μM to 49 μM. The benthic enrichment index values derived from the SPI results were similar to the geochemical monitoring method for reference sites and acceptable for most monitoring applications in soft sediments. However, SPI was thought to be less useful in benthic habitats in course sediments or rocky substrates. Sediment condition tested at a Tasmanian salmon farm was degraded with elevated sulfide levels >350 μM and organic matter enrichment directly beneath cages (Macleod et al. 2004). Effects dissipated with increasing distance from the cage and over a 36 month recovery period. Similar results were reported at a nearby farm where redox potential beneath salmon cages was significantly lower than at reference sites (Edgar et al. 2005). Impacted sediments (down to 40 mm) had redox values below -100 mV while reference sites measured around 200 mV at the sediment surface. This effect was generally limited to within 35 m of the cages. A follow up study of 42 sites (Edgar et al. (2010) confirmed that decreased sediment redox potential was measured in sediments below cages at the sediment surface (about 100 mV decrease) and down 40 mm (about 60 mV decrease). Sediment redox potential measurements at sites at least 35 m distant from the farms were intermediate between the impacted and reference sites.

Sub-Tropical Regions

The Mediterranean

Since 2000, a great deal of research has been done on benthic impacts at Mediterranean fish farms. Vita et al. (2004) measured increased organic content in the sediments from a Spanish sea bream and sea bass farm with nearly 6% beneath cages compared to 3% at control sites. Sediment sulfide was increased at the cage sites (about 750 mg/kg) compared to control sites (500 mg/kg). The experiment at this site excluded wild fish from feeding on the bottom. The results revealed that without fish feeding on the waste particulates, benthic organic matter content was about 7.5% and sulfides were 1500 mg/kg, indicating that wild fishes may be an important contributor to the degradation of farm waste. In fact, this study estimated that about 80% of the sinking particles were consumed by wild fish within 4 m below the net pen.

Another study at fish farms along the Spanish Mediterranean coast did not discern any differences in sediment organic matter levels (Maldonado et al. 2005). The authors suggest this was due to a combination of moderate farm size and siting in semi-exposed environments. In contrast, results reported by Aguado-Giménez and García-García (2004) from a one year pilot study at an offshore sea bream and sea bass farm found an increase of organic matter levels of just below 5% out to 500 m from the farm site, compared to values around 3% for control sites. An intensive study at a sea bream and meagre *Argyrosomus regius* farm in Spain compared two monitoring techniques, diver sampling versus grab sampling (Aguado-Gimenez et al. 2007). The two methods provided similar results for all but one of the geochemical parameters tested. The only exception was redox potential which was always less electronegative and had greater variability in sediments collected with the grab sampler, possibly due to inadvertent mixing of the sediment layers during collection. The diver-collected samples showed redox potential of nearly -300 mV under the cages, increasing with distance out to 500 m from the cage. Sediment sulfide levels of 0.5-0.6 ppm were found at the cage sites compared to levels near 0 ppm at 100-550 m from the cages. The cost-effectiveness analysis found that staff and equipment costs for the diver surveys were 6 and 1.2 times higher, respectively, than for the grab sampling method.

Studies in Italy have also provided a variety of information about benthic impacts in Mediterranean fish farms. For example, LaRosa et al. (2001) measured the sediment RPD below fish farms off western Italy. RPD below cages was around 2 cm at control sites, but only 0.2 cm during farm operations. Following harvest, RPD increased to 1.9 cm within a few months indicating quick recovery from enrichment. Sampling at this same farm as part of a different project found similar RPD impacts (Mirto et al. 2002). At a more northerly sea bream farm site, Vezzulli e al. (2003) found reducing conditions in surface sediments below the farm while positive redox potential values were found at control sites.

Redox potential measured at farm sites off Sicily was around +200 mV compared with control site values of +500 mV (Vezzulli et al. 2004). A study at sea bass and sea bream farms off Corsica found organic matter content of 21-24% at farm sites (to 100 m from the cages) compared to 2% at a nearby control site. Accumulation of organic farm waste was attributed to poor flushing. Sara et al. (2004) used stable isotope analysis to study sediment nutrient inputs including fish farming off the coast of Sicily. Samples were collected beneath sea bream and sea bass cages, up to 1000 m away and at control sites. The carbon ($\delta^{13}C$) signal was not significantly different along the distance gradient. However, the nitrogen ($\delta^{15}N$) isotope signature from the farm wastes was evident in the sediment and particulates up to at least 300 m away. The authors proposed that resuspension of sediment and feeding by wild fishes may play a role in the wider distribution of nutrients away from the immediate cage area. Another study off western Italy evaluated which geochemical and hydrodynamic parameters were most useful in assessing environmental impacts of fish farms (Porrello et al. 2005). Redox potential and sulfide levels were among the most useful for detecting enrichment through time and along distance gradients. The analysis also found that most geochemical effects were confined to within 50 m of cages. Vezzulli et al. (2008) reported lower redox potential values (-70 mV) in the upper (0-2 cm) sediment layer beneath a tuna fattening farm compared to samples from a control site (90 mV).

Studies in the eastern Mediterranean report findings similar to the western portion. Sediment profiling imagery data collected beneath a fish farm in the Ionian Sea was compared to geochemical analysis by Karakassis et al. (2002). The SPI results were well coordinated with the geochemical ones, validating the use of this methodology for enrichment assessments. This study did not report actual values for the geochemical parameters but the authors confirmed that enrichment, measured with both methods, was confined to within 10 m of the cages. Because of the lower cost and rapidity of image analysis associated with using SPI, it is seen as a viable alternative to traditional benthic monitoring techniques.

Belias et al. (2005) reported particulate matter sedimentation rates below Greek fish cages to be five times greater than at control sites. Despite good circulation at this site, about 30 cm of loose organic matter was observed at the sediment surface. It was reported that about half of the food at the farms is lost as particulate waste. Neofitou et al. (2010) found that organic matter concentration was higher at two semi-enclosed sea bream and sea bass farm sites in the Greek Aegean Sea (7.2 and 16.8%) than at control sites (2.6 and 10.1%). In both cases, sampling 50 m away found intermediate concentrations. Benthic sampling at seven Greek fish farms over 16 months resulted in consistently negative redox potential (around -150 mV) just beneath cages, but showed an increasing trend toward positive values within 25 m at most farms (Lampadariou et al. 2008).

As part of a waste dispersal model validation study, Jusup et al. (2009) measured sedimentation below sea bass and sea bream farms in Croatia. The sedimentation rate was around 50 g dry weight/m^2/ day directly under the cages and decreased gradually to around 9 g dry weight/m^2/day at 40 m from the cages. At a nearby Slovenian sea bass farm, Kovac et al. (2004) measured somewhat lower sedimentation rates of 31 g dry weight/m^2/day, compared with 3 g dry weight/m^2/day at local controls sites.

In contrast to the above Mediterranean studies, Apostolaki et al. (2007) analyzed sediments at farm sites off Greece and Italy finding only positive redox potential values and no difference in redox potential compared to control sites. Similarly, Matijevic et al. (2009) did not find significant changes in sediment redox potential, organic matter or sediment grain size beneath sea bass and sea bream farms in the Adriatic Sea. The use of enclosed fish to reduce benthic impacts was assessed in Israel (Katz et al. 2002). After 70 days of allowing gray mullet in small bottom pens to feed on waste below sea bream cages, several indicators of enrichment showed that the detritivores had a positive impact. Sediment sulfide levels and sediment oxygen demand decreased by 85% and 31%. The enriched sediment layer was dramatically reduced by 5 cm within the enclosures due to ingestion and resuspension by the mullet. These results have promising implications for both remediation of enriched sediments and development of commercially viable integrated multi-trophic aquaculture practices.

At gilthead sea bream farms in the Canary Islands, no difference in organic matter concentrations were detected between farm and reference sites (Molina Dominguez et al. 2001) with considerable season variability at all sites. Similar results were found by Riera et al. (2011) at other sea bream and sea bass farms. Organic matter concentrations were similarly low at both farm (0.45%) and control (0.38%) stations. Farm areas in both studies are characterized by good currents (6 and 12 cm/sec, respectively).

Asia

In Japan, Pawar et al. (2002) collected sediments from sea bream and sea bass farms to study the correlation between organic farm input, sediment quality, seasonal variation and hydrographic conditions. At all sites the organic matter (up to 12%) and sulfide levels (up to 1.8 mg dry weight/g) increased, and redox potential decreased (to <-175) with increasing organic loading from farms. This trend was consistent between the two farm areas surveyed and variation was attributed primarily to seasonal changes in feed input and temperature, rather than background sedimentation processes. Another Japanese study investigated sediment impacts at 22 sea bream and yellowtail farms (Yokoyama 2003). Elevated sulfide levels up to 2 mg/g dry weight were measured in sediments directly beneath the cages, dropping to 0.75 mg/g at 100-500 m from the cages. This level of enrichment was still below the critical value of 2.5 mg/g established as a regulatory threshold.

Tropical Regions

Pacific Islands and the Caribbean

Lee et al. (2006) measured sediment changes at Pacific threadfin *Polydactylus sexfilis* cages in Hawaii. Sediment grain size, an indication of siltation, did not appear to differ between farm and control sites, but redox values at cage sites were consistently lower than controls. Collectors discerned sulfidic odors at the cage site, suggesting reduced conditions. Benthic sampling over three years at a Hawaiian yellowtail farm documented negative redox potential readings only once (Sarver 2009). Following improvements in feeding techniques, organic matter input was reduced and redox levels rose steadily over the next year and eventually reached levels of nearby control stations.

Benthic sampling at a cobia farm in deep water with good flushing in Puerto Rico also found no difference in sediment organic matter (ranging from 4-6.2%) between cage and control sites suggesting there was no enrichment at this farm (Alston et al. 2005).

Asia

In the Philippines, organic matter from a milkfish net pen farm resulted in significantly increased sedimentation beneath the cages over a production cycle (Holmer et al. 2002). Enrichment at these shallow pens was higher compared to nearby cage farms at greater depths.

Meta Analyses

Two comprehensive reviews include analyses of the enrichment effects below marine fish cages. Kalantzi and Karakassis (2006) reviewed 41 papers covering a wide range of cultured species, habitats, site characteristics and farm management practices. This paper contains a great deal of information about the relationships between enrichment sediment variables resulting from nutrient and particle input from farms. The overarching conclusions from this work were that: 1) the most popular geochemical variables studied (e.g., organic material content, sedimentation rates, oxygen consumption and redox potential) show consistent patterns along enrichment gradients regardless of site and farm characteristics, 2) increase in water depth below the farm allows for greater dispersion of waste and improved sediment quality, 3) latitudinal differences in background productivity and sedimentation must be considered for monitoring and evaluation, and 4) the sediment type under the farm is a major contributing factor in determining the extent and severity of impacts. The authors state that applying common standards over large geographic areas is challenging due to the complex interplay of site characteristics. In general, however, their analysis suggests that the impact radius at fish farms decreases with high depth, at low latitudes and over fine sediment.

Giles (2008) developed a Bayesian network based upon 64 studies conducted at fish farms in temperate regions to quantitatively assess the relationships between environmental impact parameters, site characteristics and farm production measures. This study provides a useful analysis of the interconnectedness of farm management practices and potential impacts to the benthic environment. Sediment sulfide, redox potential, sediment oxygen consumption and nitrogen mineralization were determined to be among the most consistent and sensitive geochemical measures of benthic impacts. These parameters have the added benefit of being relatively inexpensive and applicable in multiple sediment types. The analysis found that impacts were usually confined to within 40-70 m of the farms and further confirmed that well-flushed sites in deep water tend to have significantly less impact than shallow, low-current farms. The analysis was not intended to predict the exact environmental responses, but Giles provides suggestions for its use as a decision making support tool.

Benthic enrichment impacts are often identified as the most significant and concerning environmental impact of marine aquaculture because the geochemical consequences of enrichment in the sediment can be long lasting. The trends observed worldwide reflect recurring impact patterns in benthic sediments beneath marine fish cages. In the absence of sufficient flushing or under high fish densities organic matter accumulation leads to increased microbial activity, decreasing oxygen, a shift toward reducing conditions and increase in sulfide and mineralization levels. Generally, these impacts are confined to within 100 m of the cage or farm area. A few studies are illuminating the potential far-field impacts in areas that have experienced high levels of marine aquaculture development. Research focusing upon larger scales will likely become a priority, especially in coastal areas with multiple concerned user groups. Monitoring and research to quantify downstream, far-field and long-term effects of fish farms beyond the immediate cage perimeter will continue to be important. The use of stable isotopes as tracers of farm waste output at larger spatial and temporal scales is a promising tool to help in this area. Continued effort to build upon recent work comparing different monitoring technologies and protocols to provide reliable, accurate and cost effective methods of assessing enrichment and biogeochemical impacts will be benefical.

Lacking are monitoring and assessment methods for hard bottom marine habitats, as collection of sediment may not be possible. There is little information available about marine fish farm organic discharge over hard bottom habitats. These are generally more dispersive areas with high currents and high benthic shear, and are thus less likely to exhibit accumulation of organic matter. In these cases, minimal effects may be evident immediately beneath fish cages. As marine aquaculture operations move offshore into the open ocean it is expected that organic waste will be more rapidly and broadly dispersed compared to sites in nearshore waters (Holmer 2010). Impacts in deep sediments will certainly be an area for future research as the industry expands.

SECTION 3. MARINE LIFE

The previous sections addressed a wide range of the possible physical and chemical effects of cage aquaculture on the marine environment. While some of those may have direct effects to the surrounding waters, another predominant concern is the secondary impacts to local marine life resulting from the nutrient rich effluents released from the farms. Most often, it is the benthic community directly below cages that reflects the greatest alteration. The potential impacts to other marine life near fish cages — including plants, wild fish, sharks, sea turtles, marine mammals and sea birds — are also of interest. Another area of concern is possible impacts to sensitive marine ecosystems like seagrass beds, coral reefs and mangrove forests. Research is also emerging about sites where moderate levels of nutrient discharge from fish farms may be having a beneficial effect on surrounding biodiversity particularly in nutrient limited aquatic ecosystems. In this section we review recent studies that address these issues.

The potential to cause severe and long-term damage to benthic and pelagic marine life is often identified as a concern of marine managers, scientists and regulatory entities. This topic has been identified as a priority in the U.S. (Goldburg et al. 2001, Stickney 2002, Nash et al. 2005, Pittenger et al. 2007, Johnson et al. 2008, Ocean Conservancy 2011), Canada (Hargrave 2003, Wildish et al. 2004a, Strain and Hargrave 2005), South America (Buschmann et al. 2009), Europe (Black et al. 2002, International Council for the Exploration of the Seas 2002, Huntington et al. 2006, Holmer et al. 2008, Tett 2008, Holmer 2010), the Mediterranean (International Union for Conservation of Nature 2007, Borg et al. (2011, Grigorakis and Rigos 2011) and globally (Beveridge 2004, Halwart et al. 2007, Tucker and Hargreaves 2008, Hall et al. (2011).

Primary Producers

In the **Water Quality** section recent research on the release of dissolved nutrients, predominantly nitrogen and phosphorus, from marine fish farms into the water column was summarized. Increases in nutrients have been documented at some farm sites, but not at others. At sites with increases, elevated nitrogen and phosphorus are generally measurable less than 100 m from the farm. The primary determining factor affecting such nutrient increases are farm management practices which optimize feeding and siting the farms in well flushed areas. Farms in enclosed bays and shallow water are more likely to contribute to nutrification of nearby waters. Strong currents quickly dilute the nutrients and disperse them into the surrounding environment. The potential secondary effects of these nutrients to primary production are of interest. Because marine waters are often nutrient limited, it is possible that elevated nitrogen and phosphorus may contribute to increases in phytoplankton and macroalgal production (Cloern 2001). In most marine waters, nitrogen is the limiting nutrient, but there are waters (some estuarine and tropical waters, for example) where phosphorus or abiotic seasonal factors like light are more influential on driving primary productivity.

In this section we review current research investigating the potential links between aquaculture effluent and increased primary production. Special attention is given to the issue of harmful algal blooms (HABs), both potential causes of HABs by fish farms as well as potential impacts on fish farms of HABs. Nutrient enrichment and potential eutrophication related to marine fish farm effluent are an important concern raised in the U.S. (Goldburg et al. 2001, Nash 2001, Nash et al. 2005, Pittenger et al. 2007, Johnson et al. 2008), Canada (Hargrave 2003, Strain and Hargrave 2005), South America (Costa-Pierce et al. 2007, Buschmann et al. 2009), Europe (Black et al. 2002, International Council for the Exploration of the Seas 2002, Huntington et al. 2006, Holmer et al. 2008, Olsen et al. 2008, Tett 2008, Holmer 2010), the Mediterranean (International Union for Conservation of Nature 2007, The Mediterranean Science Commission 2007, Borg et al. (2011) and globally (Beveridge 2004, Halwart et al. 2007, Tucker and Hargreaves 2008).

The rise in nutrification and eutrophication from anthropogenic sources has been documented around the world, but it is difficult to parse out the relative contributions from specific activities such as aquaculture (Wu 1995, Pearson and Black 2001). There are few long-term data series of nutrient levels and primary productivity in coastal waters, but there are data indicating that nitrogen and phosphorus fluxes in coastal waters have increased by an estimated 2-14 times the natural rates (Cloern 2001). Only a small portion of that can be attributed to marine cage culture, the dominant sources being terrestrial. While fish farms may contribute to coastal seas eutrophication, the nutrient budgets of open oceans are difficult and complex systems to quantify and there is a lack of inventory of contributing factors in most locations worldwide. Questions remain about the types and levels of risk that cage culture poses for water column enrichment and eutrophication (Olsen et al. 2008). There are cases in which primary production increases near marine cages, but many studies have not detected a chlorophyll response tied to nutrient loading from farms, perhaps due to dispersion or rapid assimilation of nitrogen and phosphorus into the food web, especially in oligotrophic waters (Braaten 2007, Holmer et al. 2008, Holmer 2010). In this section we first review studies which documented impacts to primary production, followed by cases in where no effects were found.

There is evidence that nutrification at fish farms can lead to increased primary productivity.

Temperate Regions

Robinson et al. (2005) studied the far-field effects to primary production of salmon farm effluent in the Bay of Fundy. Here, eutrophication linked to farm effluent (1 km from the intertidal zone) resulted in the growth of extensive algal mats along the shoreline, threatening a native soft clam fishery. The estimated economic direct loss of clams at one site was calculated at $168,000 (Canadian dollars).

The appearance of these algal mats in the mid- 1990s coincided with the development of salmon farming in nearby waters. Elevated zinc levels in shoreline sediment cores supported the hypothesis that nutrients from the farm were being transported near the shore. Aerial photography showed that at times the mats covered up to 40% of the beach for an estimated algal biomass production of 58.6 tons. Rensel and Forster (2007) documented over 29 species of algae growing on net pens, floats and anchor lines at a salmon farm in Puget Sound. Algal growth was linked to the farm effluent by nitrogen and carbon isotope analysis. The authors consider such bio-colonization an environmental benefit to the marine ecosystem. Some colonizing algae could have commercial value.

Photo courtesy of NOAA.

Honkanen and Helminen (2000), Nordvarg and Johansson (2002) found increased levels of planktonic chlorophyll-*a* near seven fish farms in two Finnish straits. Samples collected from the water column and growth plates showed increased algal productivity nearest the farms. Interestingly, neither dissolved nitrogen nor phosphorus varied consistently with distance from the cages. In the Baltic Sea, Nordvarg and Johansson (2002) sampled water quality and deployed algal growth plates at ten salmon farm areas and four control areas to assess farm nutrification and primary productivity response. The presence of fish farms increased both chlorophyll-a levels and periphyton growth at only a few locations. Local hydrology was thought to be the most influential factor in determining whether an effect on primary production occurred. Affected sites were semi-enclosed, shallow bays. The data were further used to develop and validate a model to predict nutrient and chlorophyll-*a* dynamics in the area.

Sub-Tropical Regions

A study in Sicily found that chlorophyll-*a* levels in water as far as 1000m from fish cages were up to 25 times higher than at five control sites (Modica et al. 2006). However, all the chlorophyll levels measured were well below the eutrophication threshold for that area and the researchers concluded that the organic enrichment from the farm was not likely to result in undesirable biological consequences. A carbon isotope analysis of seagrass and brown algae collected near the outfall of an Italian land-based fish farm and non-impacted control areas found that plants closest to the outfall sequestered significantly more $\delta^{15}N$ associated with the farm effluent than did plants up to 2 km away (Vizzini et al. 2005). However, differences in the $\delta^{13}C$ isotope signature were not as clearly evident. While this study was not conducted at a cage farm, it provides insight into nearshore nutrient processes and also provided information for a follow up study. Vizzini and Mazzola (2006) used stable isotope ratios to compare the impact of anthropogenic organic matter from onshore and offshore fish farming and a sewage outfall on seagrass and two green algae. As in the previous study, $\delta^{15}N$ was the better tracer for nutrification, showing a distinct pattern of increased isotope uptake near the fish and sewage effluent sources compared to control sites two and six km away. The effluent isotope signature was also evident at higher trophic levels, indicating local nutrient assimilation into the ecosystem. This methodology was not able to discriminate the exact anthropogenic source of enrichment in the sampled plants.

In the eastern Mediterranean, Pitta et al (2005) analyzed sites near (2-3 nautical miles) and distant from (20 nautical miles) fish farms to determine if nutrient enrichment from them caused large-scale effects on water quality and plankton assemblages. Chlorophyll-*a* was significantly and consistently increased at the near sites, but significant seasonal and regional variability was also observed. The authors suggest that there was rapid utilization of nutrients by plankton organisms, coupled with a transfer of these nutrients up the food web without leaving behind significant traces of eutrophication. There was evidence of increased chlorophyll-*a* production below the thermocline, suggesting that resuspension of nutrients, especially phosphorus in this area, was an important process. Given the large area sampled, it appears that nutrient flux out of these farms was evident at scales larger than many impact studies. Later, Pitta et al. (2009) used dialysis bag experiments to demonstrate that nutrients from sea bass and sea bream farms were being transferred up the food chain by phytoplankton grazers. In bags containing water filtered to remove grazers, chlorophyll-*a* and nitrogen levels were higher near the farms, and decreased out to 500 m. In bags with grazers, chlorophyll-a was lower, presumably due to planktivory. The authors suggest that in oligotrophic waters, the quick transfer of nutrients to higher trophic levels is possible. Such fertilization by fish farms is sometimes seen as a positive contribution of fish farms to nutrient poor marine environments.

Apostolaki et al. (2007) found varying effects of effluent on chlorophyll-*a* at three Mediterranean sea bream and sea bass farms – only one farm showed increasing productivity while the other two showed a negative influence or no effect. This suggests that the site-specific interplay of a variety of water column characteristics is important in determining ecological outcomes.

Tropical Region

A water quality study in the South China Sea found that the highest chlorophyll-*a* levels in a semi-enclosed bay were in the area of high cage aquaculture with an average of 11.74 mg/ m3 compared with levels as low as 3.81 mg/m^3 elsewhere in the bay (Song et al. 2004). In Taiwan, Huang et al. (2011) found chlorophyll-*a* levels ranging from 1.46-1.75 µg/l in a semi-enclosed bay with well-established small scale fish cage culture compared to 1.0-1.37 µg/l at a nearby reference site. Macroalgae coverage at farm sites was 2-10 times higher than at reference locations. Nutrient measurements were above eutrophication threshold values. Hydrological conditions and the use of raw fish as feed were cited as the main causative factors.

Temperate Regions

Two years of extensive sampling near salmon farms in three New Brunswick bays found no increase in chlorophyll concentrations compared with control sites (Harrison et al. 2005). This was explained by the strong tidal mixing in this area. It is possible that light, rather than nutrient loading, is the limiting factor driving primary production in this bay. Likewise, monitoring in Blue Hill Bay, Maine also did not show increased chlorophyll concentrations in proximity to fish farms (Sowles 2005). An evaluation of 43 salmon farm sites in Chile did not detect any effect of farm effluent on chlorophyll levels (Soto and Norambuena 2004) compared to control sites. This analysis included farms at nine locations of varying ages and production levels. The farms are located in deep water (15-94 m) and rapid flushing of nutrients likely explained the lack of an effect to primary production.

Water sampling at a salmon farm in Scotland detected increased nutrient levels at the farm, but no difference in chlorophyll-*a* (Navarro et al. 2008) compared to a control site 650 m away. These results supported conclusions from previous studies of no clear effect of farm nutrient enrichment on primary productivity. In fact, fish farming is considered to pose little risk of increased primary production in most Scottish waters, apart from a few heavily loaded sea lochs (Black et al. 2002). Heath et al. (2003) applied the European Regional Seas Ecosystem model (ERSEM) to identify areas of Scotland's maritime regions that may be at risk of eutrophication, with special focus on the contribution of salmon farms to nutrient loading. They concluded that the nutrients from aquaculture had no discernible eutrophication impacts on the west and north coastal and offshore waters. Recently, Tett et al. (2011) tested a computational physical- biological model, ACExR-LESV, for estimating the aquaculture carrying and assimilative capacities of Scottish fjords. Modeling output from a simulated loch was compared to water quality indicators and standards to decide how many fish farms the loch could support sustainably. The modeled carrying capacity was strongly driven by light penetration and circulation patterns that minimized chlorophyll production. This model requires further refinement before it can be reliably applied.

Sub-Tropical Regions

A study in Spain compared chlorophyll-*a* levels in the sediment below five sea bass and sea bream fish farms (Maldonado et al. 2005). Overall chlorophyll levels remained low (<2 μ/g) and, while some variability between farms was evident, there were no detectable differences between farm and reference sites. Vezzulli et al. (2008) tested chlorophyll-*a* levels in the water and sediments at a bluefin tuna farm in Italy. Only seasonal differences in concentrations were detected and the authors conclude that the water currents and depth at the farm site were adequate for preventing impacts to primary production. No eutrophication effect was found at a sea bass and sea bream farm in Greece (Belias et al. 2003), possibly due to overall low nutrient levels. However, the authors note that the alga *Ulva lactuca* has recently been found in several parts of the Astakos Gulf, possibly signaling increased nutrient loading. Demirak et al. (2006) found no difference in chlorophyll-*a* level at seven Turkish fish farm sites compared to reference sites. Basaran et al. (2010) did not detect a significant difference in chlorophyll levels at eight sea bream and sea bass farms off Turkey compared to a control station. Likewise, long-term monitoring at a tuna farm in Turkey exhibited no differences in chlorophyll-*a* at the farm site compared with reference sites (Aksu et al. (2010).

Tropical Region

In the U.S., monitoring at a cobia farm off Puerto Rico found no differences in chlorophyll-*a* concentrations at the cage versus control sites (Alston et al. 2005). Similar monitoring results were reported at an experimental cage in the Bahamas (Benetti et al. 2005) and a Hawaiian moi *Polydactylus sexfilis* farm (Helsley 2007).

Harmful Algal Blooms

Harmful algal blooms (HABs) are concentrated densities of phytoplankton that produce compounds harmful to humans or marine life. Because of the potential harm to public health and fisheries, the possibility that marine fish farm effluent could induce HABs in coastal waters has been raised. When HABs occur near fish farms, fish may die of direct poisoning, incur gill damage or show decreased growth and vigor (Beveridge 2004, Davidson et al. 2009, Borg et al. (2011). There is little research to date supporting a link between nutrient discharge from fish farms and the occurrence of HABs (Nash 2001, Silvert 2001, Black et al. 2002, Cole 2002, Huntington et al. 2006, Halwart et al. 2007). Similar to other algae species, the nutrient fluxes that influence HAB population dynamics are complex and vary for different species (Anderson et al. 2002, Anderson et al. 2008, Vargo 2009, Lewitus et al. (2012). Environmental factors other than nutrient loading also contribute to HABs. For example, in the Pacific Northwest, river discharge appears to drive HABs near many salmon farming areas (Rensel et al. (2010). As a best management practice, siting fish farms outside of nutrient sensitive habitats is encouraged (Nash 2003, Nash et al. 2005).

There are only a few studies which indicate that aquaculture effluents may contribute to occurrence of HABs. In a major aquaculture area in Guangdong province, China, the occurrence of algal blooms more than tripled from 1994-2004 compared to the previous decade (Yu et al. 2007). This included the occurrence of HABs near fish farms (Song et al. 2004), but industrial, agricultural and nuclear facilities in the same area also contributed heavily to eutrophication and warming of the semi-enclosed bay's water. In a lab study Bodennec et al. (2002) found that exposure to dead fish and fish feed elutriates could increase the growth and toxicity of ichthyotoxic algal species.

To avoid any potential negative interactions — either stimulation of HABs from aquaculture discharge or harm to cultured fish from naturally occurring HABs — it is recommended that farms be sited away from areas with a history of recurring HABs, or in marine areas with low water exchange rates and high nutrient loads where blooms may thrive (Beveridge 2004, Borg et al. (2011).

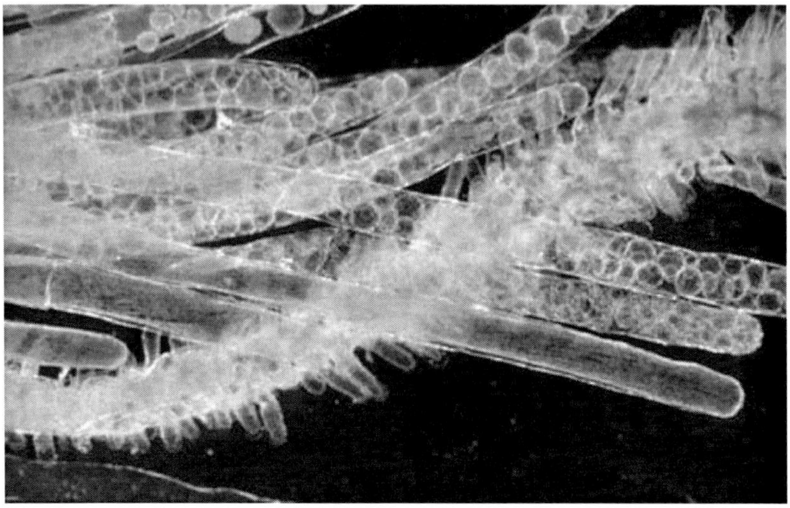

Photo courtesy of NOAA.

Table 6. Summary of primary production effects reported and modeled at fish cage sites in response to farm nutrient discharge

Impact Level	Reference	Location	Species Cultured
None Detected	Alston et al. 2005	Puerto Rico	Cobia
	Benetti et al. 2005	Hawaii	Moi
	Sowles 2005	Maine	Salmon
	Harrison et al. 2005	New Brunswick	Salmon
	Soto and Norambuena 2004	Chile	Salmon
	Navarro et al. 2008	Scotland	Salmon
	Maldonado et al. 2005	Spain	Sea bass & sea bream
	Basaran et al. (2010	Turkey	Sea bass & sea bream
	Demirak et al. 2006	Turkey	Sea bass & sea bream
	Belias et al. 2003	Greece	Sea bass & sea bream
	Vezzulli et al. 2008	Italy	Tuna
Significant	Robinson et al. 2005	Bay of Fundy	Salmon
	Rensel and Forster 2007	Puget Sound	Salmon
	Honkanen and Helminen 2000	Finland	Salmon
	Modica et al. 2006	Sicily	Sea bass & sea bream
	Vizzini and Mazzola 2006	Mediterranean	Sea bass & sea bream
	Huang et al. (2011	South China Sea	Various

In summary, while there is evidence that effluent from fish farms may result in increased primary productivity, most studies have failed to demonstrate a clear effect (Table 6). When effects are found, hydrological conditions or farm management practices may contribute. Siting farms in deep, well- flushed waters will help disperse dissolved nutrients and avoiding areas where effluent will be washed onshore will also help avoid eutrophication.

Because a change in primary productivity linked to fish farm effluents would have to be detected against the background of natural variability, it is difficult to discern effects unless they are of great magnitude (as high as a 50% increase in productivity) and duration (Huntington et al. 2006). Many farms conduct routine water sampling as part of their regulatory requirements. Given the difficulty in correlating nutrient enrichment directly to increased primary production, it is suggested that sampling include direct measurements of chlorophyll-*a*, or other metrics of productivity (Pittenger et al. 2007). Because nutrients may be flushed away from the immediate cage area and dispersed into the surrounding water body, it is difficult to assess whether far-field primary production is being affected over large areas and at longer time scales. This is further complicated by the occurrence of many anthropogenically derived nutrients in coastal marine waters, making it difficult to attribute nutrification to any one source, including aquaculture.

Benthic Community

The impact of marine finfish cage culture to the benthic community is an environmental concern. The accumulation of fish and feed waste below cages and the associated geochemical changes can induce changes to the micro and macrofauna that live on and in the sediments. Early reviews of environmental impacts by Wu (1995) and Pearson and Black (2001) address changes in benthic communities directly attributable to fish farming as an issue warranting additional attention. Benthic community impacts are identified as being one of the most critical areas requiring systematic examination and further research in the U.S. (Goldburg et al. 2001, Nash 2001, Stickney 2002, Nash 2003, Nash et al. 2005, Pittenger et al. 2007, Johnson et al. 2008, Ocean Conservancy 2011), Canada (Hargrave 2003, Wildish et al. 2004a, Hargrave 2005), Europe (Black et al. 2002, International Council for the Exploration of the Seas 2002, Huntington et al. 2006, The Mediterranean Science Commission 2007, Holmer et al. 2008, Olsen et al. 2008, Holmer 2010), the Mediterranean (International Union for Conservation of Nature 2007, Borg et al. (2011), Asia (Xue et al. 2004, Pan 2005), Australasia (Cole 2002, Crawford 2003) and globally (Beveridge 2004, Halwart et al. 2007, Tucker and Hargreaves 2008, Hall et al. (2011). Many countries require regular benthic community monitoring at fish farms as part of their regulatory process.

A great deal of recent research has increased our knowledge of how farm effluents affect benthic biogeochemistry and the resulting impacts to benthic biodiversity. We have expanded our understanding of how farm and site characteristics influence the degree, extent and duration of benthic community impacts, and work has progressed toward refining site and habitat specific indicators of perturbation (Black et al. 2008). The interplay of biotic and abiotic sediment processes is well understood and this section summarizes recent research which specifically targets the impacts of marine fish farms to the benthic community.

Microbial Communities

Excess fish food and waste can be a rich source of nutrient input to the marine environment. As organic matter accumulates on the seabed, the bacterial community is stimulated. Increased bacterial respiration rate depletes oxygen and may transform the sediment into an anoxic environment. A parallel shift in the bacterial community away from aerobic species and toward anaerobic species can occur, with mats of sulfide oxidizing bacteria like *Beggiatoa* as the only visible organism present. This process is generally well understood in marine sediments, but there has been relatively little work to more closely examine the effects of fish farm nutrients on specific components of the benthic microbial communities.

Studies conducted at Italian fish farms provide insight into the microbial processes occurring in enriched sediments. Vezzulli et al. (2002) studied the bacterial community beneath a 15 year old sea bream farm where sediment samples indicated an organically enriched benthos under reducing conditions. Bacteria levels below the farm were up to three times higher than at a reference site. There was a shift toward gram negative species at impacted locations and occurrence of pathogenic *Vibrio* (species was not reported) also increased. Danovaro et al. (2003) reported similar results from a study comparing benthic bacterial response at three fish farms. At impacted sites, the majority of culturable bacteria were facultative anaerobes, confirming hypoxic to anoxic conditions. Again, gram negative species predominated at farm sites, but gram positive species were prevalent in control site sediment. Farm sediments showed a 3-10 fold increase in bacterial biomass. Fifteen days after cage removal bacterial levels returned to levels similar to control sites. A study conducted by La Rosa et al. (2004) showed that after rapid organic accumulation following installation of a fish cage, the benthic bacterial community increased rapidly (within a few weeks) in response. As carbon levels in the sediment increased over the growout period, so did the bacterial biomass. Beneath the study cages there was a 13% decrease in eukaryotic species and a rise in cyanobacteria.

The use of commercial bacterial products to aid in the recovery of sediments beneath a fish farm was investigated by Vezzulli et al. (2004). In field trials, a blend of indigenous microbial strains (BIO-VASE) was added to enriched (high organic carbon and decreased redox potential) experimental plots below cages. This bioaugmentation stimulated carbon mobilization and enhanced extracellular enzymatic activity rates. The use of cultured bacterial products to mitigate environmental enrichment at marine fish farms is a relatively new field requiring additional research to evaluate its potential.

Meio- and Macrofauna Assemblages

The changes in the benthic faunal community resulting from marine fish farm nutrient enrichment and changes in sediment chemistry have been well documented for many years (Pearson and Rosenberg 1978). In stressed sediments, the benthic species composition and diversity shift toward tolerant generalists. The polychaete community is perhaps the most studied taxonomic group and it is widely accepted that the abundance of capitellid polychaetes and other generalist species is increased beneath impacted sites. Recent work is focused on defining additional species that can serve as consistent indicators of enrichment. There is a high cost associated with conducting benthic faunal sampling and it requires time and taxonomic expertise to process samples. Therefore, scientists and the industry are interested in identifying cheaper monitoring tools that provide reliable indication of environmental perturbation or degradation. Complex computer simulation models already used to track organic waste dispersion from fish farms are being expanded to include benthic components which predict benthic macrofaunal community response (Cromey et al. 2002).

Photo courtesy of NOAA.

The following sections reference several biodiversity indices, quantitative tools used to assess the community structure, which reflect the biological community in sediments below fish farms and at reference stations. These indices are calculated from matrices of species data to estimate biological variability and compare communities. Some of the indices mentioned here are k-dominance curves, Shannon or Shannon-Wiener Index, Pielou's Index, Infaunal Trophic Index, Simpson's Index and others. Each differs in how it is calculated and what it reflects about biodiversity, so often researchers will calculate several indices for a data set. Some indices score or rank species according to whether they are generalist species that are relatively tolerant of a range of environmental conditions or more specialized species with narrower requirements. Quantitative indices are useful for comparing two locations with different species compositions, or for assessing the community structure of a location through time as species composition shifts. Heip et al. (1998) provide a useful overview of common indices, how they are calculated and how each may be interpreted and applied.

The question of benthic community changes and how best to monitor them at fish farm sites was addressed in two chapters of *The Environmental Effects of Finfish Aquaculture* (Hargrave 2005). Holmer et al. (2005) provides a broad overview of the interconnectedness of benthic sediment conditions and the benthic macrofauna. Sediment conditions, including enrichment from farm wastes, are among the factors driving the composition of the benthic community. Likewise, feeding and burrowing behaviors can also impact sediment conditions. While there is considerable research indicating that enrichment alters benthic faunal communities, the relative roles of natural variability and site-specific determinants are less well understood. Studies which can account for these factors separate from fish farm effects are needed to be able to refine our understanding of the benthic impacts directly attributable to farm organic matter enrichment. The chapter by Wildish and Pohle (2005) reviews the benthic macrofaunal changes typically associated with marine finfish cage culture. Most of the well-documented effects pertain to near-field scales. The far-field effects of aquaculture to the ecological functionality of food webs and secondary production have not been studied and are difficult to ascertain. Many studies used the classic enrichment gradients of Pearson and Rosenburg (1978) to classify sediment and faunal conditions (Figure 2).

Building upon this older model, Wildish and Pohle (2005) developed a summary table which compares value range of several specific chemical and biological parameter ranges (Table 7). These measures were used to broadly categorize the state of the benthic sediments and community. However, review of the response of various benthic organisms, with the purpose of providing guidance for cost-effective monitoring of organic waste discharge from aquaculture. As sediments become enriched and deoxygenation ensues, sulfate metabolism becomes the major metabolic pathway driving the shifts in bacterial and macrofaunal communities. Hypoxia and sulfide toxicity inhibit the persistence of most sensitive taxa like mollusks and crustaceans. Diversity indices reflect the prevalence of as research progressed it has become evident that there are cases in which this simplistic successional stage model is inadequate. For example, the use of this model may not be adequate when the initial conditions are not 'Normal,' if one is comparing species guilds across regions, when there are seasonal macrofaunal community fluxes or if farms are located above some types of unstable marine sediments that do not adhere to the standard geochemical model. The authors suggested that additional indices should be developed to address some of these deficits.

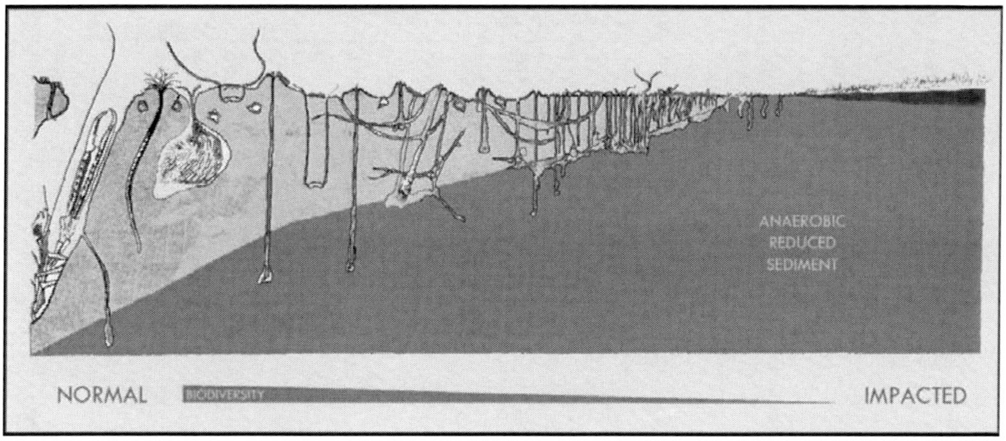

Figure 2. Enrichment gradient and biodiversity. Adapted from Pearson and Rosenburg 1978.

A tool to enhance our understanding of the interplay between biogeochemistry and benthic community structure was developed by Hargrave et al. (2008). They provides an in-depth summary of the current knowledge of the geochemical conditions resulting from organic enrichment and a opportunistic species like *C. capitata* and tolerant nematodes. Under extremely poor conditions, only tolerant species such as *Beggiatoa* (mat-forming chemoautotrophic filamentous bacteria) are found. The authors developed a nomogram which classifies marine sediments along an enrichment gradient with respect to a range of measurable biological and chemical variables (Figure 3).

This chart offers an improvement over older classifications as more chemical and biological indices with specific ranges of indicative values are included. This tool can be used by managers in developing monitoring plans at farm sites to track benthic effects. The interplay between sediment chemistry and benthic communities is widely studied (Nilsson and Rosenburg 2000, Valdemarsen et al. (2010) and research at fish farms sites will continue to add to our knowledge of these complex ecological processes.

Table 7. Value ranges of geochemical, biological and image-derived indicators of enrichment and biological effects at marine fish cages. Adapted from Wildish and Pohle 2005

Indicator	Value Range			
Geochemical	**Normal**	**Oxic**	**Hypoxic**	**Anoxic**
Redox	>100	0 – 100	-100 – 0	<-100
Sulfide	<300	300 – 1300	1300-6000	>6000
Biological				
Microbial	Normal	Oxic	Hypoxic	Anoxic
Macrofauna	Normal	Transitory	Polluted	Grossly Polluted
Sediment Profile Imaging				
OSI (Organism-sediment index)	III	II	I	Azoic
BHQ (Benthic habitat qualityindex)	>10	5 – 10	2 – 4	<2

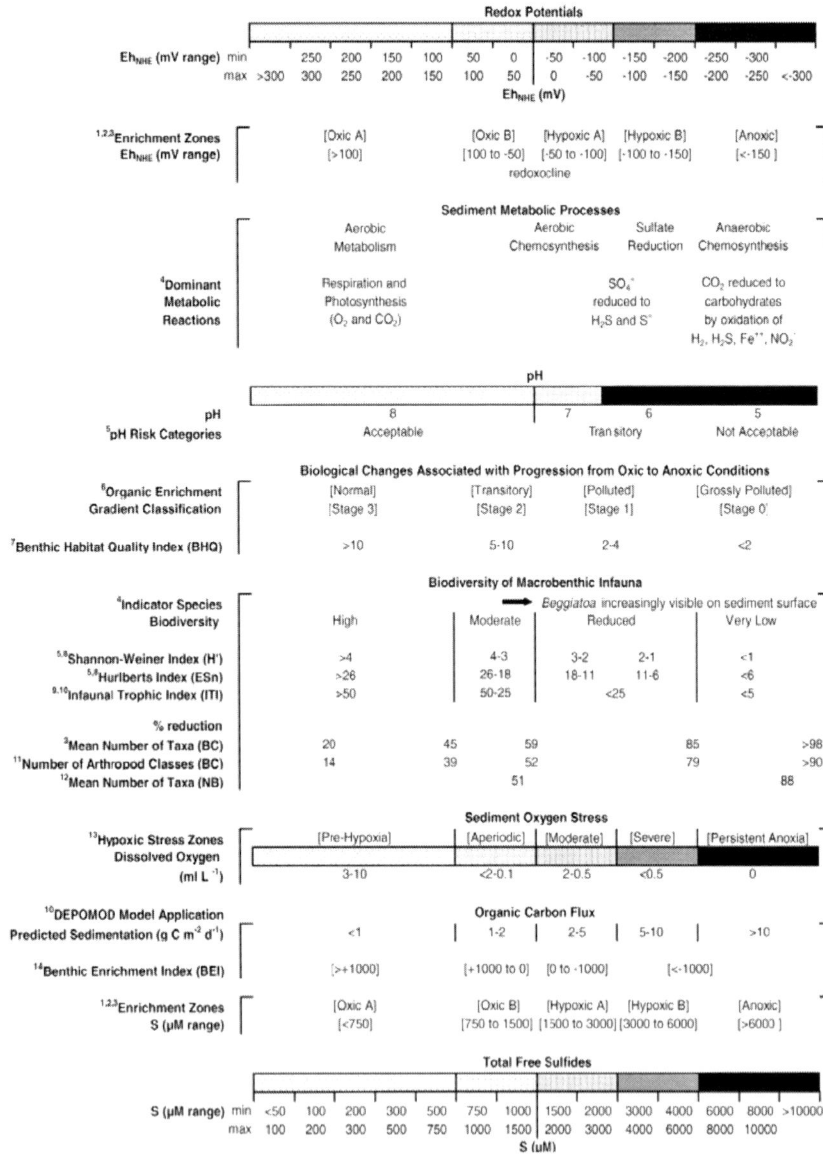

Figure 3. Nomogram depicting relationships between common geochemical and biological indicators of mari ne sediment condition below marine fish farms. Copied with permission from Hargrave et al. 2008.

Temperate Regions

North America

Much Canadian research has addressed benthic infaunal community impacts and the development of effective monitoring protocols. Wildish et al. (2001) compared the scientific results and cost effectiveness of two methods of benthic monitoring techniques — one based upon sediment geochemical analysis and one on macrofaunal analysis. The study was conducted at a salmon farm in the Bay of Fundy. Both methods found significant differences in sediment chemistry and community structure between farm and control sites, with abundance, diversity and evenness indices significantly reduced at the farm sites. *Capitella capitata* was the only species found in farm sediments. All of the geochemical tests could be conducted aboard a sampling boat, while faunal samples required extensive follow up work in the laboratory. Thus, 22 geochemical samples could be processed for every macrofaunal sampling, greatly decreasing the time cost. The training to conduct faunal identification further increases costs. Although both methods reached the same scientifically-based conclusion that the benthos was enriched, faunal sampling was found to be considerably more expensive.

A study of benthic effects at two salmon farm sites in the Letang Inlet, Bay of Fundy conducted in the late 1990's found significant enrichment (as reflected in percent organic matter) and benthic community effects compared to a nearby reference location with no aquaculture activities (Pohle et al. 2001). At the most heavily impacted site in Lime Kiln Bay, species diversity indices remained depressed and k-dominance curves (plots illustrating the relative abundance of all species in a community) elevated throughout the study period reflecting negative impacts to the benthic community. This effect persisted despite the cessation of farming one year prior to the final sampling. At the less impacted farm site, benthic community indices and k-dominance curves reflected initial degradation, but showed recovery in the last three years of surveys. Local benthic indicator species included polychaetes, bivalves and crustaceans. Enrichment tolerant species increased at impacted sites with a concomitant decrease in disturbance sensitive species.

In order to validate the use of sediment profiling imagery as a methodology for assessing benthic enrichment in Limekiln Bay, Bay of Fundy, Wildish et al. (2003) compared sediment geochemistry and faunal indicators to photographic analysis results. Both methods characterized the benthic conditions at the 20+ year old fish farming site as significantly impacted. The sediments were visually and chemically determined to be anoxic and highly reduced and only a few *C. capitata* were found. Both methods resulted in similar assessments for a nearby healthy reference site. The agreement between the two methods indicates that image technology, which is much cheaper and faster than faunal and geochemical sampling, is suitable for assessing benthic conditions beneath fish farms. Wildish et al. (2004b) also validated the use of acoustic technology to detect organic enrichment by comparing backscatter images to macrofaunal and geochemical samples collected at fish farm sites. This preliminary study confirmed that all three methods detected enriched benthic conditions at the cages. The total number of macrofaunal species was decreased and *C. capitata* numbers were elevated compared to reference sites. Additional work is needed to refine the use of acoustic mapping for benthic assessments, and this is a promising cheaper alternative to intensive sampling techniques.

Brooks et al. (2003) employed a comprehensive set of chemical and biological indicators to evaluate remediation efforts at large commercial salmon farms in British Columbia. A variety of local, benthic fauna was used in the assessment. Indices including sulfide levels, total organic carbon and redox potential followed patterns typical for an enriched benthos. Effects were greatest within 30 m of the cages, but were measurable 105-185 m downstream. Following the initiation of harvest, chemical remediation of the benthic sediments was complete prior to the cessation of harvest (about nine months) and beginning of fallowing. Changes in the macrobenthos were most pronounced within 50 m of the cages, but were evident up to 225 m from the farm during peak production. Opportunistic polychaetes thrived and increased in abundance during fish grow out, but decreased as organic loading declined. Similar patterns were observed for generalist mollusk species. Shannon's Index values increased following harvest and initiation of fallowing, with distant sampling sites (>150 m) reflecting the earliest recovery. Biological remediation was defined as a return to macrofaunal diversity comparable to reference conditions and was completed with six months of fallowing. This work also includes a description of three distinct communities of local invertebrates associated with high, moderate and low concentration of sediment organic carbon.

In contrast, a second comprehensive study at a set of farms near the previous study site found remediation periods of several years (Brooks et al. 2004). Again, chemical and biological indices were used to evaluate the recovery of heavily enriched and degraded benthic conditions following harvest in 1997. By 2002, chemical remediation was nearly complete out to distances greater than 80 m from the cages. However, biological remediation did not progress as quickly and, even after four years of fallowing, community structure beneath the cages had not fully recovered. Recruitment of benthic invertebrates was slow and included few species. The authors note that many previous studies have documented a coupled chemical and biological benthic recovery process. As burrowing infauna and epifauna recolonize enriched sediments, their burrowing and feeding activities resuspend organic matter and oxygenate the benthos, aiding in chemical remediation. This process was not, however, observed during this study. This farm site represented a worst case scenario for the authors who note that such a lengthy remediation is uncharacteristic of salmon farms in that area. Siting over depositional sediments in an area with poor dispersion was the likely explanatory factor.

Hargrave (2003) summarized the current research on the far field effects to benthic communities and food webs from aquaculture enrichment. These are more difficult to assess as there are multiple sources of enrichment in most water bodies with aquaculture operations and because there are few large-scale, long-term data sets. In New Brunswick, studies have found indications of far field impacts in terms of benthic community structure which may take greater than five years to manifest themselves. The potential food web disruptions at larger scales could be significant and could negatively impact Canada's demersal commercial fisheries, including groundfishes, scallops, lobsters and sea urchins (Wildish et al. 2004a).

Benthic faunal impacts are identified as an environmental issue of concern for the U.S. aquaculture industry (Nash 2001, Tlusty 2001, Nash et al. 2005, Pittenger et al. 2007). Nash (2003) identifies the potential effects to benthic faunal abundance and diversity as a high risk issue for salmon aquaculture in the U.S., especially if cages are placed in poorly flushed areas. He suggests that proper siting, appropriate production levels and fallowing will likely restrict harmful impacts to within less than 15 m of the cages. At well-flushed sites where accumulation of organic matter is not a factor, there may even be a net increase in benthic biomass from slight increases of carbon.

Nash (2001) summarized the benthic community effects from studies in the 1990s at North American salmon farms. Invertebrate infauna consistently responded to enrichment in terms of abundance and taxonomic diversity. Enhanced production was sometimes documented in the early phases of production, but usually adverse impacts were associated with increase in farm biomass. Polychaete, crustacean and mollusk abundance and diversity tended to decrease within 75 m of fish cages, with tolerant generalist species prevailing. Nash et al. (2005) later suggested that as the number of farms increases, there will be greater potential for far-field benthic effects from cumulative nutrient loading. They note that the scale of benthic impact will depend upon the degree of flushing in and around a facility and the biomass of the animals cultured. Appropriate siting of intensive aquaculture facilities is critical to the management of benthic effects. The high cost of monitoring of benthic community impacts may be reduced by using video techniques in combination with less expensive assessments of total volatile solids, redox potential and free sulfides.

Heinig (2001) reported on the results of long-term monitoring during the 1990s at salmon farms in Maine as part of that state's Finfish Aquaculture Monitoring Program (FAMP). Infaunal analysis indicated that net pen effects on the abundance and diversity of benthic community tended to be most pronounced directly beneath fish cages and within a 30 m radius. Impacts were typically not measurable beyond 60-80 m. Abundance and dominance by *C. capitata* was highly variable, ranging from 0-100%, but generally decreased with distances from the cages. This report urged the use of multiple indices to assess the benthic community because site- specific conditions and variability may be significant factors.

Similar conclusions were reached following nematode sampling conducted in the Gulf of Maine at two fish cage demonstration sites to collect baseline monitoring data (Abebe et al. 2004). A survey of the pre-stocking population status found low similarity in the nematode composition between the sites, even though sediment characteristics were similar. These results highlight the need to document site-specific changes in biodiversity before and after deployment of cages to be able to understand long-term effects. Goldburg and Triplett (1997) summarized reports from the 1990s of degraded benthic community conditions at Maine salmon farms. However, voluntarily fallowing and other mitigation measures were successfully implemented.

A study by Rensel and Forster (2007) points to the beneficial effect of nutrient loading to the colonizing organisms found at marine fish farms. This study did not examine the benthic community, but did find over 100 algae and invertebrate species populating the cages, lines and floats at a salmon farm in Puget Sound. Marine birds, fish, shrimp and crab were observed near the farms but not at reference areas nearby without farms. Such data supports the idea that limited nutrient input from fish farms in well flushed areas can have an overall enhancing effect on marine biomass (especially substrate colonizing organisms) and food webs.

South America

Sediment profile imaging (SPI) was used to examine sediment cores collected beneath salmon farms in two Chilean fjords (Mulsow et al. 2006). Image analysis and geochemical parameters reflected high levels of ecological degradation in the more enclosed Pillan Fjord where the benthos was largely azoic. The organism sediment index (OSI) values ranged from -4 to -2 at the most impacted sampling areas and the benthic habitat quality (BHQ) index from 0.67 to 1.3 at the most degraded sites and from 3.6 to 10 at the least impacted sites. In the more exposed fjord, there was less perturbation, but still evidence of impacts. Though no sites were devoid of benthic fauna, some areas had low OSI (2.7) and BHQ (5) scores compared to reference locations where OSI was 11 and BHQ was up to 12.7. The researchers concluded that SPI alone might have underestimated the impacts of fish farming and recommended this technology to be coupled with biogeochemical analysis for environmental assessments.

A large scale comparative study in Chile found impacts of 29 salmon farms on the benthic communities (Soto and Norambuena 2004). Shannon index and evenness values (how the individuals are spread among the species present) were significantly higher at reference sites compared to farm sites. Hardy, tolerant species were more common at impacted sites. Significant latitudinal differences in taxonomic representation were also found reflecting the need for regional knowledge about benthic fauna and identification of local species to serve as indicators of disturbance. A consistent relationship of decreasing species richness with high sediment phosphorus concentrations was detected. As the aquaculture industry expands in Chile, the potential environmental impact to benthic communities has been identified as a concern (Buschmann et al. 2009).

Northern Europe

A great deal of research has been conducted in Europe to understand effects of fish farming on benthic communities. Shifts in community dynamics beneath cages were already being documented in the 1990s (Lu and Wu 1998, Mazzola et al. 1999). Work has continued to further understand these impacts at multiple scales and develop effective monitoring and remediation strategies and techniques (Holmer 2010). A study in Finland was not able to detect consistent patterns of eutrophication effects to the benthic community with respect to geographical location or distance from fish cages (Honkanen and Helminen 2000). Here, plankton chlorophyll-*a* was a better indicator of enrichment, likely due to water circulation and mixing or high degrees of spatial variability in benthic community structure. In Norway, benthic sampling is a core component of the monitoring system (MOM) used to track environmental impacts of marine fish farms (Hansen et al. 2001). Impacts are assessed at the cage site, in the surrounding waters and regionally. Investigations of environmental conditions include monitoring of organic sedimentation rates, geochemical and sediment condition parameters and benthic macrofauna. Combined results are used to score overall impacts at farm sites. A trial test of this monitoring program at 44 farms found that sediment chemistry analysis was as definitive as more expensive and time consuming faunal sampling in detecting unacceptable sediment conditions. Most samples were categorized as showing little impacts. The polychaete *Malcoceros fuliginosus* was identified as an indicator species in enriched sediments.

Kutti et al. (2007) examined the temporal and spatial variations in infaunal communities for two years near a Norwegian salmon farm sited in very deep water (230 m). At peak farm production, infaunal abundance was 10 times and biomass was 35 times higher within 250 m of the farm, compared to a reference site 3 km away. Species richness was highest 550-900 m from the farm. Sediment parameters (visual observations and redox) indicated degraded sediment conditions in the immediate farm footprint (< 250 m) but the researchers concluded that overall the benthic capacity to decompose farm waste was not exceeded at this site. The organic wastes released from the farm served as a readily utilized food source for opportunistic polychaetes, bivalves and echinoderms. The researchers noted that effects at deep water sites may be spread over larger spatial scales than farms in shallower areas.

Benthic community structure was used to assess the recovery of sediments in two farming areas in the North Baltic Sea impacted by long term organic pollution (Kraufvelin et al. 2001). Historical monitoring data (1982-1991) showed that 12-18 species of opportunistic chironomid, oligochaete and bivalve species characterized early samples. Samples collected over seven years following pollution abatement found varying results. At one site, increase in abundance and species diversity indicated slow, but partial recovery. The other farm site, however, reflected continuing deterioration. Differences in topography and water exchange patterns were suggested as explanatory factors. Carroll et al. (2003) compared four common monitoring techniques — diver surveys, faunal analysis, sediment chemistry and sediment profile imagery — for evaluating sediment enrichment at five Norwegian salmon cage farm sites. They found that beneath the cages, faunal abundance and diversity indices were negatively impacted with effects decreasing out to 50 m. The four monitoring methods generally gave similar results with respect to assigning sediment condition. While faunal analysis is a very sensitive indicator of ecological impacts, it is time consuming and costly. Image analysis and sediment chemistry provide cheaper alternatives, but may underestimate the spatial extent of impacts.

A benthic recovery project in Scotland compared recovery time predicted using DEPOMOD (a model which tracks waste nutrient dispersion from fish farms) to actual recovery at five salmon farms (Black et al. 2012). The sites either had high initial impacts and recovered within a year or had low impacts but were further from recovery after two years, thus reflecting the site-specific nature of benthic recovery processes. At both sites, a suite of biodiversity indices, including Shannon's H′, AMBI, Pielou J′ and ITI, were compiled from benthic faunal samples used to compare site conditions over time. At peak biomass, all sites showed decreased indices values out to 50 m, with recovery times of 251-774 days. A new index, the Brooks Recovery Index (BRI), was also developed which evaluated the persistence of individual benthic taxa from grow out through recovery. A recent study by Wilding (2011) focused on the abundance of the sea pen *Pennatula phophorea* near Scottish salmon farms finding that most transects within 30 m contained no sea pens. Sea pen numbers increased at intermediate distances (to 50 m). Interestingly, peripheral transects were highly variable with some transects containing no sea pens while others had relatively high densities. This variability was potentially explained as a protective effect of sea cages at some sites where the presence of cages deterred trawling activity.

Recently, a large-scale study was undertaken by Borja et al. (2009) to: 1) identify the suitability of selected quantitative indicators to assess the effects of aquaculture on benthic communities, 2) assess their applicability over a range of ecosystems and production systems, and 3) investigate the factors to which the indicators respond across Europe. Ten study sites from differing latitudes, water depths, sediment types and cultured species were included in the analysis. Macrofaunal abundance, biomass, species diversity indices and species richness were inversely related to the sampling distance from the farm. The influence of and relationships with other environmental variables such as depth and current speed were also analyzed and found to account for 53.2% of the variability in macrofaunal parameters. Thus, this study concludes that the response of benthic communities to marine aquaculture enrichment is a function of site-specific environmental factors, which should be considered when studying the cumulative impacts of existing farms and for selection of future farm locations.

Sub-Tropical Regions

North America

The impacts of a tuna sea cage operation in Baja California, Mexico, on the local benthic fauna (including mollusk, crustaceans, polychaetes and echinoderms) were investigated by

Diaz-Castaneda and Valenzuela-Solano (2009). Sampling was conducted over two years at 18 stations in the bay surrounding the farm with the closest samples being taken 250 m from the cages. Shannon diversity index values varied only slightly over the study (2.26-3.40) and were distributed around the bay. An analysis of environmental condition base upon diversity and evenness index scores suggested that most sites were in favorable or stable condition. Polychaete diversity remained high and *C. capitata* represented only about 9% of polychaete abundance. Overall, the authors concluded that this area hosted a rich benthic invertebrate community, but because sampling at the cages was not possible they were unable to test for near-field impacts.

The Mediterranean

Many studies have been conducted throughout the Mediterranean in the last decade to investigate the impacts of marine fish culture, primarily of sea bass and sea bream, to benthic biodiversity. A study conducted at five farms in Spain (Maldonado et al. 2005) was able to detect significant benthic community impacts at only one site, though trends indicated biodiversity reduction at farm sites. The authors reported that poor statistical power was likely an experimental factor, but even so, the effect of fish farming seemed to be relatively weak based upon the results of clustering and ordination analyses. Opportunistic capitellid polychaetes and nassarid gastropods were the most common taxa collected at the impacted farm. Another Spanish study compared chemical and biological parameters tested in sediments from a single fish farm (Aguado-Gimenez et al. 2007). Here, only sulfide levels correlated well with the effects to the indices of macrobenthic community condition. Two methods of grab sampling, Van Veen grabs and diver sampling, were also compared. Both sampling methods found that effects to biotic indices were most evident directly beneath the cage and within 100 m of the cage. Cost effectiveness analysis concluded that at this farm, faunal sampling was necessary to assess ecological impacts and that grab sampling was less costly than using divers.

In Italy, La Rosa et al. (2001) studied the response of meoifauna to fish farm sediment impacts and subsequent cage removal. Both total abundance and biomass of meiofauna and nematode were decreased under the cages compared to a control site in response to organic loading from the farm. Within a few months of cage removal, the nematode population began to increase suggesting a relatively quick onset of sediment and benthic community recovery. A follow up study at the same farm (Mirto et al. 2002) focused on the nematode assemblage response to organic farm deposition. Again, nematode abundance and biomass were decreased below the cage compared to a control site. Individual nematode body weight was higher at the cage site, however. Community structure of the nematode assemblage differed beneath the cage. For example, four of the dominant genera found below the cage were negligible in control sediments. *Setosabateria* characterized the control site and *Pierrickia* the cage station. Within a few months of farm deployment, k-dominance curves from cage samples were above control sites.

Impacts to meiofauna were examined by Vezzuli et al. (2003) at an established (15 years of continuous production) fish farm. Total meiofauna abundance and density increased below cages, with nematodes accounting for more than 85% of the abundance at cages, compared to 67% at the control site. Copepods were more abundant (26%) at control sites, compared to the cages (4%). Several minor taxa were absent below cages. Sampling of macrofauna at an intensive sea bream and sea bass farm in Trieste (Aleffi et al. 2005) also found increased biomass beneath the cages. However, the number of species declined under cages with opportunistic polychaetes *Neanthes caudata* and the mollusk *Mytilus galloprovincialis* dominating. K-dominance curves reflected disturbed communities beneath the cages. Overall, the severity of impact was considered low compared with other studies and impacts were confined to within 100m of the cages. Vezzulli et al. (2008) investigated the meiofaunal assemblage below a bluefin tuna fattening farm off the southwest coast of Italy. Due to good flushing at this site, environmental impacts were generally not detectable, but slight differences in the meiofaunal community were evident. The total number of species decreased and nematode composition differed beneath cage stations. Limited sampling and seasonal flux may have masked farm effects. A more intensive study at an Italian sea bass and meagre farm found significant differences in the benthic (Aguado-Gimenez et al. 2007) community. The overall number of species was lower beneath cages and increased with distance. Shannon diversity index was lower and opportunistic polychaetes prevailed beneath and near cages. Cluster analysis clearly differentiated between cage and control sites. AMBI (AZTI Marine Biotic Index) values were also calculated and confirmed progressively degraded conditions below the cages over the 18 months of increasing biomass during grow out.

Italian sea grass meadows near fish cages have also demonstrated changes in their macrofaunal assemblages. Terlizzi et al. (2010) found that seagrass beds 100-500 m downstream of a sea bass and sea bream farm showed increased organic loading compared to control sites. Benthic communities closest to the farms had fewer individuals, decreased evenness and diversity index scores, and increased k-dominance curves. The 500 m site values were intermediate between the 100 m site and controls, but statistically similar to control sites. At the impacted sites, a few opportunistic species tolerant of stressed habitats with high organic matter represented most of the faunal abundance. Impacts to Mediterranean seagrass meadows macrofauna were compared for three farm sites in Spain, Italy and Greece (Apostolaki et al. 2007). Enrichment was negligible at the seagrass meadows according to geochemical analysis. Species number and abundance showed little variability between the farm and reference stations at all sites. Interestingly, macrofaunal biomass peaked at the intermediate sampling locations between the farm and control stations for all sites. Overall, the meadows displayed high diversity and contained the expected species arrays, resulting in the conclusion that effects to the seagrass at these farms were mild. These two studies highlight the need to examine benthic effects of marine aquaculture in sensitive habitats.

In the last ten years many studies were conducted in Greece to understand the effects of fish farms on benthic communities. For example, Karakassis et al. (2002) used sediment profiling image analysis to characterize macrofaunal assemblages for comparison with traditional core samples and geochemical analysis. The two methods showed a high degree of correlation, validating the use of image analysis as a cheap monitoring alternative. The stations directly below the stations became heavily degraded over the 10 month study, with stations out to 50 m from cages more closely resembling conditions at a control site. Lampadariou et al. (2005) found increased meiofaunal abundance 25 m from cages at three Greek fish farms compared to control sites, mostly due to increased nematode and copepod densities. Diversity indices decreased significantly at only one of the farms. Meiofaunal response was in agreement with redox and organic matter levels which indicated enriched sediment conditions. Similarly, Klaoudatos et al. (2006) found significant increases in abundance, but decreased species number, diversity and richness during a year of sampling at floating cages off eastern Greece. The polychaetes *Nereis diversicolor* dominated (35%

abundance) at farm sites compared to controls (12%). Other opportunistic polychaete species were also abundant at farm sites. K-dominance curves indicated that the highest impacts to benthic assemblages occurred in October, reflecting the importance of seasonal variability in faunal abundance.

Lampadariou et al. (2008) developed a biomass fractionation index (BFI) to rapidly assess benthic organic enrichment. A quick sieving method is used to quantify the biomass fraction >0.5 mm and <1.0 mm as a proportion of total biomass in each sample. This technique was tested at seven Greek farm sites sited in varying conditions by comparing assessments to geochemical analysis and full macrofaunal identification. Impacted sediments and changes in macrofaunal assemblages, primarily due to the prevalence of *C. capitata*, were evident up to 25 m from the cages. The BFI values were elevated (up to 0.8) below and near cages (out to 10-25 m), and decreased (down to < 0.1) out to 50 m from the cages. This was largely due to increased total biomass contributed by large macrofauna such as mollusks and echinoderms further away from the cage areas. Recently, Neofitou et al. (2010) studied the spatial and temporal benthic community effects at two Greek fish farms in a semi-enclosed gulf. Spatial, but not seasonal, differences in enrichment were detected; the greatest impacts were within 10 m of the cages and decreased closer to control levels by 50 m. Total abundance and biomass were variable with regard to distance from the cage, season and farm. However, the number of species, species richness, evenness and the Shannon-Wiener index were all lower within 10 m of the cages, with an increasing trend out to the control station 300 m away. The 50 m sampling stations were often intermediate between the high impact area under the cage and the controls. K-dominance curves followed a similar pattern and, as in other studies, opportunistic polychaetes *C. capitata* principally accounted for the dissimilarities in benthic communities between farm and control sites.

A recent study used six sea bream and sea bass sites in the eastern Mediterranean to test a computer simulation model, MERAMOD, which tracks waste nutrient particles and predicts benthic impacts from fish farm discharge (Cromey et al. 2012). The abundance and biomass of opportunistic species including *C. capitata* (representing over 85% of total abundance near the cages) increased significantly beneath cages. The Shannon-Wiener diversity index and species richness decreased 10-25 m from the cage and increased 50 m away and at the references stations. Statistical comparison indicated that the relationship between the field data and the model's predicted values for benthic biodiversity impacts (abundance, Shannon-Wiener H′ and Simpson's index) were classified as excellent, demonstrating the usefulness of tools like MERAMOD for understanding the ecological impacts of fish farms in the marine environment. In this study, 25 m was the greatest distance at which benthic impacts were detected according to modeled and observed measures of biodiversity and sediment flux.

In the Adriatic Sea, Kovac et al. (2004) reported meiofaunal changes beneath a sea bass farm where species diversity was impoverished and several groups (Gastropoda, Acarina, Ostracda and Ophiuroidea) were completely absent. Effects were evident up to 200 m from the cages. Najdek et al. (2007) reported a localized (20 m) impact to benthic communities in enriched sediments below a sea bass and sea bream farm site. Relative abundance of nematodes at the cages was up to 92%, compared to 69% at a control site 1 km away, while total density and species diversity decreased. Stable isotope analysis was used to detect fish farm derived nitrogen in sessile benthic invertebrates (Dolenec et al. 2007). Enrichment of $\delta^{15}N$ up to 5.2-6.8 ‰ in organisms collected from the perimeter of fish farms compared to reference locations with values close to zero. This study was able to distinguish invertebrate nitrogen uptake in sponges, anemones and barnacles from farm and municipal waste sources, providing insight into the use of this tool for assessing near and far field impacts in areas with multiple sources of enrichment.

In Israel, the benthic foraminiferal community was studied in relation to organic enrichment gradients associated with a commercial sea bream farm (Angel et al. 2000). These taxa are used to study anthropogenic effects in marine sediments as their abundance, calcium carbonate test formation and community composition may be altered in response to perturbations. Over 50 species of foraminifera were found around the farms with highest abundance in the enriched sediments associated with the farms compared to control sites. Species richness was generally low in the vicinity of the farms, but lack of historical data in this area made it impossible to draw a conclusion about changes in diversity due to enrichment. Eden et al. (2003) studied the response of the mud snail *Nassarius sinusigerus* to changes in sediment biogeochemistry beneath an Israeli sea bream farm. The peak in snail abundance occurred with moderately impacted sediments 20-80 m from the farm's center. Highly deteriorated sediment conditions caused the snails to move further away, but they returned following recovery.

A broad scale, comparative study looked at the influence of marine cage culture on the development of sublittoral fouling communities on identical artificial structures in Scotland, Crete, Slovenia and Israel (Cook et al. 2006, Tsemel et al. 2006). The results provide information about the nutrient availability to food webs and epibiotic recruitment around fish farms compared to reference locations. After six months, biomass was higher at the Scottish and Israeli fish farm structures, but the opposite was true for the site in Slovenia. Community composition varied between the sites with 26-73 colonizing species recorded and algal production also differing among the sites. It is likely that in oligotrophic waters, nutrients released from fish farms may increase the biomass of fouling organisms, possibly reducing other environmental impacts of cage culture.

Recently, the AZTI Marine Biotic Index (AMBI) was developed to provide a tool to assess the benthic biotic condition in a wide range of European coastal sediments (Muxika et al. 2005). This index can be used to determine the ecological quality of benthic habitat impacted by a wide range of human activities including marine fish aquaculture. The proportions of benthic species, ranging from very sensitive to disturbance tolerant to opportunistic, are used to calculate a location's AMBI value, with higher values reflecting heavily disturbed conditions. This type of index is valuable in making comparisons of impacts across wide geographical areas. A case study at three Greek fish farms found AMBI scores increased directly below the cages, but decreasing out to 25 m. This index has been applied in Atlantic and Mediterranean regions.

Studies in the Canary Islands have found significant changes in benthic communities below fish farms. Boyra et al. (2004) used image analysis to show an increase in sea anemone coverage at a sea bream and sea bass farm (1.1-2.3% seafloor coverage) compared to reference sites (0.3-1.2%). A six year study of meiofaunal community dynamics under fish farms in Tenerife found significant shifts in impacted sediments toward nematode dominated communities. Overall meiofaunal abundance was greatly increased below cage sites compared to control and intermediate stations, and seasonal and yearly effects were also significant. Harpacticoid copepod abundance and presence were also affected by the presence of farms, while polychaete abundance was low (less than 5%) in all samples.

Photo courtesy of NOAA.

Asia

Japanese regulatory requirements for sustainable aquaculture include benthic community monitoring. Yokoyama (2003) describes the relationships between macrofaunal communities, degree of embayment of farm sites, farm production and sediment chemical parameters to provide additional information for evaluating environmental criteria. Six indicator groups of macrofaunal species were derived based upon sampling at 22 fish farms. Results suggested that topography (depth and width of a bay) is the primary factor affecting the benthic communities under cages, with farm production as a secondary driver. This information was combined into a simple graph which can be used to guide production or siting decisions based upon predicted benthic community impacts. The relationship between benthic fauna, topography and current velocity as indicators to assess the assimilative capacity of fish farms was further examined by Yokoyama et al. (2007). An index based upon water depth and current velocity was refined and a graph similar to the one described above was developed to guide siting and production. They also included a table of critical threshold values of biotic and abiotic parameters typically collected during monitoring. Similar research was conducted in Korea (Kim et al. 1998, Park et al. 2000) and other Asian countries, but it is often difficult to obtain translated research articles.

Australasia

A benthic fish exclusion study conducted by Felsing et al. (2005) additionally analyzed changes in the benthic community. Sedimentation and enrichment were highest under cages with restricted wild fish access, as feeding activity greatly ameliorated farm effects. Macrobenthic community structure was also most affected when fish were excluded. Capitellid and spionid polychaete abundances were predominantly present in the more degraded sediment inside the exclusion nets. Shannon diversity index values were lower beneath cages with and without fish exclusion nets compared to reference site values.

Tropical Regions

Caribbean

Several U.S. marine fish farms have conducted benthic fauna sampling as part of their monitoring regimes. Benthic macroinvertebrate sampling at a mutton snapper and cobia farm in Puerto Rico found no differences for the Shannon diversity index, species evenness index, or species richness index at the cage site versus the control site (Alston et al. 2005). Contrast analysis, however, showed that abundance of macroinvertebrates in the sediments at the control site was significantly higher when compared to stations beneath the cages, especially at the snapper cage. Seasonal differences in abundance related to maximum feeding loads were also noted. Overall, impacts were confined to the area directly beneath the cages.

Pacific Islands

Benthic sampling at a Hawaiian yellowtail farm found a high proportion of gastropods and few bivalve species (Family Lucinidae) known to occur in anaerobic and high sulfide conditions (Sarver 2009). The report concluded that two years of sampling indicated only a low impact of the farm to the benthic fauna. An increase in the number of dorvilleid polychaetes was related to increased feed amounts and fish size at a moi farm in Hawaii (Bybee 2001). This opportunistic group of polychaetes was possibly an indicator of enrichment, even though crustacean and nematode abundance were not affected. After harvest, the dorvelleid polychaete numbers returned to pre-stocking values. A follow up study sampled the polychaete infaunal community at the same sites a few years later under increased production (Lee et al. 2006). This study found chemical measurements indicative of an enriched benthos, and Shannon-Weiner diversity index values of the polychaete community decreased both over time and in relation to distance from the cages. Pielou's evenness levels were lower at cage than control sites and opportunistic species *C. capitata* and *Ophryotrocha adherens* were more abundant at impacted sites. Polychaete communities at control sites were unchanged over time. Following a six month fallowing period, polychaete diversity and community structure at the affected sites began to resemble communities at reference sites (Lin and Bailey-Brock 2008). Enrichment indicator species declined sharply, but the study concluded that full recovery had not yet been achieved as significant differences in community structure still existed.

Asia

Research is being conducted in India to assess benthic impacts as mariculture expands in that country. A survey of the macrobenthos in the Bay of Bengal, India found significant changes in the assemblages there compared to surveys conducted 50 years ago (Raut et al. 2005). Yet, two taxonomic distinctness indices still reflected an overall healthy level of biodiversity and a taxonomically stable community structure. Increased human activities, including aquaculture, were identified as factors affecting this region. A pilot study of the culture of sea bass in small open cages (hapa) on the west coast of India monitored benthic macrofauna (Imelda- Joseph et al. 2010). During the four month culture period the macrofaunal abundance increased nearly fourfold due to high organic loading from the trash fish used as feed.

In China, Gao et al. (2005) report that fish farm activities near Hong Kong had adverse impacts to the macroinvertebrate assemblage. A one year field study was conducted in a semi-enclosed bay with grouper, snapper and sea bream mariculture operations. Changes in polychaete, brittle star and bivalve abundance were evident beneath fish cages and at intermediate (100 m away) stations compared to controls (600 m away). Shannon-Wiener index value significantly decreased beneath fish cages and at intermediate sites, and K-dominance curves reflected similar patterns of perturbation.

Australasia

An Australian study examined sediment and macrobenthos condition to assess recovery at a high production salmon farm in operation for 14 years (Macleod et al. 2004). Sulfide levels at the cage and out to 35 m returned to reference site values within six months and sediment organic matter at the cage site had decreased by 30-40% after two years. Benthic faunal recovery, however, was slower. Even after three years, Shannon diversity values remained lower at the cage site (1.6) than at the reference station (2.0) although stations beyond 10 m were recovered within a few months. The farm was situated in a depositional environment, which tended to localize impacts to within 10m but also slowed recovery.

A comparison of sediments and macrobenthos at 20 fish farms in Tasmania showed an increase in opportunistic polychaetes and decreased species diversity correlated to the degree of farm waste (Edgar et al. 2005). Species richness of crustaceans and gastropods and biomass of bivalves were lower at cage than less-impacted sites. Sites 20 m from the cages and 35 m from farm boundaries showed community effects intermediate between cage sites and reference sites 1-2 km distant. A recent follow up study in the same area (Edgar et al. 2010) found similar results. Redox potential, the proportional abundance of capitellid polychaetes, and bivalve/ mollusk ratio were determined to be the most useful environmental indicators of fish farm impacts for this region. The study also attempted to discern large scale, regional environmental impacts from fish farming. However, even though 42 farm sites were sampled, there was insufficient statistical replication and power to make definitive inferences about long-term impacts. Nonetheless, background changes through time suggested to the authors that organic matter, faunal abundance and proportion of introduced taxa increased in Tasmania during the six year monitoring period, while redox potential, sediment particle size and abundance of capitellids decreased.

A recent study used 12 years of sediment data collected at five New Zealand salmon farms to evaluate five benthic indicators (abundance, number of taxa, redox potential, total sulfides and total organic matter) and 10 biotic indices (Margalef 's d, Peilou's J□, Shannon H□, AMBI, M-AMBI, MEDOCC, BENTIX, BOPA, ITI and BQI) under low and high flow regimes to identify those that best define organic enrichment gradients (Keeley et al. 2012). Though most of the indices performed well, none was able to consistently discriminate biotic impacts over the full range of enrichment gradient for both flow regimes. Overall, redox potential and the M-AMBI index performed well for both flow regimes across the enrichment range. At the upper end of the enrichment scale, however, the M-AMBI showed increased variability, while redox had lower correlation with enrichment at higher flows. At very high enrichment, redox, sulfides, number of taxa and abundance were relatively clear, but some of the more complex biotic indices were less discerning. Of the ten indices, BQI (Benthic Quality Index) performed best for higher enrichment levels. At the lower end of the enrichment scale geochemical parameters were less sensitive, and biotic indices were better at reflecting early impacts to the community. Perhaps one of the most important findings of this study was that it was difficult to discriminate biotic impacts at high levels of enrichment especially when trying to determine the threshold point at which opportunistic macrofauna can no longer survive and the benthos become azoic. The researchers concluded that a combination of geochemical parameters and regionally validated indexes will be useful to assess benthic community conditions at fish farms, but caution that expert judgment must be applied to weight indicator variables appropriately, identify spurious results and to provide an integrated approach to the assessment process.

In his summary, Phillips (2005) reports that benthic community impacts are generally confined to within 50 m of fish cages. Flushing and feed management are two factors which determine the radius of impact. Highly variable fallowing periods range from a few months to greater than 10 years. Two other studies have summarized global trends in benthic impacts of fish farms. Kalantzi and Karakassis (2006) conducted a meta-analysis of benthic impacts drawing on data from 41 studies covering a range of farmed species, geographic areas, management practices and site characteristics. Benthic faunal parameters included as dependent variables were Shannon diversity index, Pielou's evenness index, number of species, abundance and biomass. Stepwise regression found that most biological variables were determined by distance from the farm, water depth and latitude. For example, 68% of the variance associated with Shannon- Wiener diversity index was explained by these three factors. Results were also presented for analyses done with respect to three sediment types, providing further insight into site- specific characteristics which may influence the severity of benthic impacts from fish waste. Giles (2008) accessed data from 64 studies and used Bayesian networks to quantitatively assess the relationships between impact parameters and farm characteristics. Macrofaunal abundance and biomass, and Shannon diversity index were used to quantify benthic diversity effects. These were significantly correlated to sediment nitrogen, sediment organic matter and redox potential, as well as to farm characteristics including farm volume, current speed and sediment type.

These types of comprehensive data analyses are possible because of increased monitoring and recent research incorporating repeated measurement of biological and geochemical parameters at farm and reference sites. Such tools will be useful to the marine aquaculture industry and regulatory agencies for making decisions regarding siting of new farms and setting environmental impact thresholds. The need for improving the power of benthic surveys to detect trends in offshore marine sediments impacted by aquaculture or other human activities is addressed by Rogers et al. (2008). This work describes the resource and sampling requirements necessary to effectively assess ecosystem status in benthic sediments, concluding that large numbers of relatively expensive repeated samples will generally be required to detect impacts. Overall, megafaunal sampling provided the best tradeoff between sufficient power of detection and costs. Research has continued to document previously reported effects to the benthic community and build upon this research in terms of improving our knowledge of the interactions of biotic and abiotic components of the benthic environment (Table 8). Improvements in farm management, proper siting and fallowing are consistently reported to substantially decrease and minimize the impacts of marine fish farm waste to benthic communities. Advances in image analyses and acoustic technology are being evaluated to develop quicker and cheaper sampling methods. Continued monitoring of benthic communities at farm sites is imperative, as is site specific reference data establishing baseline faunal conditions. Future research to gain understanding about regional and far-field impacts of marine fish farming to benthic communities is needed. This will require long-term data sets and repeated sampling over large areas and refinement of techniques that can differentiate between aquaculture waste and other natural and human sources of nutrients. Methods to measure impacts to hard bottom sites are needed to assess impacts to those habitats.

Table 8. Effects to the benthic community at marine fish farms

Community Effect	Reference	Location
Increase in anaerobic microbes in sediment	Vezzulli et al. 2002	Mediterranean
	Danovaro et al. 2003	Mediterranean
	La Rosa et al. 2004	Mediterranean
Macrofauna		
Decreased abundance or species diversity and/or increase in opportunistic species	Nash 2001	North America
	Wildish et al. 2001	Bay of Fundy
	Pohle et al. 2001	Bay of Fundy
	Lee et al. 2006	Hawaii
	Brooks et al. 2003, 2004	British Columbia
	Hargrave 2005	Canada
	Soto and Norambuena 2004	Chile
	Kraufvelin et al. 2001	Baltic Sea
	Carroll et al. 2003	Baltic Sea
	Borja et al. 2009	Europe
	Aguado-Giminez et al. 2007	Spain
	Vezzuli et al. 2003	Italy
	Terlizzi et al. 2010	Italy
	Neofitou et al. 2010	Greece
	Gao et al. 2005	China
	Imelda-Joseph et al. 2010	India
	Macleod et al. 2004	Australia
	Edgar et al. 2005	Tasmania
	Kalantzi and Karakassis 2006	Global
	Giles 2008	Global
Minimal effects on species diversity; increased abundance or biomass	Diaz-Castenada and Valenzuela-Solano 2009	Mexico
	Alston et al. 2005	Puerto Rico
	Sarver 2009	Hawaii
	Rensel and Forster 2007	Puget Sound

Fish

The excess food and waste released from fish cages may be a food source for wild fishes, especially benthic feeders (International Council for the Exploration of the Seas 2002, International Union for Conservation of Nature 2007, Grigorakis and Rigos 2011). Cages may also provide shelter and foraging habitat for fish. These characteristics may be considered a benefit to the local and regional environment because of increased production of local fish and potential benefits to the benthic environment.

On-Site Effects

A study in Israel (Katz et al. 2002) found that bottom-feeding gray mullet stocked in enclosures beneath a sea bream farm increased sediment oxygen levels and decreased organic loading and sulfide levels as a result of increased bioturbation during mullet feeding. In this study, wild demersal fishes were excluded from portions of the sediment resulting in degraded conditions demonstrating the positive effects of sediment resuspension. Planktivorous damselfish and snails were also attracted to the area beneath fish cages, with the latter being present in higher numbers in the mullet cages (3970 snails/m^2) compared to unstocked enclosures (457 snails/m^2).

Similarly, Vita et al. (2004) found a beneficial effect of fish at sea bream and sea bass farms in Spain during an exclusion experiment. When wild fish were excluded from areas of the sediment below the cages the organic content, nitrogen and sulfide levels, and *Capitellid* polychaete and macrofauna numbers increased significantly. Analysis of sedimentation showed about 80% of the organic waste was consumed within 4 m of the cage bottom, thus reducing its accumulation on the bottom. Additional bioturbation of sediment during fish foraging activity greatly improved sediment quality. Likewise, wild fish exclusion experiments at a marine rainbow trout farm in Australia estimated that wild fish consumed 40-60% of the sedimented nutrients from the fish cages (Felsing et al. 2005). Carbon deposition under cages without fish was 4.5 g carbon/m^2/day compared to only 0.7 g carbon/ m^2/day under cages open to benthic foraging. After 62 days, conditions beneath cages excluding fish had deteriorated as reflected by both geochemical measurements of nutrient enrichment and a shift in the benthic community towards *Capitellid* and *Spionid* polychaetes.

Fish aggregation devices (FADs) are structures deployed in aquatic environments to attract fish communities, and marine cages have been called 'super-FADs' because of the large numbers of wild fish attracted to these structures (Dempster et al. 2005).

Temperate Regions

Three fish species were recorded being attracted to a salmon farm in Puget Sound (Rensel and Forster 2007), but this fish data was incidental within the larger context of the study which focused on invertebrate fouling organisms. Nash (2001) also states that salmon cages act as fish attractants.

Sub-Tropical Regions

From diver surveys, Oakes and Pondella (2009) found much higher fish abundance, density and diversity below cages stocked with white sea bass *Atractoscion nobilis* off Catalina Island (10,234 fish, 142 fish/100m^2, and H□= 2.29) compared to adjacent (8452 fishes, 117 fish/100m^2, and H□= 1.45) and distant (500 m away, 8958 fishes, 124 fish/100m^2, and H□= 1.13) reference reefs. Tuya et al. (2006) found that cages continued to attract wild fish even after the cages had been emptied and farm feeding activity ceased. Prior to harvest, wild fish abundance was about 50 times greater at cages than at control sites; after harvest wild fish abundance was double control levels. Fish that fed on particulate organic matter, large benthic chondrichthyid rays and *Pagellus* spp. declined at the fish farm after the cessation of farming, but herbivores and mid-sized benthic carnivores remained. Dempster et al. (2005) concluded it is difficult to predict the fish community structure response to the presences of fish cages. Their comparison among fish assemblages at cages in Spain and the Canary Islands, found no pattern of fish distribution vertically in the water column around the cages or with respect to size classes present. Their results suggest that it may be difficult to predict *a priori* at which sites demersal fish foraging and bioturbation might be expected to contribute to assimilating or dispersing farm waste. However, wild fish abundance and species richness were ubiquitously high near cages, with planktivores and pellet eaters consistently being numerically dominant. Recently, a computer model, MERAMOD, which tracks waste particle flux from fish farms was used to assess environmental impacts in the eastern Mediterranean Sea (Cromey et al. 2012). The researchers incorporated a module to account for site-specific rates of wild fish feeding on waste particulates and found this improved model performance. This was confirmed by direct diver observations which also indicated that removal of particulates by wild fish can be an important part of the nutrient flux process. Most nutrient dispersal models do not explicitly incorporate the ecological roles of wild fish into their simulations, either as direct consumers of particulate waste or as ultimate consumers benefitting from farm nutrient transfer up the food web (Pitta et al. 2009). This approach appears to be a valuable addition to these quantitative tools, especially for modeling areas with fish communities rich in demersal species.

Tropical Region

Alston et al. (2005) monitored a diverse fish assemblage around a cobia sea cage in Puerto Rico. While species richness indices fluctuated throughout the year with no definitive pattern, the fish community at the cage site was always more diverse and about 40 times more abundant than at control sites. Both reef and pelagic species were present and Carangids were the most numerous fish representing 92% of the fish censused. In Indonesia, Sudirman et al. (2009) found 29 species of mostly small reef fishes aggregating around small marine cage farms, only five of which were observed directly feeding on waste pellets, consuming around 27%. The fish appeared to be permanent residents using the cages as shelter and for foraging on the fouling communities. In Queensland, researchers made incidental observations of wild fish aggregating near a barramundi farm, and concluded that most of the fish there were Siganid herbivores seeking shelter and feeding on fouling organisms (McKinnon et al. 2008).

Negative interactions with wild fishes have also been observed. For example, bluefish are reported to seasonally aggregate around and invade fish cages in the Mediterranean (Sanchez-Jerez et al. 2008). Of the 23 farms surveyed, bluefish aggregations were detected at 16, but only four farms reported significant impacts to the cultured fish. In addition to direct predation on the cultured fish, these four farms reported decreased productivity due to stress and additional costs associated with removal of bluefish and net repair. While marine cages primarily provide a habitat and foraging opportunity for fish, Nash et al. (2005) suggest that lights used at salmon cages to extend the photoperiod for growth enhancement may attract fish at night, possibly interfering with juvenile migratory fishes. Other documented negative interactions include entanglement of wild fish (Huntington et al. 2006) and exposure to antibiotics and other chemicals (Fortt and Buschmann 2007, for example). The potential for wild fish to consume medicated feed and then be captured for human consumption, and the possibility of disease transfer from farmed fish (Braaten 2007) may also be important issues to consider.

Fish Community Effects

The effect to fish communities has been investigated at larger scales. Machias et al. (2004) studied the species composition of demersal fish assemblages in the Aegean Sea prior to and 12 years after the deployment of commercial fish cages. Fish abundance increased by a factor of four within the bay, the number of species caught increased from 37 to 42 and the trophic level value increased from 3.59 to 3.79 after the onset of marine cage culture. Traditional diversity indices showed that despite some differences in species composition, the overall fish community structure after the establishment of fish farming was not phylogenetically impoverished. The average lengths and weights of several fish species were also compared and fish were found to be either similar in size or larger after the farm was established. The results were thought to reflect an overall benefit to the local fish community at a regional scale, most likely due to nutrient driven increases in primary production. In another study, Machias et al. (2005) conducted trawls near (within three nautical miles) and far from (> 20 nautical miles) Greek fish farms. They found that the abundance and biomass of wild fish was greatest close to the fish farms compared to nearby reference sites without cages and at the distant sites. Although seasonal and substrate differences in fish abundance were evident, the Shannon-Wiener diversity indices showed no deleterious effect due to fish farms. Increased abundance of several important commercial fish species was also documented. These observations were bolstered by an analysis of the relationships between 18 years of data fish farming activity and fish landings (Machias et al. 2006) throughout Greece. The researchers found no negative correlations between farming activity and fisheries landings, and there were strong indications of increased fisheries production in areas with farms (presumably as a result of nutrient discharge).

The secondary role of fish cages as FADs, especially near reef habitat and in the open ocean, warrants further research (Holmer 2010) and keeping track of wild fish aggregates at open ocean cage facilities is recommended as part of standard environmental impact monitoring procedures (Lee and O'Bryen 2007, Holmer et al. 2008). Dempster et al. (2006) point out that the unintentional role of sea cages as FADs may have significant conservation impact to marine fisheries and they encourage banning commercial and recreational fishing activities around farms and the designation of these areas as Marine Protected Areas. A similar recommendation was made by researchers in Turkey (Akyol and Ertosluk 2010) as a result of their study of fish farmers who set traps near fish cages. This lucrative, but illegal, harvest of aggregating fish has been the basis of conflict with local artisanal fishers who are not allowed to fish near the cages.

Sharks

There is little published information about the interactions of sharks and marine cage farms, but they have been documented as being attracted to fish cages in the Pacific Northwest (Nash et al. 2005), Puerto Rico (Alston et al. 2005), The Bahamas (Benetti et al. 2005), Latin America (Rojas and Wadsworth 2007) and Australia (Australian Government 2009). Because sharks pose a threat to the stocked fish and potentially divers, dangerous species may be destroyed. In Australia, an estimated 20 great white sharks a year are killed at marine aquaculture farms (Australian Government 2009). Siting of a salmon farm off South Africa within an ecologically significant great white shark congregation area and eco-tourist destination elicited major negative public response (Scholl and Pade 2005) and the farm was later closed. A recent telemetry study of sand and tiger sharks near fish cages off Hawaii found that sharks did aggregate near the cages with some individuals being recorded for the entire term of the 2.5 year study (Papastimatiou et al. 2010). These animals were considered to pose minimal threat to humans. The economic and ecological potential risk of large scale fish releases due to sharks tearing nets may be a concern as the industry moves into offshore sites (Holmer 2010) depending on the types of nets and locations used.

Deterring shark predation at marine cage sites can likely be accomplished by the use of tear-resistant nets. Sharks guards are small rigid mesh nets installed at the bottom of a fish cage to prevent sharks from damaging nets while attempting to feed on dead fish that have fallen to the bottom (Jamieson and Olesiuk 2002). Good husbandry practices such as removing sick or dead fish promptly from cages is also an effective predator deterrent. Given the recent global interest in shark population declines and the need to implement conservation efforts, the potential impacts of marine cage culture to sharks is likely a fruitful area for research.

Marine Mammals

The interactions of marine mammals with marine fish cages and efforts to minimize potential problems are recognized, but there is little recent published, peer-reviewed literature that specifically addresses the issue. Marine mammals such as seals, sea lions, cetaceans and otters at fish cages can represent a threat to cultured fish of direct predation, injury, stress mediated increased susceptibility to disease, decreased growth due to stress, and escapement loss through torn nets (Nash et al. 2000, Jamieson and Olesiuk 2002, Würsig and Gailey 2002, Rojas and Wadsworth 2007, Belle and Nash 2008). Reciprocally, marine aquaculture operations may displace marine mammals from their foraging habitats (Markowitz et al. 2004, Cañadas and Hammond 2008) or cause other disruptions to their behavior (Early 2001). Entanglement in nets or lines around fish farms may cause injury, stress or death to marine mammals.

Photo courtesy of NOAA.

Nash et al. (2000) provide a summary of information to assess the risk associated with aquaculture and marine mammal interactions in the Pacific Northwest salmon industry. Loss due to direct predation, fish injury or stress and escapement can account for losses of up to 10% in terms of fish and represents significant financial loss. Pinniped attacks on cage divers have also been reported. The authors conclude that physical barriers including rigid netting around cages are the best management options to decrease harm along with siting of cages offshore far away from haul out sites and rookeries.

A report by Jamieson and Olesiuk (2002) provides a thorough review of pinniped interactions with salmon farms in Canada, the financial impacts to the industry, methods for non-lethal intervention and the ecological implications of lethal deterrents to the seal and sea lion populations. The authors summarize estimates from the 1980-90s for damages caused by pinnipeds at salmon farms around the world. Losses range from a few thousand fish up to 10% of the stocked fish. Damages may be only a few thousand dollars for an individual farm, but can total millions of dollars for a single country in a year. The growth of the fish farming industry and concomitant expansion of pinniped populations has tended to increase the number of interactions, but lethal control methods are less viable than previously due to ecosystem conservation objectives and regulatory protection. Typically, only single individuals may be killed and only after multiple forays into the farm with repeated attempts to deter the animal. They note that the U.S. has even stricter regulations with respect to lethal removal. Nonlethal interventions include harassment by boat or with noise (such as underwater seal firecrackers), aversive conditioning, predator (killer whales) models or sounds, and the use of acoustic devices and relocation. Often, harassment techniques are effective in the short term, but may be less efficacious over time as animals become habituated. Acoustic deterrent devices (ADD and AHDs) are designed to cause auditory discomfort to pinnipeds by emitting sound underwater at a range of frequencies. However, these devices have also been shown to deleteriously impact non-target marine mammals.

Würsig and Gailey (2002) provide useful information on the conflicts between aquaculture and marine mammals and potential resolutions. They report on the damage and financial loss that marine mammals, especially pinnipeds, may inflict on commercial fish farms. The need for non-lethal management options to reduce conflicts is recognized, with the goal of decreasing impacts to non-target animals and preventing the killing, both licensed and illegal, of pinnipeds. Six options for reducing marine mammal impacts are discussed: harassment, aversive condition, exclusion, non- lethal removal, lethal removal and population control. Harassment by chasing, explosives, and ADDs has been found to be only somewhat effective and generally only in the short term until animals become habituated. In fact it is possible that over time noise harassment devices may actually become attractants to habituated individuals who come to recognize the sound as an unpleasant dinner bell. Predator models and sound devices (imitating killer whales for example) are also not very effective.

The dangers that these harassment techniques pose toward target and non-target marine life are discussed. Aversive conditioning refers to feeding with poisoned (with lithium chloride for example), but not deadly, bait to sicken the offending animals. This has also proved to be only ephemerally useful. Non-lethal capture and relocation of problematic individual animals is feasible, but very expensive, time-consuming and minimally effective. Relocated animals often return quickly to the farm area. Lethal removal and large scale population control (or culling) are generally not very effective, popular or legal options.

Removing problem animals may help in some instances where an individual is causing damage, but typically more animals will just move in.

Furthermore, there is a segment of the public which opposes the killing of marine mammals, especially for private gain. Likewise, large-scale population control methods like culling are unlikely to be supported. Wursig and Gailey (2002) conclude that exclusion is the most effective measure. Also, siting is noted as being an important tool. For example, farms located distantly (> 20 km) from haul out sites tend to have fewer interactions with pinnipeds.

Research results support the views and conclusions in the foregoing three review papers. At 11 out of 25 sea bass and sea bream fish farms surveyed in the Turkish Aegean, individual monk seals were documented taking fish and damaging nets, mostly at nighttime feedings during the winter months (Guecluesoy and Savas 2003). A range of non-lethal deterrents was ineffective and only the installation of anti-predator nets was successful in avoiding fish losses. Aerial and ship surveys conducted in New Brunswick by Jacobs and Terhune (2000) suggested that harbor seals do not congregate in salmon farming areas, but nor do the farms seem to disrupt the mammals normal movement patterns. A later study by Terhune et al. (2002) found that ADDs near aquaculture facilities in the Bay of Fundy did not elicit startle responses, measurable avoidance behavior or changes in haul out behavior in pinnipeds that had been exposed to ADDs for many years. Surveys of salmon farm managers in Scotland (Northridge et al. 2010) indicated that seal predation has declined over the past decade and that less than a quarter of salmon farms reported major problems with seals despite nearly daily siting of seals near farms. Rogue individuals were thought to cause the most damage and individual recognition techniques are being improved as a potential management tool. ADDs were not in use at all farms and they were not thought to be very effective, while farm management strategies including net tensioning, removing mortalities, lower stocking densities and seal blinds at the bottom of the nets deterred predation.

Photo courtesy of NOAA.

The most damaging marine mammal interactions are with pinnipeds while dolphins, porpoises and whales are viewed as a minor threat to fish cages. Dolphins have been documented feeding on wild fish attracted to marine fish farms off Italy, but were not reported to predate caged fish (Diaz Lopez et al. 2005). In a recent five year study at Italian sea bass, sea bream and meagre cages Díaz López (2012) observed individually identified dolphins to assess patterns of habitat use and farm fidelity. Dolphin occurrence near the farm varied with time of day, season and year. Individual animals fell within four farm fidelity categories: farmers (occurrence rates > 50%; 20% of individuals), and frequent (occurrence rates 0.25-49%; 10% of individuals), occasional (seasonal occurrence rates < 25%, yearly occurrence > 0.25; 20% of individuals) or sporadic visitors (occurrence rates < 25%; 50% of individuals). Dolphins near farms were typically foraging on wild fish concentrated in the farm, but also fed on discarded or escaping fish during harvesting operations. Annual dolphin mortality was 1.5 per year and five animals were found entangled in nets during the study period. The potential for marine mammals to become entangled and drown in farm structures or lines is a predominant concern because many are protected in the U.S. under the Marine Mammal Protection Act (Würsig and Gailey 2002). From surveys at marine fish farms off Italy, Diaz Lopez and Shirai (2007) estimated one bottlenose dolphin mortality per month due to entanglement with farm nets. This risk can be minimized by siting farms in areas away from known migration routes, using rigid net materials or secondary rigid anti-predator nets, and keeping mooring lines taut. In Scotland detectors were placed at salmon farms and reference sights to monitor porpoise activity and response to ADDs (Northridge et al. 2010). Generally porpoises avoided farm areas when ADDs were turned on but returned quickly when they were deactivated. Some animals were observed foraging near farms with active ADDs, especially in areas where the devices had been deployed for some time.

Concerns are raised about the impacts of the noise pollution caused by ADDs to non-target marine mammals which pose no predation threat. For example, in British Columbia, harbor porpoises avoided areas during times when ADDs were activated (Olesiuk et al. 2002). A study in New Zealand (Stone et al. 2000) found that Hector's dolphins, a rare species, avoided acoustic gillnet pingers, suggesting that use of similar devices at salmon farms to deter pinnipeds could also impact non-target mammals. Early (2001) notes that killer whales in British Columbia will avoid marine farm areas where ADDs are in use. This is confirmed by another study in the Broughton Archipelago where killer whales avoided marine areas near salmon farms with ADDs installed to deter pinnipeds (Morton 2002). Following removal of the devices six years after deployment, the whale numbers rose to levels similar to previous levels. Killer whale numbers in a nearby farm area without ADDs remained stable during this same time period. To date, there are no available reports on the impacts of marine fish cage culture to manatees and dugongs, yet potential impacts to these animals should be considered at sites within their habitat range (Würsig and Gailey 2002).

The greatest success in deterring pinniped predation is the use of rigid net materials for fish cages or the installation of rigid exclusionary nets around salmon farms. These may be expensive to install, require follow up maintenance and cleaning and, in the case of secondary nets, may decrease water flow through the fish cage. Exclusion nets must be strong enough to resist chewing or tearing. Lines made of stiff materials will help prevent entanglement. Best management practices to deter predators include siting away from marine mammal aggregations, installing predator nets and other barriers, varying farm routines, using olfactory deterrents, and dogs where appropriate (Belle and Nash 2008), and installing electrical fencing around cage perimeters (Rojas and Wadsworth 2007). In the U.S., marine mammals are protected by either or both the Marine Mammal Protection Act of 1972 or the Endangered Species Act of 1973. Permitting to authorize fish farms in U.S. waters must include impact assessments to evaluate the threat, if any, posed by a marine aquaculture operation. These measures determine the allowable impacts to marine mammals and regulators may impose monitoring requirements or other farm operation and management measures to eliminate or reduce negative interactions.

Sea Birds and Turtles

At marine fish farms, entanglement in the cage nets poses the biggest threat to sea birds, especially those that may dive to feed on fish or fouling organisms (Belle and Nash 2008). Sea birds are reported to congregate near marine fish farms but are typically considered a low risk in terms of predatory threat, though they may scavenge mortalities or pick off fish during transfer or harvest (Pearson and Black 2001, Nash et al. 2005, Huntington et al. 2006, Rensel and Forster 2007). In contrast, the often significant impacts to freshwater aquaculture (Goldburg and Triplett 1997, Belant et al. 2000, Snow et al. 2005) and fisheries (Karpouzi et al. 2007) by piscivorous birds like cormorants and pelicans are better understood.

Permits are available to implement non-lethal predator controls to frighten birds away from cages and, because birds become habituated to noise harassment, farms often use overhead netting or screens to exclude sea birds from cage areas (Nash 2001, Huntington et al. 2006, Halwart et al. 2007). Siting of fish farms away from important sea bird habitats is encouraged or required in many countries (Bridger and Neal 2004, Borg et al. 2011) to minimize conflicts. Overviews of environmental impacts of marine aquaculture often refer to sea birds as species of concern, but contain few specific examples of measures implemented to aid in sea bird conservation (Halwart et al. 2007, International Union for Conservation of Nature 2007).

Photo courtesy of NOAA.

Like sea birds, sea turtles are generally perceived as incidental visitors at sea cages and not as predatory threats (Nash et al. 2005, Helsley 2007). Because these animals are protected in the U.S. and elsewhere as threatened or endangered species, potential impacts to sea turtles are an environmental concern associated with marine cage culture (Bridger and Neal 2004, Huntington et al. 2006, International Union for Conservation of Nature 2007, Borg et al. 2011), yet relatively little is known about how sea turtles may be impacted by such facilities. The primary concern with respect to these animals and marine cage culture tends to be the threat to the animals of entanglement with nets, mooring lines or other floating equipment. Management recommendations to reduce negative interactions include the use of rigid netting material for the cage, keeping mooring lines taut and removing any loose lines or floating equipment around the farm. Lines made of stiff materials will help prevent entanglement. Additionally, the proper disposal of all trash will reduce the risk that sea turtles will ingest plastic or other trash associated with farm operations. A recent study investigating hearing capabilities in sea turtles indicates they hear best at frequencies <1,000 Hz (Piniak et al. 2012), which is outside the range typically used for marine mammal ADDs.

Sensitive Habitats

The potential impacts of marine cage culture to sensitive habitats like corals, seagrass beds and mangrove forests are of concern to resource managers and scientists. Because they are nutrient sensitive, siting of fish farms near these habitats may have long-term consequences.

Corals

Eutrophication and sedimentation are known stressors on corals and reef communities, including those in U.S. marine waters (Torres 2001, Smith et al. 2008). There is interest regarding the potential impacts of marine fish farm effluent on sensitive coral reefs (Holmer et al. 2005) and within the last ten years research efforts have been focused upon this issue. A series of reports published in 2003 provided conflicting conclusions about the impacts of sea bream farms on the north shore of the Gulf of Eilat in the Red Sea. Bongiorni et al. (2003a, 2003b) and Rinkevich et al. (2003) presented data suggesting that coral exposed to fish farm effluent showed enhanced growth and reproduction. However, their study design and conclusions were criticized by Loya and Kramarsky-Winter (2003), who presented their own data which indicated the opposite effect – that fish farm effluent had deleterious impacts to coral growth and reproduction. Additional studies seem largely to bear out the latter conclusion. Villanueva et al. (2006) examined the survival, growth, physiology and reproduction of different life stages of reef-building coral in the Philippines. After 81 days of exposure to enriched conditions at a milkfish farm site, the survival, growth rate, photosynthesis to respiration ratio, and larval output were significantly lower than corals at intermediate and reference sites. Kramarsky-Winter et al. (2009) examined 15 cellular markers to assess potential physiological differences in corals and their algal symbionts growing near fish farms (the same Gulf of Eilat site as above) or at reference locations. Their results indicated significant differences in the physiological status between corals at impacted versus control sites for many of the markers tested. Two heat shock proteins (indicators of stress) were elevated in both corals and symbionts at the farm sites; five oxidative stress proteins (important for maintaining oxidation-reduction reactions and cellular signaling) were decreased in corals but increased for symbionts; three metabolic proteins (indicators of metabolic condition) showed varying differences; two cytochrome P450 (metabolic and xenobiotic response proteins) values differed, one being elevated and the other decreased. For stress proteins several of the biomarker showed no site- specific significant differences. These studies indicate that fish farm effluents can affect coral physiology and highlight the need for a better understanding of the effects to coral communities at multiple scales. A study of corals in the Philippines (Garren et al. 2009) revealed that the microbial communities on transplanted corals were altered within five days of exposure to milkfish farm effluents. This included increases of microorganisms pathogenic to coral, as well as humans. After 22 days of exposure, the microbial communities on the corals at effluent exposed sites, intermediate and references sites appeared to be transitioning back to pre-study composition. However, levels of the coral pathogen *Desulfovibrio* remained high at the sites nearest the fish farm. Stable carbon isotope analysis has been used to quantify the effect to coral skeletal growth from long-term exposure to the heightened nutrient loads associated with two decades of mariculture operations (Levy et al. 2010). Corals growing near fish farms showed a growth rate of 9.28 mm/year while samples collected at two sites away from the farms grew at slightly higher rates of 10.28-12.2 mm/year.

The impacts of marine fish farming to sensitive red corraline algae gravel habitats, also known as maerl, are being investigated. Hall-Spencer et al. (2006) documented major effects of Scottish salmon farms on this habitat of high conservation importance. Diver surveys found a build-up of organic waste beneath the farms and out to 100 m. Scavenging fauna increased 10-100 fold and live maerl cover was significantly reduced beneath the cages. The benthic infana community was shifted toward enrichment tolerant species. Similar results have been found at other maerl sites in the Mediterranean (Borg et al. 2011). Maerl (Wilding 2011) and reef (Tett 2008) habitats are sensitive to salmon farming, but additional investigation was recommended.

The European Union made the following recommendations with respect to environmental quality standards regarding corals: deviation from mean ambient nitrogen concentrations should not exceed 5%; deviation from mean ambient phosphorus concentrations should not exceed 5%; no increase in mean ambient levels of suspended solids; changes in salinity levels from seasonal ambient state not to exceed 5 ppt (Huntington et al. 2006). The potential impacts of aquaculture operations to sensitive reefs in U.S. waters has been identified as a concern, especially for nearshore reefs which already experience considerable stress from anthropogenic sources including terrigenous sediments and nutrients (Torres 2001, Smith et al. 2008, Otero 2009), and sewage outfall (Kaczmarsky et al. 2005, Nagelkerken 2006, Sutherland et al. 2011).

Seagrasses

Seagrasses are an important marine habitat and provide a range of ecological services, and are under threat from a variety of anthropogenic influences like sedimentation and nutrient loading (Orth et al. 2006). The effects of marine fish farming on seagrasses have been investigated, notably in the Mediterranean, where seagrasses are considered a sensitive habitat. Seagrass beds are often found in clear water with high rates of advective exchange, which are also good areas for farming fish. Potential negative impacts of fish farms on seagrasses include reduced water clarity due to sedimentation and nutrient loading (Cancemi et al. 2003, Dolenec et al. 2006). Organic loading can result in anaerobic sediment conditions and the accumulation of sulfides in the root zones, both of which may be toxic to the seagrass (Holmer et al. 2003, 2005). Increased primary production in seagrasses may occur at low levels of nutrient input, but a possible secondary increase in herbivore (such as sea urchins) pressure may lead to an overall decrease in seagrass biomass.

Changes in the macrofaunal assemblages in seagrass meadows due to fish farm effluent were recently investigated by Terlizzi et al. (2010). Meadows heavily impacted by the study farm saw a shift toward mollusk, gastropod amphipod and polychaete species associated with muddy and high organic content sediment and stressed habitats. A review of impacts to seagrass beds by Pergent-Martini et al. (2006) found that fish farms contributed to a decrease in seagrass meadow surface area and cover, shoot density and size, leaf and rhizome growth and photosynthetic capacity, as well as increases in epiphyte biomass and leaf size. Their recommendations to minimize such impacts include siting at least 200 m away in deep, well-flushed waters and careful monitoring of the community to enable early response. This recommendation is somewhat supported by a recent study which found only minimal community responses in a seagrass meadow to effluent from a large offshore fish farm 3 km away (Ruiz et al. 2010).

The European Union made the following recommendations with respect to environmental quality standards regarding seagrass: no increase in mean seasonal levels of suspended solids; light levels at 2 m depth should not normally fall below 10% of surface incident light; total Kjeldahl N not to exceed 140 μg/L; mean total N not to exceed 500 μg/L (Huntington et al. 2006).

Mangroves

Alongi (2002) summarized the status and threats to mangrove habitats, including aquaculture operations. The harmful effects of marine aquaculture to mangrove habitat are primarily associated with near or onshore operations including pond construction for growing milkfish and shrimp (Barbier 2003). However, this tropical habitat could be impacted by fish net pen operations in shallow coastal waters with onshore currents. For example, a preliminary study at a sea bass farm in Malaysia suggests that nutrient enrichment from farm effluent may affect primary production in mangroves (Alongi et al. 2003), but more research is needed to fully understand the impacts of aquaculture within the larger context of human impacts.

SECTION 4. CHEMICALS

Chemicals used at marine cage culture sites may be released into the surrounding environment. Some of these chemicals pose only a short-term hazard because they are relatively benign, are diluted before being released into ocean waters, degrade quickly in water or do not accumulate in sediments. Other chemicals may persist and pose short or long-term risks to marine biodiversity near farms. The potential effects of antibiotics, therapeutants and antifoulants are often identified as being a concern. Antibiotics and therapeutants are administered to recover sick fish. Antibiotic and chemical use has declined in marine aquaculture in many countries and is limited in the U.S. Antifoulants are chemical treatments used on nets and other marine farm equipment to prevent the settling of marine fouling organisms. Heavy metals, especially zinc and copper, may be released from fish farms in feed or antifouling treatments. In this section we review current literature addressing the effects that these chemicals may have on the surrounding marine environment.

Antibiotics

The environmental impact of antibiotics used in marine aquaculture has been identified as an area of concern and was addressed in early review documents (Wu 1995, Stickney 2002). More recently the known and potential effects of antibiotics have been considered in summary reports for marine aquaculture globally (Beveridge 2004, Nash et al. 2005, Halwart et al. 2007), in the Pacific Northwest (Nash 2001, Nash and Waknitz 2003), in the Mediterranean (International Council for the Exploration of the Seas 2002, The Mediterranean Science Commission 2007), in Europe (International Council for the Exploration of the Seas 2002, Huntington et al. 2006, Braaten 2007), in the U.S. (Goldburg et al. 2001, Nash 2001, Benbrook 2002, Boyd et al. 2005, Lee and O'Bryen 2007, Pittenger et al. 2007), for Canadian salmon farming (Scott 2004, Phillips 2005), and for salmon farming in general (Burridge 2003, Armstrong et al. 2005).

Antibiotics are administered in medicated feed, by injection or by immersion, with the latter resulting in the largest amounts released into natural waters. Commercial medicated feed is readily available and is commonly used at farms in response to disease outbreaks. However, the amount of antibiotics released depends upon the fish species, amount of feeding activity and absorption in the fish digestive tract. Studies estimate that as much as 75-99% of the antibiotic administered is released into the environment (Goldburg and Triplett 1997, Scott 2004, Armstrong et al. 2005, Pittenger et al. 2007).

The most common antibiotics in use around the world in marine aquaculture operations include oxytetracycline, sulfamerazine, amoxicillin, florfenicol, sulphonamides, quinolones, nitrofurans, and erythromycin (Benbrook 2002, International Council for the Exploration of the Seas 2002, Burridge 2003, Armstrong et al. 2005, Huntington et al. 2006). ICES (2002) summarized which antibiotics are used in the UK, Norway, Ireland and Canada. Burridge (2008) includes a recent list of antibiotics in use in Norway, Chile, the U.K. and Canada. According to the Food and Drug Administration (FDA) website antibiotics are infrequently used in U.S. marine fish farms, in part because only three antibiotics are approved for use in the U.S. according to the—oxytetracycline, florfenicol, and sulfadimethoxine/ormetoprim —

and these are only allowed for specific indications in fresh water (www.fda.gov/ AnimalVeterinary/DevelopmentApprovalProcess/Aquaculture/default. htm; visited 07 February 2011). Application in marine aquaculture requires extra-label approval by a licensed veterinarian or under an investigational new animal drug (INAD) approval through the FDA, generally with direct oversight by a veterinarian. Benbrook (2002) has compiled a summary of antibiotic drug use in U.S. aquaculture, but this report focuses largely upon catfish production.

Antibiotics which are administered, but not assimilated by the fish, are released into the environment where they either become dissolved in the water column or settle to the sea floor and accumulate in the sediment (Capone et al. 1996, Lalumera 2004, Rigos et al. 2004). Some antibiotics have relatively short residence times in marine sediment, while others may remain at measurable levels for longer periods. Laboratory and field studies have found that antibiotic persistence in sediment ranges from a few days to years depending on the drug in question and the geophysical properties (including light level, oxygen levels, pH, temperature, and sediment type) of the water or sediment (Scott 2004, Armstrong et al. 2005, Rigos and Troisi 2005).

Rigos and Troisi (2005) have developed a model to illustrate the transport and environmental fate of orally administered antibiotics (Figure 4) at Mediterranean fish farms. This model is useful for visualizing pathways by which antibiotics may become available for uptake in the water column or sediment, and how they may impact marine food webs. The authors review environmental concerns about the potential impacts from antibiotics released from cage aquaculture including the emergence of resistant bacteria, impacts to marine biodiversity and risks to human health.

Exposure to antibiotics in the environment allows bacteria to adapt and become resistant to them (Kummerer 2004). This is true for both the targeted disease pathogens and for other microbes occurring naturally in the ecosystem. Both have been affected by exposure to antibiotics. The occurrence of resistant strains of fish pathogens has been reported in Spain (Zorrilla 2003), Washington state (Nash 2001), Finland and Chile (as cited in Armstrong et al. 2005), and Denmark (Schmidt et al. 2000, Bruun et al. 2003). Questions remain about the relative risk posed to the marine ecosystem by antibiotic use at marine fish farms due to the relatively low doses that may actually be entering the ecosystem, the high solubility and short half-life of many antibiotics, and uncertainty about the level of selective pressure that is exerted on target pathogens or non-target resident microbes (Alday et al. 2006).

Non-target sedimentary microbes may be exposed to antibiotics released from marine farms, raising concerns about the impacts to benthic microbial biodiversity (Burridge 2003). Chelossi et al. (2003) found a higher fraction of antibiotic resistant bacteria isolated from sediments beneath Mediterranean fish farms than from control sites. Similarly, microcosm studies conducted by Kerry et al. (1996) found that oxytetracyline-resistant bacterial colonies were present in 6-90% of replicates depending upon dosage. Review papers by Armstrong et al. (2005) and Scott (2004) reference numerous studies in Canada, Chile, Denmark, Asia, Norway, and the U.S., which report antibiotic resistance in bacteria from sediments underneath freshwater and marine fish farms. The authors suggest that the antibiotics released may create sedimentary reservoirs for resistant bacteria.

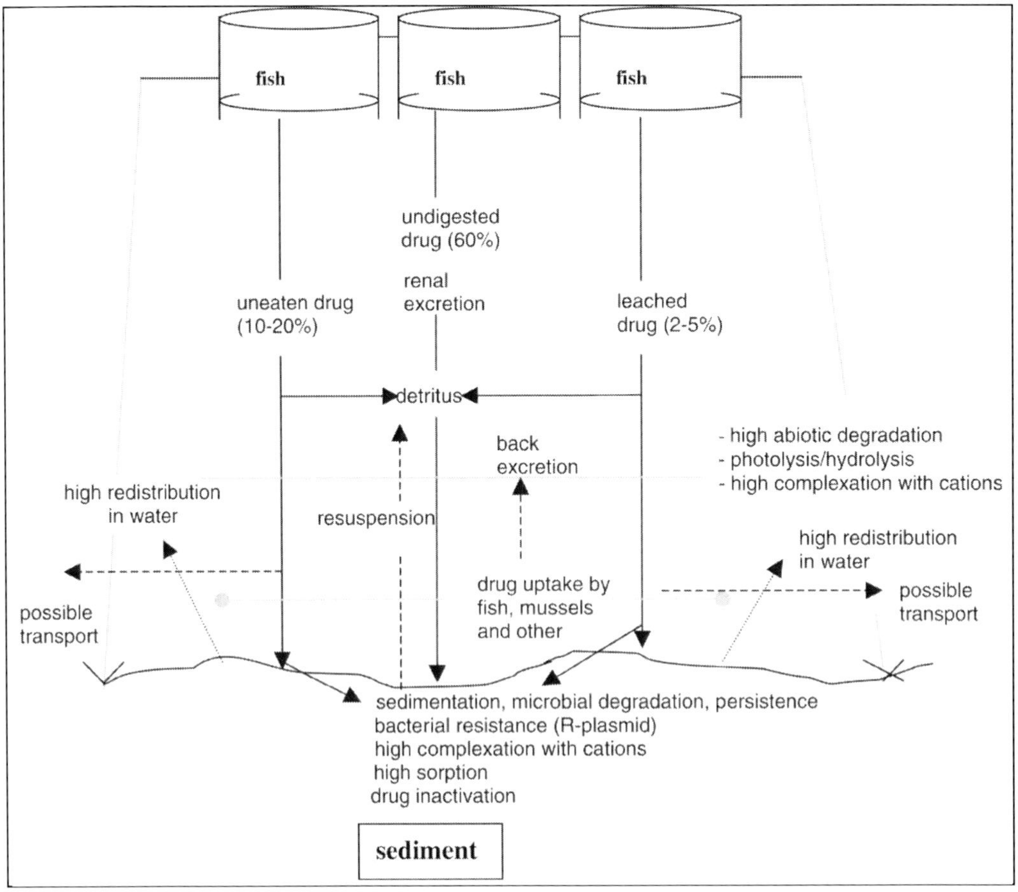

Figure 4. The transport and environmental fate of orally administered antibiotics. Copied with permission from Rigos and Troisi (2005).

The toxicity of antibiotics to marine plants is a concern as they are known to be toxic to some phytoplankton (Halling-Sorensen et al. 1998). For example, when Oh et al. (2005) tested the effects of six antibiotics commonly used in finfish aquaculture on seaweed, they found a dose-dependent decrease in photosynthetic efficiency and ammonium uptake.

Antibiotics thought to originate from marine aquaculture operations have been isolated in wild fauna collected near farm sites (Beveridge 2004). Capone et al. (1996) found oxytetracycline in Dungeness and red rock crabs collected near salmon farms in Puget Sound during and up to 12 days after treatment, but samples collected 41 and 75 days after treatment showed only trace residues. Migliore et al. (1997) compared the toxicity of five antibiotics to *Artemia* nauplii. Mortality rates ranged from 0–100 % depending upon exposure time and dose. The most toxic antibiotic, bacitracin, was found to significantly decrease hatching rates, and flumequine altered nauplii pigmentation. Additional studies in the field would be useful to determine if antibiotic exposure doses that might be measured at a fish farm are sufficient to cause ecological effects.

Studies have found antibiotics present in wild fish feeding on feces and food originating from marine farms (Fortt and Buschmann 2007). This may serve as a pathway for development of antibacterial resistance within wild populations (Rigos et al. 2004). Armstrong et al. (2005) and Scott (2004) document the accumulation of antibiotics in marine fish and invertebrates near farms, but these were studies conducted during years of higher antibiotic use. In their assessment of the environmental impacts of antibiotics to the Mediterranean, Rigos and Troisi (2005) reviewed a range of potential impacts of antibiotic exposure to marine biodiversity, but state that there is a need for studies that directly address the levels of and environmental impacts of antibiotics under sea cages. Burridge et al. (2008) suggest that continued research into the accumulation of antibiotics from fish farms to the flesh of other organisms, including humans, is needed. To protect human health, most governments require a withdrawal period of days to months before fish treated with antibiotics may be harvested (Burridge 2003). He also notes that aquaculture workers may be exposed to antibiotics in the dust aerosols associated with feed production and distribution.

In general, improved husbandry in marine cage culture over the last 10-20 years has resulted in a tremendous decline in the use of antibiotics in Scandinavia, Canada, the U.S. and Europe (International Council for the Exploration of the Seas 2002, Cardia and Lovatelli 2007, Pittenger et al. 2007, Holmer et al. 2008). For example, data from Norway (Tveterås 2002) and British Columbia, are representative of the general decline in antibiotic use in the marine aquaculture industry since the 1980s (Figure 5). Nash (2001) concludes that antibiotics used in marine aquaculture are safe and effective, stating that over 40 years of salmon culture in the Pacific Northwest has not resulted in adverse impacts to wild salmon. Phillips (2005) also suggests there is little risk to native salmon from antibiotic use in the Puget Sound. In other countries, however, considerable concern remains about the levels of antibiotics used in marine aquaculture (Cardia and Lovatelli 2007, Fortt and Buschmann 2007).

Recent interest in the use of probiotics to promote fish health and decrease the need for antibiotics has opened a new avenue for immunological and husbandry research. Probiotics are nutritional supplements generally comprising yeast, algae and bacteria thought to be beneficial for health, growth, immunocompetence and survival (Das et al. 2008). Early studies have resulted in promising results for cultured species (Aguilar-Macías et al. 2010, Dharmaraj and Kandasamy 2010). Probiotics have been in use in China since the 1980's (Qi et al. 2009) and are being investigated for use in U.S. aquaculture as well.

While some research to understand the effects of antibiotics used in marine finfish culture has been conducted, there are many unanswered questions. For example, additional research to investigate the role of the sediment beneath farms as a reservoir for resistant bacterial or the development of resistant strains of infectious diseases is warranted. Because antibiotics may persist in the marine environment, questions remain about the long-term impacts to biodiversity, ecosystem function and human health. Although their use in marine fish farming is limited in the U.S. (e.g., no antibiotics have been applied at Maine salmon farms in the last 6+ years, Jon Lewis, personal communication), and the environmental impacts of antibiotics may be minimized at coastal and offshore sites through dilution (Goldburg et al. 2001, Holmer 2010), it will continue to be important to consider the ecological effects that antibiotics can have in the marine environment. An Integrated Pest Management approach, which decreases stress to the fish, employs stocking densities to keep fish healthy and uses preventative vaccination, is used to maintain fish health while minimizing or eliminating the use of antibiotic drugs.

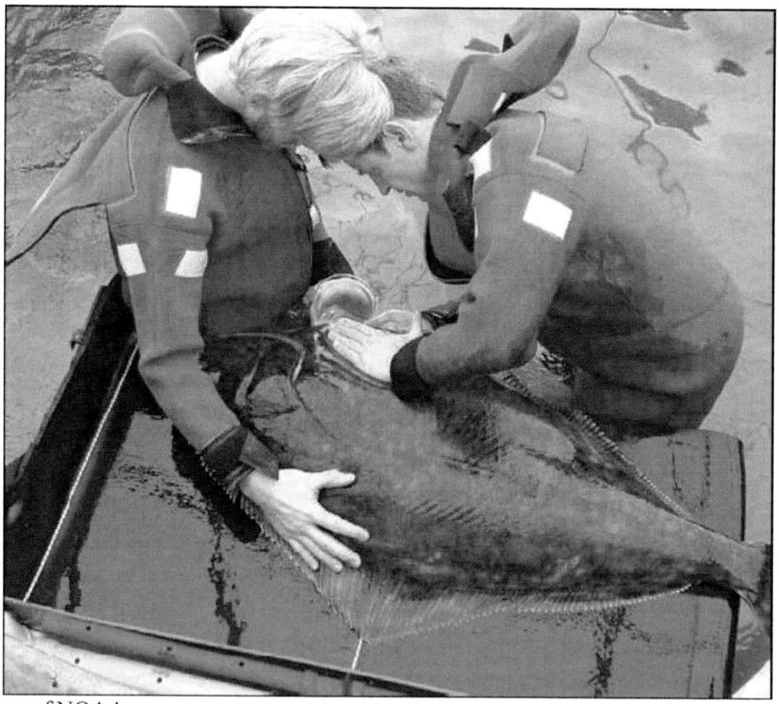

Photo courtesy of NOAA.

Therapeutants

The environmental effects of therapeutic chemicals used in marine finfish aquaculture are consistently identified worldwide as a concern (Cardia and Lovatelli 2007, Pittenger et al. 2007). Therapeutants are used to treat parasitic, viral, fungal and bacterial infections and to treat aquaculture facilities for disease causing agents. These drugs may be administered in medicated feed or by immersing the fish. Therapeutants are released into the environment from uneaten food, the feces if not absorbed in the gut, or direct release into the water. Therapeutants may be present in the water column or may accumulate in the sediment below cages. Assessments of the therapeutants being used in aquaculture have been conducted in the U.S. (Nash 2001, Phillips 2005), Scotland (Black et al. 2002), Canada (Burridge 2003, Scott 2004, Burridge et al. 2008), Europe (International Council for the Exploration of the Seas 2002, Huntington et al. 2006, Burridge et al. 2010), Asia (Graslund and Bengtsson 2001), the Mediterranean (The Mediterranean Science Commission 2007) and by the United Nations (Joint Group of Experts on the Scientific Aspects of Marine Environmental Protection 1997, Cardia and Lovatelli 2007). These reports review known and potential environmental impacts of therapeutants to the environment in order to promote their safe and effective use. In the U.S., the FDA maintains the list of therapeutants approved for use in aquaculture, which is currently limited to hydrogen peroxide (applied as an extra label use with veterinarian approval) and formalin (which is not currently used in marine fish farming operations) (www.fda.gov/AnimalVeterinary/DevelopmentApprovalProcess/Aquaculture/default.htm; visited 14 February 2011).

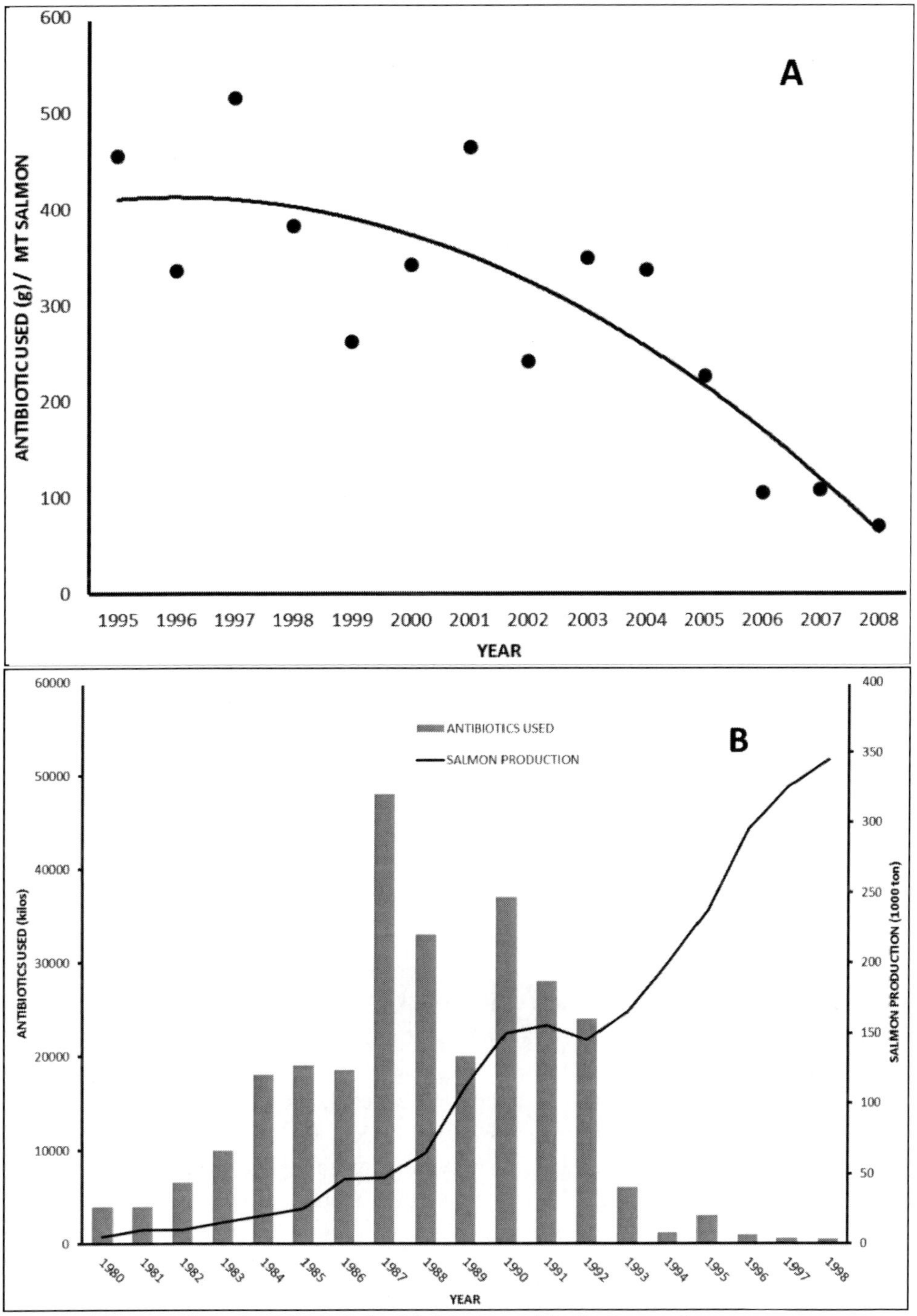

(A) Adapted from graph on the British Columbia Ministry of Agriculture webpage. (http://www.agf.gov.bc.ca/ahc/fish_health/antibiotics.htm; visited.

Figure 5. Antibiotic use in salmon aquaculture.

Therapeutants pose several environmental risks including the evolution of resistant strains of pathogenic organisms, nonlethal toxicity, direct mortality, and bioaccumulation in the food chain. Summary reports by Hallig-Sorensen et al. (1998) and GESAMP (1997) provide earlier compilations of research done on the environmental effects of therapeutants, documenting acute and sublethal impacts of drugs to non-target organisms in the marine environment. Hallig-Sorensen et al. addresses a very wide range of pharmaceuticals that are present in terrestrial and aquatic environments, but with no focus upon aquaculture. However, many of the compounds covered are used in fish farming. The GESAMP report is specifically geared toward chemicals used in coastal aquaculture. While much has been added to the body of knowledge since its publication, this report remains a valuable source of information.

More recently, summaries have been prepared by Gräslund and Bengtsson (2001) for shrimp aquaculture in Asia, by Burridge (2003) for Canadian marine finfish aquaculture, by Scott (2004) for Canadian freshwater aquaculture, by Haya et al. (2005) for sea lice therapeutants, and by Burridge et al. (2008) for chemicals used in salmon aquaculture. Collectively, these reports review a wide range of chemicals used topically or in feed at aquaculture operations around the world. They address the applications of therapeutants, the dosages used, and known or potential effects these chemicals have on plants and animals in the aquatic environment. Most of the effects that are documented in these reports are LC_{50} values from laboratory toxicity trials. Less information is available about sublethal impacts or cumulative impacts resulting from accumulation of therapeutants in marine sediments.

Persistence of therapeutants in the environment is highly variable. Half-lives in the water column may be only a few hours. In sediments, however, therapeutants may accumulate and remain high for months after application (Telfer et al. 2006) and it is the effects of residual therapeutants in sediments that are generally believed to pose the greatest environmental risk (Black et al. 2002, Huntington et al. 2006, Burridge et al. 2008).

Some of the most common therapeutants in use in North America and Europe are insecticides including pyrethroids, emamectin benzoate, teflubenzuron, ivermectin. In the U.S. only emamectin benzoate is approved for limited use under an INAD and is typically administered using a well boat to avoid dumping the treatment bath at sea. Therapeutants are administered either by immersion or in medicated feed, and are often used to treat sea lice and other external parasites. The insecticides can have toxic effects on molting marine organisms like crustaceans, amphipods and zooplankton (Grant and Briggs 1998, Beveridge 2004, Tett 2008), including commercially important species such as crabs, lobsters and shrimp which may reside or forage in sediments near fish farms (Haya et al. 2001).

In addition to direct mortality in crustaceans, insecticides are found to inhibit chitin production, induce behavioral changes, affect swimming ability, decrease spawning, cause premature molting (Burridge 2003, Haya et al. 2005) and changes in enzyme activity (Gowland et al. 2002). Field and laboratory studies reporting little or no effects of insecticide treatment on non-target organisms are also reported (Graslund and Bengtsson 2001, Burridge 2003, Haya et al. 2005). For example, a study in Scotland determined that the sea lice medicines cypermethrin, hydrogen peroxide, azamethiphos and emamectin benzoate had no measurable impact to phytoplankton, zooplankton, benthic macrofauna or barnacle settling rates near four active salmon farm sites (Black et al. 2005). Copepod and nematode communities were negatively impacted by sea lice treatments at one site, though these changes were also related to an organic enrichment gradient. The cypermethrin and emamectin benzoate concentrations were elevated around the cages when treatments occurred, yet these levels were generally lower than what laboratory studies found to be toxic to copepods (Willis et al. 2005).

Currently, hydrogen peroxide can be used as a parasite treatment in aquaculture in the U.S. Hydrogen peroxide is commonly used as an immersion treatment for sea lice and other external parasites, bacteria and fungi. Because it dissolves quickly in water, this chemical is perceived to pose very little environmental risk (Haya et al. 2005, Schmidt et al. 2006, Yanong 2008). Hydrogen peroxide was found to have no significant side effects on walleye blood chemistry (Tort et al. 2003) suggesting that it may pose low risk to non-target fishes that may be exposed in treated water released into the ocean. Mansell et al. (2005) measured acute secondary hematological effects of hydrogen peroxide to infected cultured fish (kingfish), but concluded they were outweighed by overall health benefits to the fish. The use of cleaner fish like wrasse is suggested as an ecological alternative to hydrogen peroxide for the treatment of external parasites (Costello et al. 2001).

Formalin, the other therapeutant approved for use in the U.S., is used as a fungicide and to control ectoparasites (Joint Group of Experts on the Scientific Aspects of Marine Environmental Protection 1997) and as a disinfectant (Burridge et al. 2008). Acute toxicity trials of formalin have been conducted for fish (Graslund and Bengtsson 2001, Scott 2004), but sublethal effects in marine organisms are not well studied. In the U.S., formalin is not specifically approved for use in the marine environment and is not applied in open ocean fish farms.

Many types of chemicals are used in aquaculture facilities as disinfectants for pens, facilities and gear including iodophores, chlorine derivatives, benzalkonium chloride, malachite green (Joint Group of Experts on the Scientific Aspects of Marine Environmental Protection 1997, Burridge 2003, Burridge et al. 2008). Most of these are not approved for use in U.S. marine fish farms by the EPA. Acute toxicity trials have been conducted for some of these chemicals, and they are recognized as environmental toxins, but little is known about their potential impacts in the marine environment (Burridge 2003). These chemicals are used effectively to control a wide range of unwanted or harmful organisms in aquaculture facilities, and it is their non-specific toxicity which poses a threat to non-target marine organisms. Persistence in marine sediments needs to be evaluated as degradation of these compounds is highly variable and dependent upon environmental factors like temperature, pH, the level of dissolved oxygen, light intensity and the presence of micro-organisms (Graslund and Bengtsson 2001).

Because improvements in fish husbandry have decreased the use of therapeutants and the few therapeutants approved for use in the U.S. are considered to be relatively non-toxic in the marine environment, some believe that these chemicals pose low risk to the environment (Nash 2001, Fairgrieve and Rust 2003, Nash 2003). More research is needed to understand cumulative, long- term and sublethal effects that therapeutants may have in the marine environment. Environmental risk should be evaluated for each chemical and take into consideration site characteristics like flushing rate and sediment type. Costello et al. (2001) provide a list of best practice recommendations to guide the use of chemicals in the environment. The illegal application of non-approved therapeutants, the misuse of approved drugs or the use of INADs are avenues by which additional drugs may be introduced to the environment (Love et al. 2011, Ocean Conservancy 2011).

Antifoulants

Antifoulants are used to control or eliminate the growth of marine organisms which attach to aquaculture cages, ropes, and structures. Heavy and persistent biofouling impedes water flow through cages, increases biological oxygen demand in cages, causes net drag and can shorten the useful life of nets and ropes (Swift et al. 2006, Braithwaite et al. 2007, Belle and Nash 2008, Burridge et al. 2010). Chemical antifouling treatments on nets and other farm structures are effective at reducing biofouling because their toxicity to many marine attaching organisms prevents settlement of larvae on nets and equipment. For this same reason, their toxic effects on other organisms in the water and sediments around fish farms are a concern (Wu 1995, Burridge 2003, International Union for Conservation of Nature 2007, Tett 2008, Burridge et al. 2010).

In the past, tin compounds (e.g., tributyltin) were used in antifouling treatments, but were banned in many countries for use in aquaculture because of their harmful environmental effects (Davies et al. 1988, Minchin et al. 1996, Joint Group of Experts on the Scientific Aspects of Marine Environmental Protection 1997, Kan-atireklap et al. 1997, Minchin et al. 1997, Nash 2001). Tin compounds are currently not in use on fish farms in the U.S., Canada, much of Europe and other countries. However, tin is still used for other marine applications — in antifouling paint for boats, for example. The accumulation of anthropogenic tin in coastal waters (Sudaryanto et al. 2002, Ščančar et al. 2006) and its toxic effects continue to be documented for marine life including mollusks (Meng et al. 2005) and fish (Dimitriou et al. 2003). Elevated tin levels were found in some marine aquaculture products in Japan six years after the use of tin was banned (Ueno et al. 1999) and in pearl culture areas following reported illegal use of tin compounds (Ramaswamy et al. 2004).

Currently, copper-based antifoulants are the most common and effective chemical net treatments in use worldwide (Burridge 2003, Braithwaite et al. 2007, Guenther et al. 2009). Burridge et al. (2010) provide a thorough overview of the effectiveness and biological impacts of copper used in aquaculture. They report copper has been found to leach out of nets and accumulate to levels above regulatory water quality guidelines in sediments below fish farms in Canada and Scotland. Copper tends to bind strongly with sediments, so its bioavailability and long-term toxicity warrant further investigation. Two reviewed studies found that tissue from fish inside copper-treated net pens did not show copper accumulation. Likewise, Solberg et al. (2002) detected no differences in copper levels in muscle and liver tissue in both farmed salmon or wild fish collected at farm sites using copper net treatments versus a reference farm that did not use copper.

Burridge et al. (2010) provides a general summary of copper's toxicity to marine organisms and documents lethal and sublethal effects to phytoplankton, bacteria, crustaceans, crabs, mollusks and fish. With respect to research conducted specifically at fish farming sites, Chou et al. (2002a) found that lobsters collected near aquaculture sites had elevated copper levels in their digestive glands. Similarly, Chou et al. (2003) found elevated copper levels in sea urchins collected within 75 m of salmon aquaculture sites. Additional reviews summarizing the use and toxic effects of copper in aquaculture are available (Graslund and Bengtsson 2001, Burridge 2003, Burridge et al. 2008).

Little research has assessed the persistence of copper in sediments near marine fish cages and its accumulation in benthic marine organisms. In a study to predict benthic recovery at farm sites in New Zealand, Morrisey et al. (2000) suggest that accumulation of copper and other metals in the sediments may impair recolonization of benthic organisms. Nash et al. (2005) conclude that best management practices to reduce copper contamination has greatly reduced risk to the marine environment around fish cages, and that copper is limited to non-toxic levels when combined with sulfides in organic sediments. Rensel and Forster (2007) report no measurable copper increase has been found in sediments monitored at net pen sites in Puget Sound. The IUCN (2007) reports that current environmental effects of antifoulants are less than in the past.

Work is underway to develop and implement alternative chemicals and methods to reduce biofouling at marine fish farms. For example, the pan-European Collective Research on Aquaculture Biofouling project (CRAB, www.crabproject.com) is working to develop and implement effective strategies to manage biofouling. CRAB estimates that more effective control of fouling will result in saving of 5-10% of the market value in European aquaculture (Willemsen 2005). Nets may be air dried to kill biofouling organisms by lifting the top area above the water line, or manually cleaned (preferably on land to eliminate bioloading to the sediment) using high pressure spray or scraping to remove encrusted organisms (Belle and Nash 2008). Air drying of nets is phasing out the use of copper in antifouling agents (Braaten 2007, Holmer et al. 2008). Other, less toxic chemicals like acetic acid (Forrest et al. 2007) are being investigated as immersion treatments. The use of non-toxic biofilms as alternative net coatings is also being researched (Bazes et al. 2006, Sarà 2007). Biofilms can be derived from marine organisms including plants, bacteria and sponges that naturally repel the larvae of sessile marine organisms (Qian et al. 2007). This technology is still in its early stages, but may yield promising results in the development of less or non-toxic antifoulants. Using grazing animals or mechanical robots inside fish cages to remove attaching organisms have also been proposed as alternatives to copper treatment (Willemsen 2005, International Union for Conservation of Nature 2007). Sala and Luchetti (2008) describe a low cost, effective Wave Brush® prototype that prevented biofouling in oyster farms and which may be adapted to other aquaculture operations. One study (Lander et al. 2009) has reported positive results of recycling the mussel encrusted nets from fish farming for spat collection in integrated multi-trophic aquaculture (IMTA).

Heavy Metals

Heavy metals are present in trace amounts as nutrients in feed used at marine fish farms and over time may accumulate in benthic sediments beneath cages. Both direct toxicity of heavy metals and accumulation in the benthic food chain have been identified as potential environmental impacts of marine aquaculture in the U.S. (Goldburg and Triplett 1997, Pittenger et al. 2007), Canada (Johannessen et al. 2007), Chile (Buschmann et al. 2009), the European Union (Black et al. 2002, Huntington et al. 2006), the Mediterranean (International Union for Conservation of Nature 2007, The Mediterranean Science Commission 2007), Australia (Cole 2002) and globally (Cardia and Lovatelli 2007, Burridge et al. 2010). Much research has been done in the last ten years to measure the release, accumulation and persistence of heavy metals at marine fish farms.

In the U.S., the accumulation of heavy metals such as zinc and copper (also see antifoulants section) was identified as an issue of concern in salmon net pen farming (Nash 2001). However, reduction in the bioavailability of metals due to soil geochemistry and improvements in feed formulation were identified as factors which could decrease environmental risk. Zinc does not tend to accumulate readily in muscle tissue and therefore the risk to humans is thought to be low (Nash et al. 2005). Nash (2003) states that in long-term studies background levels of metals are achieved after fallowing, but no specific work was cited. Phillips (2005) compiled a summary of recent contaminant research on farmed salmon and other fish. The potential effects of these contaminants – including metals, PCBs, organochlorine and dioxin – range from negligible to very serious.

Many heavy metal studies have been conducted in the last ten years at Canadian farm sites providing valuable insight for potential U.S. farm operations (Scott 2004). Copper (up to 55 μg/g compared to 13 μg/g) and zinc (253 μg/g compared to 49 μg/g) levels were elevated beneath salmon cages at 14 sites in New Brunswick compared to reference stations, but iron and manganese were not (Chou et al. 2002b). There was a decreasing trend of copper and zinc enrichment 50 m from the cages, with anoxic sediments showing the highest levels of accumulation. In this same study, wild lobsters *Homarus americanus* were captured around farm sites. Although copper levels were higher at one heavily farmed site, all tissue concentrations were comparable to other North American lobsters. These same results were used in a later paper (Chou et al. 2004) to develop regression models for understanding sediment dynamics for environmental monitoring. In their study of sea urchin tissue collected under and up to 100 m from salmon aquaculture cage sites, Chou et al. (2003) found that at normoxic sampling sites the tissue levels of copper, zinc, iron, manganese and cadmium levels were comparable to reference locations. At hypoxic and anoxic sites, varying elevations of heavy metals in the urchins were correlated with body size and distance from the cage site. Sampling at a salmon farm in British Columbia over a fallowing cycle found that elevated zinc sediment levels (up to 200 μg/g) returned to background concentrations within six months (Brooks et al. 2003), but never exceeded the apparent effects threshold of 260 μg/g. In British Columbia, sampling along a single transect (0-300 m from a salmon farm) found some zinc and copper levels above the sediment quality guidelines, but this was not the case for iron, aluminum and manganese (Schendel et al. 2004). A sediment geochronology study in the Bay of Fundy showed elevated levels of copper and zinc in cores taken in an area with 20 years of salmon farming (Smith et al. 2005). Enrichment was highest in the immediate vicinity of cage sites, with decreasing levels out to 200 m. Even five years after cage removal, copper and zinc levels remained high indicating that bioremediation of these metals may be slow. Zinc concentrations were higher at 16 out of 22 sampling stations near fish farms in British Columbia compared to reference sites (Sutherland et al. 2007). Copper was also increased at near field (0-30 m) sites, especially in fine-grained sediments. This study's objective was to determine if metals like copper and zinc could be used as waste tracer elements to differentiate between natural and anthropogenic sources of contamination at larger scales. Yeats et al. (2005) similarly found that zinc and copper could be used in sediment analysis to identify waste directly linked to aquaculture cage sites.

Burridge et al. (2010 and references therein) provide a comprehensive review of heavy metal release and accumulation at fish farms in Chile, Norway, Canada and the U.K. Research in all of these countries has detected metal accumulation in sediments and fish tissue. In Chile, as in other countries, sampling generally found elevated copper in farm sediments, but Burridge concludes that copper likely is not readily bioavailable in carbon-rich, oxygen-poor sediments which tend to bind metals. Similarly, they conclude that zinc, also found in elevated levels beneath Chilean cages, is also bound in the sediments minimizing its availability. This conclusion is supported by non-specific ecotoxicological bioassays in which mesocosms were stocked with test organisms and sediments collected from aquaculture sites (Rudolph et al. 2009). No toxic effects were observed, but the authors recommended monitoring to detect changes and effects at larger spatial and temporal scales. A Chilean study investigating the use of fish farm sediments as agricultural manure measured nutrient and heavy metal concentrations in sediment samples from below salmon farms (Salazar and Saldana 2007). Generally, manure from marine salmon farms contained lower heavy metal concentrations than freshwater lake or trout farm manure. Only copper concentrations were consistently higher at 89 mg/kg dry weight compared with 45 and 33.4 mg/ kg, respectively. All concentrations were well below environmental limit values.

A Scottish review of the environmental impacts of aquaculture reports that elevated levels of zinc and copper from fish feeds and antifoulants have been found at farm sites (Black et al. 2002). One study found that sediments beneath and within 30 m of farms were contaminated with zinc and copper and adverse effects to the benthic community were predicted. While immediate bioavailability may be minimal, remobilization of the metals by strong currents or trawling was identified as a potential concern. Heavy metal contamination may be a barrier to recolonization during fallowing. Further research addressing the toxicology, biological impacts and management was recommended. Dean et al. (2007) sampled 70 stations near a Scottish fish farm and found maximum sediment concentrations of 921, 805 and 3.5 μg/g of zinc, copper and cadmium, respectively. The calculated losses from the farm (feed input minus fish output) were 87.0%, 4.3% and 14.0% of the background-corrected inventories for Zn, Cu and Cd, respectively, indicating that for Cu and Cd at least, the feed is not the only source. Metal concentrations decreased away from the farm and reached background levels about 300 m from the farm. Background heavy metal concentrations must be accounted for by using standard methods such as lithium normalization appropriate for the region being sampled (Yeats et al. 2005). Huntington et al. (2006) determined that the release and persistence of metals into the water and sediments is an important environmental effect to consider, but note that information about bioavailability and long-term ecological implications is lacking.

The accumulation of heavy metals in marine sediments due to fish farming has been reported in Spain. Compared to upstream control sites, Mendiguchia et al. (2006) found increased levels of zinc (140% above levels), copper (362%) and lead (97%) in several marine aquaculture facilities. Macroalgae cultivated in an intensive Spanish sea bream growout facility did not accumulate heavy metals during a normal growth period and were considered safe for the human food industry (Hernandez et al. 2005). An analysis of a variety of commercial fish feeds used in Italian aquaculture found no mercury present in any samples and only low levels of arsenic, cadmium, chromium and lead (Abete et al. 2004). The researchers concluded that although the feeds used complied with governmental standards, but raised the potential for cumulative impacts to the cultured fish and the food chain after long term application of feeds.

In the eastern Mediterranean, heavy metal enrichment of the water column and sediments around three fish farms was measured and attributed directly to the feeds used (Belias et al. 2003). The greatest increases in sediment metal concentrations relative to control sites were for iron (up to 80 ppb compared to 28 ppb), zinc (up to 9 ppb, compared to 4 ppb) and copper (up to 1.3 ppb compared to 0.5 ppb). The three farms sampled had been in operation for 7-9 years. Similarly, Aksu et al. (2010) sampled sediments off Turkey at eight sea bream and sea bass farms. Although the farm sites generally had elevated levels of zinc, iron and copper, the concentrations detected were below guideline limits or probable effect levels. The authors concluded that the sediments had not reached metal concentrations that were ecologically harmful.

Sediment samples from three marine aquaculture sites in Hong Kong found elevated levels of zinc and copper (Wong et al. 2001). The farm sited in an isolated area with low human population generally had lower heavy metal levels indicating that anthropogenic sources of metal other than aquaculture also contributed to sediment pollution.

Sampling at a farm site in New Zealand found sediment zinc levels of 665 μg/g dry weight (Morrisey et al. 2000). This was about 38 times the level measured at nearby (100 m away) control sites, suggesting that metal accumulation was localized. No differences in copper concentrations were detected.

Recent research indicates that heavy metals do tend to accumulate in the sediments below fish farms. However, most studies have found that concentrations are within acceptable environmental guidelines even at farms that have been in production for many years. Because the metals are often bound in the sediments, they are generally perceived to be of low risk in terms of environmental effects. Improvement in feed formulation is expected to decrease zinc loading to the marine environment, as many manufacturers are adding lower amounts of a more available form, zinc methionine (Burridge et al. 2010). Long term monitoring is usually recommended, particularly if fish farm manure is going to be used for agricultural fertilization.

In addition to heavy metals, other contaminants have been measured around Canadian fish farms. For example, Hellou et al. (2005) analyzed feed pellets and sediment samples for polycyclic aromatic hydrocarbons (PAHs), polychlorinated biphenyls (PCBs) and chlorinated pesticides including DDT. PAHs and pesticides were found in three of the five feeds tested and in fish oil. PCBs were found in two of the feeds and the oil. In sediment samples, PAHs concentrations ranged from 90-10,234 ng/g compared to 4ng/g at reference sites. PCBs were generally highest under cage sites, decreasing with distance. Of the 12 pesticides analyzed, only DDE (a DDT breakdown product) was consistently detected with the same trend as PCBs.

A recent study reported increased mercury bioaccumulation in rockfish *Sebastes spp.* around aquaculture facilities in British Columbia as the rockfish switched to feeding at higher trophic levels by including overall more fish prey in their diet (Debruyn et al. 2006). Mercury in the prey fish was thought to originate from foraging on waste feed and fish feces and the mobilization of native and added mercury in sediment due to farm-induced anoxia, resulting in increased bioaccumulation in the rockfish predators.

SECTION 5. MANAGEMENT TOOLS

Perhaps the most effective marine fish farm management tool is good decision making when selecting sites. This approach is emphasized in the preceeding sections, with recurring recommendations to locate farms in well-flushed waters, including deep offshore areas over erosional sediments. Additionally, computer simulation models (Cromey et al. 2002, Nordvarg and Hakanson 2002, Doglioli et al. 2004, Rensel et al. 2007, Jusup et al. 2009, Tett et al. 2011, Cromey et al. 2012) that track nutrient discharge and predict environmental impacts are tools for farm managers and government regulatory agencies to understand interactions of fish farms and the marine environment and to guide decision making about siting, monitoring and management. However, this topic is not covered here as it was beyond the scope of this report to provide a comprehensive analysis and comparison of such models.

The next sections cover two areas of marine fish farm management that have received attention due to their potential to reduce environmental impacts. Fallowing is recommended and implemented to allow the benthic sediments and communities below fish cages to recover from negative effects of nutrient loading. We reviewed research that aims to establish how long it takes for the benthic habitat to recover and whether fallowing is effective at re-establishing a normal faunal assemblage. Integrated multi- trophic aquaculture, or IMTA, is a tool that can both broaden the economic base of the mariculture industry while at the same time adding species to the farm facilities that can decrease nutrient loading, improve water quality and decrease sedimentation.

Fallowing

Fallowing refers to the practice of relocating marine fish cages or delaying restocking of cages to allow the sediment below to undergo natural recovery, both geochemically and ecologically, from the impacts of nutrient loading. This management tool is implemented around the world for preventing long-lasting damage to the benthic environment and for decreasing risks to farms of pathogens and parasites. This section provides an overview of the duration, extent and degree of recovery associated with the practice of fallowing.

Nash (2001, 2003) summarizes information collected in Pacific Northwest salmon farms showing that geochemical recovery may take as little as a few weeks or months at some sites, but up to two or three years at others. Biological remediation typically lags behind chemical recovery depending upon recruitment of new fauna. Typically, as organic material is dispersed or degraded and sediment redox potential increases recovery of the benthic fauna progresses with generalists and bottom feeders recolonizing rapidly, and rare taxa reappearing more slowly. A study at a Hawaiian Pacific threadfin farm reported on the effectiveness of a six month fallowing period, as reflected by faunal community recovery, following five years of commercial farm activity (Lee et al. 2006). Changes in the community had been evident up to 80 m from the cage site compared to controls almost 400 m away. Although improvement was evident during the study period, comparisons of community structure indicated that complete recovery was not yet achieved. As part of an extensive completion report on a mutton snapper and cobia farming operation, fallowing was recommended by Alston et al. (2005) as a management practice for aquaculture operations in Puerto Rico.

A comprehensive analysis of fallowing practices at salmon farms was prepared for Fisheries and Oceans Canada in the late 1990s (Stewart 1998), but was driven primarily by fish health rather than environmental impact concerns. It concluded that fallowing of at least three months, in combination with good farm practices such as adequate distance between cages, was beneficial in reducing the spread of infectious pathogens. Brooks et al. (2003) conducted an 18 month study to determine the extents of benthic chemical and biological impacts at high production salmon farms in British Columbia. Impacts increased progressively from stocking through the production cycle and were evident to at least 50 m from the cages. Chemical remediation of the sediments was evident immediately following the onset of harvest in August and was considered complete nine months later when harvest ended in April. However, it took another six months of fallowing before biological indices suggested that the benthic community was also recovered. In contrast, a seven year study of remediation conducted at a nearby (5 nm away) set of salmon farms reported much longer recovery times (Brooks et al. 2004). After eight years in production, there was extensive waste accumulation and sediment degradation up to 200 m away from the cages. Twelve surveys over five years indicated steady, but very slow and incomplete remediation at this site. Chemical improvements were evident after the first year and continued to progress over time. The biological indices reflected an even slower pattern of improvement and after five years there was minimal recovery within the first 80 m impact zone. Biodiversity indices from samples within 100 m of the farm also remained below reference values and the taxa were primarily common or generalist species. The highest impact area shifted 50 m down current of the cage site after four years. The researchers believe that siting the farm in a net depositional hydrodynamic regime was responsible for this abnormally extended recovery time. A review by Wildish and Pohle (2005) supports the idea that shorter fallowing cycles of 6-24 months are generally adequate for restoring benthic macrofauna, depending upon site specific conditions. Tlusty et al. (2005) reports that Canadian salmon farm sites without fallowing protocols displayed higher organic loading and impacts to nematode diversity compared to fallowed farms.

Studies in Europe also support the effectiveness of fallowing and point to the influence of local conditions on the extent of the recovery period. Kraufvelin et al. (2001) compared recovery at two farming sites in Finland 4-7 years after decreased nutrient loading and cessation of farming. The two sites varied greatly in topography, water circulation and hydrodynamic conditions. At the shallow sheltered site, there was no recovery of the infaunal community and even after seven years there were sampling stations with no organisms present. In contrast, the well flushed site showed faunal colonization and recovery even with ongoing fish farming in the area. Sediment sampling at Norwegian salmon farms found severe chemical and biological benthic impacts within 50 m of cages, but detectable effects up to several hundred meters away (Carroll et al. 2003). Fallowed farms exhibited significantly less environmental impacts, so fallowing was recommended as a management tool for sustainable fish farming. When Schaaning and Hansen (2005) sampled sediment beneath abandoned farm locations they found evidence of chemical remediation approaching reference site levels. Sediments became normalized within 0.5-3 years after farming stopped, thus demonstrating the positive results that might be expected from fallowing.

A similar study was conducted at 10 salmon cages in Scotland (Pereira et al. 2004) after the cessation of farming. Chemical and biological samples were collected over 15 months. Within this time the sites 30 and 55 m from the farm indicated a recovered benthic assemblage, but at the station adjacent to the cage significantly reduced and hypoxic conditions persisted. Species abundance and diversity also remained in degraded conditions at the cage site compared to the other two sites where biological recovery steadily increased over the study period. This farm had been under constant production for several years prior to fallowing, and the results of the study suggest that additional time was required to fully recover benthic conditions at this farm.

A research project conducted in the Canary Islands quantified the use by wild fish assemblages of the area below sea bream and sea bass cages before and after the cessation of farming (Tuya et al. 2006). Over the course of fallowing the abundance of large mugilids, rays and *Pagellus* species decreased drastically, while the numbers of herbivores and benthic carnivores remained relatively stable. Large carnivorous fishes increased in numbers. The authors concluded that the lack of farm feed input was the driving factor in shifting the fish assemblage.

Investigations in the Mediterranean support fallowing as a management option to reduce environmental impacts. La Rosa et al. (2001) sampled microbes and meiofauna at a sea bass farm off western Italy just prior to harvest and following the removal of the cage. Within four months significant chemical and biological recovery was evident, but not complete. Microbial response was quite rapid — in only 15 days, microbial components returned to control values. In a eutrophic lagoon heavily impacted by fish farming effluents, Lardicci et al. (2001) found an impoverished benthic community even six years after the implementation of environmental restoration measures. While not a fallowing study *per se*, these results are useful in understanding the interplay between nutrification from aquaculture waste, geochemical processes and benthic community response in enclosed, poorly flushed marine habitats. Porrello et al. (2005) recommended fallowing as a management tool at sea cages to minimize environmental impacts. Their study concluded that the negative impacts to the benthic chemistry were confined to within 50 m of the open ocean sea bass farm they evaluated. Because the bottom currents were slow and the cages were placed in shallow water, fallowing was suggested to allow recovery of the sediments. A comparison of benthic impacts at seven Greek fish farms in the eastern Mediterranean found low organic carbon concentrations, high redox and high diversity index at the single operational farm which implemented fallowing (Lampadariou et al. 2008). In fact, the fallowed farm was the only one at which there were no differences between samples taken beneath the farm and at distances out to 50 m. Additional comparative studies would be useful in developing fallowing guidelines for the Mediterranean.

Vezzulli et al. (2004) compared the effectiveness of two bioremediation products for mobilizing carbon in organic rich sediments below fish cages. Biovase, a proprietary blend of indigenous microorganisms, stimulated carbon mobilization (up to 23% increase) and enhanced extra-cellular enzymatic activity in the sediment. A commercial product containing oxygen releasing compounds was less effective, but did result in a slight increase in carbon mobilization. The widespread use of such remediation products is likely cost prohibitive, but continued development of technology to aid in farm waste management is warranted (Chavez-Crooker and Obreque-Contreras 2010).

Morrisey et al. (2000) conducted a study at New Zealand marine salmon farms, including a site fallowed after more than a decade of commercial mariculture, to validate a modeling approach to predict benthic impacts. The observed levels of carbon sedimentation were in agreement with values generated by the model, and the model accurately estimated one-year recovery rates for the fallowed sediments. Additional findings suggested that zinc and copper in farm sediments may impair the process of chemical and biological remediation. In Tasmania, sediments below a salmon farm recovered very quickly after removal of the cage (Macleod et al. 2004). After only two months geochemical parameters indicated marked improvement. However, the macrobenthos remained unrecovered for 36 months. This study stresses the importance of defining recovery benchmarks to make certain the correct environmental indicators are used to establishing fallowing guidelines. Fallowing was strongly recommended based upon an investigation of nutrient fluxes in the sediments below bluefin tuna farms in Australia (Lauer et al. 2009). High sedimentation rates and up to ten-fold increases in nitrogen and phosphorus were reversed following a four month fallowing period. The high potential for organic sediment enrichment in this industry makes the implementation of fallowing an important aspect of ensuring long term environmental sustainability.

Lin and Bailey-Brock (2008) constructed a summary table of ten recent (since 1999) fallowing studies from seven countries with five cultured species. Fallow time ranged broadly from 3-48 months. Current flow, depth, substrate and biological response information are included. Additional work to investigate the correlations between farm site characteristics and the required duration and effectiveness of recovery would aid development of fallowing guidance for the mariculture industry. Fallowing is being implemented or suggested as a valuable component of modern marine fish farm management in the U.S. (Nash 2001, Nash et al. 2005, Tucker and Hargreaves 2008), Canada (Environment Canada 2001, Sutherland 2004), the Mediterranean (International Union for Conservation of Nature 2007, Borg et al. 2011), Europe (Black et al. 2002, International Council for the Exploration of the Seas 2002, Huntington et al. 2006), Australia (Crawford 2003), globally (Stewart 1998, Halwart et al. 2007) and by the Global Aquaculture Alliance (Global Aquaculture Alliance 2011).

IMTA

Integrated multi-trophic aquaculture, or IMTA, is the practice of culturing finfish in combination with other species that utilize waste particulates and dissolved nutrients, thereby reducing nutrient discharge and expanding the economic base of a farming operation (Chopin 2006). The Food and Agriculture Organization of the United Nations (FAO) produced a comprehensive report on the current state and the requirements for expansion of IMTA in fish and shellfish aquaculture in temperate and tropical waters, and the Mediterranean (Soto 2009). In addition to nutrient uptake benefits and production enhancement, there may be mitigation of pathogens (Pang et al. 2006) and improved social acceptance (Barrington et al. 2010). The organisms most commonly integrated with finfish culture are seaweeds and shellfish. Polyculture has been an integral part of aquaculture for hundreds of years, but IMTA differs in two important ways. First, IMTA always employs organisms from different trophic levels as biofilters, while polyculture may refer more broadly to the growing of different fish species together without a biofiltering benefit. The IMTA approach strives to emulate natural nutrient cycling processes. Also, polyculture systems generally grow animals in the same enclosure, while IMTA systems may strategically position the different trophic components nearby the fish cages to take advantage of current-driven nutrient plumes.

The commercial feasibility of IMTA is under investigation around the world and has gained momentum in the past decade (Troell et al. 2009). In their summaries of the environmental impacts of marine cage culture, Wu (1995) and Pearson and Black (2001) review studies of nutrient uptake by algae, seaweeds and shellfish, and promote expanded research on the use of other organisms like polychaetes for filtering and bioremediation. As the marine finfish aquaculture industry expands, there may be increased demand for farm management protocols emphasizing nutrient uptake or other potential advantages of IMTA. IMTA is a remediation and filtering system with the potential to recoup its implementation costs (in the form of revenue from additional seafood products), as all other approaches incur costs without the potential for generating revenue (Troell et al. 2009).

Research at marine fish farms has demonstrated the potential benefits of IMTA with shellfish. For example, early work by Stirling and Okumus (1995) found that mussels cultured on rafts and lines at Scottish salmon farms showed augmented growth compared to those grown at a nearby shellfish farm. Local primary productivity and seasonal mussel growth cycles were noted as significant factors to be taken into account when considering mussel culture. Changes in cage design to increase buoyancy and the potential for shellfish to harbor organisms, chemicals or therapeutants which affect humans or fish were also identified as important factors.

A demonstration farm in the Gulf of Maine grew mussels on long lines deployed 200 m from submersible fish cages (Langan 2004). The amount of nitrogen discharged from the cages over a three year growout cycle of halibut and haddock was calculated as was the amount removed by bivalve harvest. The total discharge exceeded assimilation by mussels, and installation of another six lines was estimated to be needed to compensate for all the nitrogen waste. Ultimately, the scale of the mussel culture would need to be greater than that of the fish farm to achieve no net nitrogen increase to the environment. However, the study found that improved feed formulation or addition of seaweed culture could also be incorporated to increase nutrient removal. Another study in Maine found that sea scallops *Placopecten magellanicus* cultured in suspension at salmon farms grew to sizes comparable to those grown at a nearby scallop aquaculture site (Parsons et al. 2002).

A lab and field study in Canada measured the absorption efficiency of blue mussels cultured adjacent to open water salmon cages (Reid et al. 2010). In the lab study, mussels consuming salmon feed and fecal particulates showed the same organic matter absorption efficiency as those on commercial algal bivalve feeds. The mussels grown at the cage sites appeared to rely on both farm waste and natural seston. Hydrodynamics of particulate waste is important for deciding the placement of mussel growing structures within a constant nutrient plume. A successful IMTA pilot project at salmon farms in the Bay of Fundy is growing seaweed and mussels with the fish (Ridler et al. 2007). The profitability of this approach was economically evaluated. The additional profits from the mussel and seaweeds compensated for the added investment needs and there was a 24% increase in net present value. When different risk scenarios were run, IMTA was more economically resistant than salmon monoculture to catastrophic events or decrease in market value of salmon. Social acceptability surveys found an increased approval rating for IMTA compared to monoculture, reflecting an additional public perception benefit when IMTA practices are implemented.

The economic potential for salmon and mussel IMTA in the United Kingdom was assessed by Whitmarsh et al. (2006). They found that under current economic conditions appropriately sited mussel lines could assist in removing a proportion of the organic waste produced by fish farms while also offering financial benefits to the farm owner. However, the future market price for both cultured species and integration benefit are important when considering long-term economic opportunities. A study in Italy (Sarà et al. 2009) tested the potential of integrating mussel culture at a sea bass and sea bream farm. Mussels grown downstream of fish cages showed greater growth after one year compared to those at control sites. Results suggested that bivalve culture could be helpful in recycling organic waste from marine fish farms. However, implementing IMTA in the Mediterranean may be a challenge because the predominantly oligotrophic conditions mean the baseline productivity may be insufficient to support cultivation of other organisms, even when farms are discharging nutrients (Angel and Freeman 2009).

In contrast to the above studies Navarrete-Mier et al. (2010) concluded that oysters and mussels grown for three months along an 1800 m transect downstream of a sea bream and sea bass farm in Spain did not show enhanced growth from exposure to farm waste. Stable isotope and heavy metal bioaccumulation analysis indicated that farm waste was not a significant part of the bivalve diet. Other studies have found no growth enhancement of mussels grown in proximity to fish cages (Cheshuk et al. 2003), yet recent work on feeding behavior of mussels near salmon farms (MacDonald et al. 2011) and the use of stable isotopes to trace assimilation of fish feed in mussels (Redmond et al. 2010) indicate that these shellfish can be successfully used to capture and absorb farm nutrients.

In addition to shellfish, other invertebrates are being considered for IMTA. Voluntary recruitment of spiny lobster post-larvae was observed at sea cages in Puerto Rico (Davis et al. 2006). This pilot study confirmed the feasibility of collecting larval and juvenile lobsters for commercial grow out at the sea cage. Ahlgren (1998) showed that sea cucumbers stocked into salmon net pens in Alaska effectively cleared organic debris from the pens while also displaying increased growth compared to sea cucumbers feeding naturally. While not a common food species in the U.S., sea cucumbers are a commercially harvested detritivore with the potential for use in IMTA. Sea urchins *Paracentrotus lividus* cultivated in nets suspended in Scottish salmon cages grew faster than urchins grown 2.5 km away. Though tests diameter was similar, gonad maturation – which determine marketability – was higher for urchins cultured with the fish. Urchin survivorship in the fish pens was 98% compared to 57% at the reference station, resulting in much higher final total biomass.

The culture of primary producers for human consumption at fish farms is common in land-based and nearshore aquaculture (Neori et al. 2004), but the technology to make it possible at high energy off shore sites is still being developed. Buck and Buchholz (2004) describe an offshore ring structure successfully used to cultivate *Laminaria saccharaina*, a kelp used for food and biofiltration, in the North Sea. Such IMTA initiatives are being integrated with offshore wind farm installations (also see Michler-Cieluch et al. 2009) in this part of the world, but require structures rigid enough to withstand the intense wave energy while being easily handled during harvest. The ring system performed better compared to long-line, ladder and grid structures that were simultaneously tested. Technical problems were associated with all structures due to rough sea conditions, but the ring structure allowed for easier sampling and could be towed to shore intact. This study was not conducted near a fish farm, so it remains to be seen how this technology might be integrated with sea cage operations.

Efforts in Chile to incorporate IMTA approaches are reviewed by (Buschmann et al. 2009), but primarily in land-based, freshwater systems. The profitability of a seaweed market must be proven, but there is interest in developing the technological capacity for coastal IMTA. It is estimated that culturing 50-60 hectares of algae downstream of a salmon farm producing 1500 tons yearly would result in an 80% reduction of nitrogen entering the environment. This could also be integrated with existing mussel production.

The growth of *Porphyra* (nori) at salmon farms in the North Atlantic Ocean was compared to seaweed growing away from farm influence (Chopin et al. 1999). Seaweed nitrogen and phosphorus levels varied with location and seawater nutrient concentrations, and growth varied by species and seasonally. The use of local cultivars and a solid understanding of seaweed biology are needed to make its cultivation commercially attractive and useful as a bioremediation tool. There is still only limited knowledge about a few species that may be useful for marine finfish IMTA endeavors and include *Ulva*, *Porphyra, Gracilaria* and *Laminaria* (Neori et al. 2004). Challenges facing this industry are epiphyte growth on seaweed monocultures or biofilters, grazing by herbivores, managing plant tendrils in oscillating and strong currents, the physiological dynamics of nutrient uptake. While much can be learned from land-based operations, the turbulence and remoteness of the open ocean will require innovative solutions.

The application of IMTA primarily as a filtration or remediation technique, rather than for food production, is also being investigated (Chavez- Crooker and Obreque-Contreras 2010) and improvements in culture technology are being made for the open ocean environment. Chung et al. (2002) evaluated several species of seaweed in Korea to determine which would be most effective at nutrient uptake to reduce the eutrophication in coastal fish farming areas. *Porphyra* and *Ulva* had up to six times higher rates of short term ammonia uptake than the other algal species analyzed. A simple model was developed to predict the overall nitrogen scrubbing that could be expected from integrating algal culture into farming areas.

Photo courtesy of NOAA.

Photo courtesy of NOAA.

As part of a study in Japan, three species of seaweeds were cultured year round at fish farms to improve the water quality (Kitadai and Kadowaki 2007). The nitrogen and phosphorus uptake, oxygen production and growth rates of the seaweeds were calculated. Growth rates of up to 4.2 cm/day were reported and oxygen production was 8-11 times higher than consumption. Nitrogen and phosphorus uptake were 2.9-3.6 mg/m^2/day and 0.19-0.54 mg/m^2/day, respectively. These results were used to determine the seaweed biomass needed to clear the nitrogen output from the fish farms. A similar effort was made at sea bass and croaker farms and macroalgae culture areas in Nansha Bay, China (Jiang et al. 2010). Water nutrient levels, eutrophication status and the fish nitrogen excretion rates were determined in order to estimate the optimal co-culture proportions of fish to macroalgae needed to maintain environmental quality. The authors calculated that for each fish cage 450 m^2 of *Laminaria* and 690 m^2 of *Gracilaria* were required to maintain water quality and prevent eutrophication. At an intensive sea bream growout farm, *Ulva* and *Gracilaria* tanks were added to treat the farm effluent and remove nitrogen and phosphorus (Hernandez et al. 2005). After less than three months the algal growth exceeded the capacity of the tanks. The *Ulva* removed 8.9% of the phosphorus input and 24% of nitrogen. The *Gracilaria* removed 3.2% of the phosphorus and 19% of the nitrogen input. Heavy metal analysis of the seaweeds determined there was no contamination making the tissues eligible for use in the food industry. Asian and U.S. species of *Porphyra* have also been tested in the laboratory by Kraemer et al. (2004) and appear to be an excellent choice for commercial and bioremediation applications based upon nutrient uptake rates. Lab experiments in Korea with *Codium*, a seaweed with food and medicinal applications, suggest this species may be useful for IMTA in fish farming areas with high water temperatures (Kang et al. 2008).

Mathematical models designed to predict waste dispersion and carrying capacity relative to multi- species invertebrate mariculture are being developed (Duarte et al. 2003, Reid et al. 2009, Reid et al. 2011), and may be useful in designing IMTA systems that balance the nutrient inputs of the finfish with optimized culture of filtering organisms. Recently (Sarà et al. 2012) applied Dynamic Energy Budget modeling to examine growth of blue mussels *Mytilus galloprovincialis* and oysters *Crassostrea gigas* grown around fish cages in the Mediterranean. The model predicted that both species would exhibit greater growth (nearly doubled for oysters) in water enriched by farm effluent, compared to sites away from cages. The modeling results correlated well with results from field experiments with mussels. Development of advanced models to integrate IMTA into nutrient discharge models will be especially relevant in areas with high fish production demands and moderate flushing capacity to help manage nutrient inputs by harnessing the assimilative capacity of IMTA biofiltering components.

Troell et al. (2009) provide a useful summary of the current status of IMTA knowledge, research and engineering for offshore marine systems. Generally, there is great interest in pursuing the expansion of IMTA to marine cage culture, and successful pilot projects indicate that it is feasible (Blouin et al. 2007, Robinson et al. 2011) and profitable. The high energy environment of the open ocean poses great challenges to all aspects of mariculture in that environment, but significant advances are being made. For seaweed culture, open ocean harvesting techniques and estimating the capacity to remove nutrients in an open water system are two areas requiring further work. The integration of bivalves and other filter feeding organisms into sea cages also needs further research and technological innovation. For both seaweeds and filter feeders, economic cost and benefits analyses are still needed.

Photo courtesy of NOAA.

IMTA has been attempted for many years, but due to significant knowledge gaps, technological challenges and economic viability, it is not yet a widely-implemented management tool. The incorporation of IMTA into marine aquaculture is of interest for both commercial and environmental reasons, and it is identified as a best management practice (Stickney 2002, Belle and Nash 2008, Johnson et al. 2008). IMTA systems are thought to be near commercial scales in the U.S. (Barrington et al. 2009). As this industry expands domestically, continued research and development efforts to identify the right species and technologies for co- culture will advance IMTA from an experimental to a profitable aspect of marine fish aquaculture. The integration of IMTA into marine monoculture may also increase societal acceptance of marine aquaculture if it reduces potential negative impacts such as nutrient discharge (Barrington et al. 2010).

CONCLUSION

This report provides a comprehensive review and summary of some predominant environmental effects of marine finfish aquaculture. We compiled over 420 research papers, studies and reports to gather current information as a tool for industry, coastal managers, scientists and the public. Our work intends to provide stakeholders a scientific basis for decision making about siting, monitoring, regulating and managing a marine aquaculture industry in the U.S.

There is opportunity to proactively apply environmental lessons learned in the U.S. and other countries to existing and new locations in U.S. waters. As the scope of human activities in the coastal and marine environment expands, there is a need to understand the individual and collective, short and long-term impacts that may result, and how to prevent, minimize or mitigate these effects. Marine cage aquaculture is only one such activity. The NOAA Aquaculture Policy presents national guidance for marine aquaculture which provides food security and economic opportunity for the U.S., while also conserving our ocean resources. This policy supports the joint goals of environmental and economic sustainability, and implementation of the policy will be aided by having the best available science to guide decision-making.

The scope of our report does not allow for a quantitative or comprehensive analysis of the environmental effects of marine finfish aquaculture. Yet, we are able to discern certain trends in the research and draw some general conclusions about the risks of the industry to marine ecosystems. We also highlight aspects of the industry that would benefit from further investigation or further technical advancement.

Water quality is the first issue covered in this report. The trend over the last 20 years reflects great improvements in feed formulation and management, which are largely credited for the reduction in water quality impacts. Proper siting in areas with sufficient flushing is important for eliminating nutrient enhancement, turbidity, and decreased dissolved oxygen. In the open ocean, water quality impacts are not likely to pose a great environmental threat when farms are properly sited and appropriately managed. Protection of water quality will be best achieved by siting farms in deep, well-flushed waters. Impaired water quality is typically observed around farms in nearshore or intertidal habitats where flushing is minimal and at farms using feeds that include unprocessed raw fish rather than formulated feeds.

Elevated levels of nitrogen and phosphorus are the most common impacts to water quality, but are typically not measurable 30 m beyond farm perimeters. Generally, both nutrients follow the same trend in terms of measurable changes in concentration, although sometimes nitrogen increase is found without a concomitant rise in phosphorus. Nutrient elevations will be most evident following feeding and just prior to harvest when the farm stock biomass is at its greatest. Farm management and monitoring protocols should take this into account.

The trophic status and background nutrient flux of the receiving waters, as well as other sources of nutrient loading, should be considered when assessing the relative contribution of marine fish farm discharge to the environment. Because nutrients tend to get flushed away from the farm area faster than they can be assimilated into the local food web, it is difficult to measure direct impacts to local phytoplankton production. It is uncommon for phytoplankton productivity and blooms, including harmful algae blooms, resulting from eutrophication to be attributed to fish farms. Continued research to understand and model the complex array of forces driving nutrient dispersion in and around fish farms will provide additional tools to effectively manage fish farming practices. Few comparative analyses have investigated correlations between farm site characteristics (e.g., depth, latitude, current profile), farm management factors (e.g., species cultured, volume of cages, biomass, feeding rate), and observed water quality impacts so additional research in this area would be valuable.

Operating farms in well-flushed areas over erosional sediments and improved feeding practices are factors which also reduce geochemical impacts to the sediments below farms. Organic carbon, nitrogen and phosphorus discharge from fish farms continues to be an environmental concern because benthic enrichment will occur if nutrient discharge and sedimentation exceed the environment's assimilative capacity. Because much of the nutrient input to farms is lost through excretion and excess feed, it is important to continue optimizing feed formulation and closely monitor feeding to reduce waste. The accumulation of some organic matter below farms is to be expected, especially toward the end of a grow-out period when farm biomass is at its peak. Benthic environments typically recover to baseline condition in less than one to three years after a farm is removed from a site. Visual observations of benthic conditions below farms are a valuable tool throughout a crop cycle for assessing whether operations are within the capacity of the ecosystem. Farms located in deep water with continuous or episodic benthic scouring of organic waste will be less likely to exhibit sediment degradation. As with water quality, benthic geochemical impacts are most pronounced at enclosed, nearshore or coastal farm sites with insufficient depth and flow to disperse organic wastes.

Further research is needed to determine the best indicators of biogeochemical perturbation which also incorporate seasonal or regional variability. Total organic carbon, sulfides and redox potential are consistently the most reliable parameters that reflect the geochemical condition of sediments below farms. Of course, these must be assessed relative to background levels of the surrounding sea floor. Local pre-farming conditions should be determined for comparative analyses and sediment management should contribute to regulatory decision-making, developing monitoring protocols and informing site management. Where changes in the sediment are documented, they are generally confined to within 500 m of the cages and often recovery progresses quickly after harvest. At well-managed farms this footprint may be reduced to 100 m.

Effective monitoring protocols should be implemented that ensure early detection of geochemical perturbations that signal potential ecological consequences with adaptive management response options to minimize impacts. Ongoing work continues to identify new benthic monitoring methods resulting in reliable, accurate and cost-effective data which provide early indications of biogeochemical alterations. Such methods will also prove useful for surveying benthic conditions in very deep open ocean sites inaccessible for diver surveys or at farms over hard bottom habitats. There is a need for long-term data to evaluate cumulative impacts of organic matter accumulation from multiple farms over many production cycles to the benthic processes at large, regional scales. Analytical methods to quantify the relative contribution of fish farm discharge are critical for understanding and modeling the full scope of impacts within marine ecosystems under nutrient loading from many anthropogenic and natural sources.

Early detection and intervention will be important to minimizing ecological benthic impacts. Once sediment chemistry is deleteriously affected by farm operations, the benthic community responds by shifting toward generalist species tolerant of perturbed conditions. Often these effects are of short duration and the benthic community recovers within months following harvest and fallowing. At farms with high fish biomass production and nutrient enrichment, fallowing for several years may be required before recovery is evident. Effort should be made to prevent the need for extended fallowing periods due to severe enrichment. Proactive, careful site selection and modeling to predict likely impacts based on production parameters and site characteristics can optimize farm locations to minimize or eliminate most effects. Once a farm is operational, best management practices should be followed, environmental monitoring should be conducted for early detection of environmental impacts, and adaptive farm management strategies should be implemented if warranted. Cost effective, quick sampling methods to assess the condition of the benthic community and site-specific indicator species that reflect impaired sediment conditions are tools that will enhance on-site evaluation and management.

As with water quality and sediment condition, improvements in farm siting and management have tended to decrease impacts to benthic communities. Effects to the infaunal communities are usually confined to within 100 m of a farm. Most studies of marine fish farm impacts to benthic communities highlight site-specific factors that affect the response and stress that the availability of good historical records of local biodiversity is a key factor in being able to assess the changes due to enrichment. Opportunistic polychaetes are the most common indicator species of degraded benthic conditions. However, other local benthic species or assemblages that best reflect sediment condition and quality should be identified for optimal monitoring. Biodiversity indices should be coupled with geochemical assessment to fully ascertain the degree of sediment and community perturbation. An interesting finding at fish farms is that moderate organic enrichment may stimulate benthic community productivity and even enhance biodiversity. As the relationships between sediment geochemistry and the benthic community response are better understood, computer models are beginning to incorporate benthic community response to farm nutrient loading. This approach will enhance our ability to make decisions about fish farming within an ecological context.

The broader ecological role of aquaculture operations within the marine environment should also be considered. Nitrogen and phosphorus inputs to the water column may increase primary productivity, especially in oligotrophic waters. However, rapid uptake by phytoplankton and zooplankton, coupled with dispersive water currents make it difficult to track or measure such impacts over large spatial scales. Dissolved nutrients may contribute to increasing biomass of the fouling community on farm structures. Hydrology of farms located near shore or in semi- enclosed water bodies must be carefully examined to prevent eutrophication and increased primary productivity in coastal areas and habitats. One knowledge gap continues to be how dissolved nutrients are dispersed and assimilated over large marine areas, and how productivity may be affected under increasing production from more farms. The cumulative effects of long-term nutrient loading to marine waters should remain a research priority. It is especially important that as marine fish farming expands to larger scales in a variety of geographic regions and marine habitats, monitoring protocols incorporate a framework for discerning cumulative, far-field and long-term nutrient responses in phyto- and zooplankton productivity and the mobilization of energy from fish farms into marine food webs.

Scavenging fish also serve as a route for mobilizing farm nutrients into the marine food webs, thus providing a natural approach for decreasing nutrient accumulation and organic waste impacts on the sea floor. Wild fish often aggregate and forage around fish cages and this is often considered a beneficial impact to marine life. As fish are attracted to farms, the potential for interactions with human fishers may increase and farm management or regulatory steps should be considered to minimize such conflicts. Likewise, marine fish and mammalian predators may also be attracted to farms. Little research has documented the extent to which marine predators target wild fish around farms, but this would be useful for understanding ecological interactions between farming and marine life. In contrast, impacts to predatory sharks and marine mammals are being minimized with improved net technologies that prevent predation on cultured fish. Proactive siting away from areas known to harbor dense populations of protected marine organisms is an effective strategy for minimizing negative interactions. Impacts to cetaceans, sea turtles and sea birds are typically minimal. Acoustic deterrent devices should be avoided as they have not proven to be an effective long-term solution and may have negative secondary consequences to non-target species.

Concerns about siting marine cage farms near corals, seagrass meadows and mangrove forests support the need for monitoring to detect impacts to these habitats. Because of the ecological roles these habitats play, caution should be used when siting farms near them. If farms are sited upstream of sensitive habitats, careful monitoring should be in place for early detection of any nutrient or sedimentation impacts directly to corals, seagrasses or mangroves, or the ecological functions and biodiversity associated with them. Although the potential for impacts to deep-water coral habitats is unknown, potential effects should be considered in siting farms in deep water locations with known corals.

The use of antibiotics, therapeutants and antifoulants at marine farms has greatly decreased in the last 20 years, resulting in decreased potential harmful effects on the marine environment. The misuse of approved chemicals or the application of unauthorized chemicals has the potential to cause environmental damage. In the U.S and other nations implementing modern fish farming, vaccination has largely replaced the use of antibiotics and less toxic, soluble therapeutants are more commonly approved for use. Better husbandry techniques to improve fish health are also being implemented. Best management practices such as onshore net de-fouling and mechanical alternatives for net cleaning reduce the release of antifoulant chemicals into the ocean. Ideally, few chemicals will be used on a regular basis at farms by implementing best management practices. Continued efforts should be made to determine the persistence and accumulation of antibiotics, therapeutants and other farm chemicals in the marine environment, and the extent to which they affect wild organisms.

However, the most effective way to avoid negative impacts to marine life is to continue use and refinement of non- toxic therapeutic measures along with effective farm management and husbandry practices to safeguard health of cultured fish. Mechanical methods for de-fouling cages contribute greatly to the decline in heavy metal use at farms and should be continued. Though heavy metals are a small component of fish feed, they can build up in sediments after many years of operation and should be monitored to detect long-term accumulation.

Fallowing and IMTA are two management tools which may mitigate or reduce environmental effects of some marine aquaculture operations. Siting and operating farms appropriately may eliminate the need for fallowing altogether. However, where benthic impacts are observed, fallowing should be implemented to avoid long-term damage to benthic ecosystems and reduce risks to farm animals from pathogens. Farm lease sites should be large enough and cage spacing appropriate to allow for such rotation of production cages. Benthic sediment and faunal community recovery generally occur within months of initiating fallowing. Because biological recovery may lag behind chemical recovery care should be taken to select appropriate, site-specific indicators of ecosystem health to evaluate the remediation process. Ideally, farms managed in equilibrium with the ecosystem's capacity to assimilate nutrient waste will not require fallowing to offset organic enrichment. Fallowing for the purposes of reducing the presence of pathogens, and thus the risk of disease outbreaks, may be appropriate even if no benthic impacts are measured.

IMTA technology and methods are improving. This approach integrates the culture of additional marketable marine organisms and provides benefits in terms of a reduction in nutrient enrichment and economic profitability. Though still largely experimental, IMTA is becoming more viable and offers economic as well as social acceptance benefits, sometimes at little extra cost. There is a long tradition of integrated culture in freshwater aquaculture, particularly at smaller community or family farms. Making use of farm effluent as the energy input for additional crops at other trophic levels provides efficiency and reduces harmful impacts. Globally there is an increased focus on fish production that reduces waste discharge. IMTA in the turbulent open ocean faces technological challenges, yet there is considerable interest in making it a viable component of ocean aquaculture.

Research to shed light on far-field and regional processes, especially in intensively farmed areas and over longer time scales, should continue to examine the ecological role of fish farms in marine environments. The research in this report was conducted at farms in temperate, subtropical and tropical waters. Geographical determinants of the type and extent of environmental impacts from marine fish farms include latitude, temperature and current regimes, sediment type and trophic status of the water. Generally, the extent to which such geographical aspects of farms interact with local hydrography and bathymetry is a driving factor behind the risk associated with farming in marine environments remains unclear. The combination of these is likely important in trying to understand and predict how fish farms will affect and interact with ocean ecosystems. The correlation of latitude, geographic area and trophic status of the receiving waters with the degree of biological and geochemical response to farm discharge is an area for further investigation. Comparative meta-analyses of environmental impacts would provide additional information, but are challenging to conduct because of the different monitoring protocols used, inconsistency in reporting of farm site (latitude, depth, current regime, sediment type, background nutrient levels, biodiversity) and management characteristics (farm size, loading density, stock biomass, feed type, feeding rate), and language barriers.

One pattern that does emerge is that decreasing environmental risk from aquaculture appears to be driven by prudent siting of operations outside of shallow, enclosed, coastal and nearshore waters lacking dispersive current regimes, coupled with modern feed and farm management. This observation is important as it suggests that fish farming with minimal environmental effects is possible in many ecosystems as long as proper safeguards are in place to minimize nutrient and chemical discharge and to manage its immediate and cumulative impacts. These safeguards may be in the form of regulatory oversight or industry-developed best management practices. Ideally, a combination of the two approaches is most beneficial.

This report provides a broad perspective on a range of potential environmental impacts and their relative intensity, which should be coupled with detailed, site-specific information to make good management decisions about a proposed or operational farm site during its lifetime. If marine fish farming expands, cumulative impacts may or may not become more apparent, but robust monitoring protocols are necessary and should be proactively designed to be able to discern both near and far-field environmental impacts. A standard guide and protocols for environmental monitoring at marine fish farms may be useful in the U.S. for monitoring and comparing results at local, regional, and national levels.

The scientific information reviewed in this report provides current information to NOAA and other stakeholders about the environmental effects of marine finfish cage culture. Development, regulation and operations of marine finfish cage farming should continue to be supported by a strong underpinning of scientific knowledge and expertise. Additional national, regional and local decision- making tools, including computer simulation modeling, to aid in siting and refining monitoring protocols would be useful to government agencies and fish farm operators. Continued assessments of the effects of marine fish farming in the ocean has on marine environments should be undertaken. Our understanding of nutrient flux and accumulation in the water and sediments around farms, and resulting impacts to local biodiversity has greatly improved in the last decades. However, questions remain about far-field effects over large time scales.

The rising world population is becoming more reliant on aquaculture for food production. In the U.S., the regulatory process for permitting offshore aquaculture facilities is moving forward and industry growth is expected to follow. Domestic production of seafood can aid in decreasing U.S. reliance on imported products, provide jobs and food security, and meet the rising demand for seafood. Every effort should be made to ensure that industry growth occurs within a framework of environmental responsibility and ocean stewardship.

REFERENCES

Abebe, E., Grizzle, R. E., Hope, D. & Thomas, W. K. (2004). Nematode diversity in the Gulf of Maine, USA, and a web-accessible, relational database. *Journal of the Marine Biological Association of the United Kingdom*, *84*, 1159-1167.

Abete, M. C., Prearo, M., Andruetto, S., Pavino, D., Colussi, S., Tarasco, R., Agnetti, F. & Ghittino, C. (2004). A preliminary note on residues in aquacultural feed: Arsenic, cadmium, chromium, mercury and lead research. *Ittiopatologia*, *1*, 68-76.

Aguado-Gimenez, F. & Garcia-Garcia, B. (2004). Assessment of some chemical parameters in marine sediments exposed to offshore cage fish farming influence: A pilot study. *Aquaculture*, *242*, 283-296.

Aguado-Gimenez, F., Garcia-Garcia, B., Hernandez-Lorente, M. D. & Cerezo-Valverde, J. (2006). Gross metabolic waste output estimates using a nutritional approach in Atlantic bluefin tuna (*Thunnus thynnus*) under intensive fattening conditions in western Mediterranean Sea. *Aquaculture Research*, *37*, 1254-1258.

Aguado-Gimenez, F., Garcia-Garcia, B., Hernandez-Lorente, M. D. & Cerezo-Valverde, J. (2006). Gross metabolic waste output estimates using a nutritional approach in Atlantic bluefin tuna (*Thunnus thynnus*) under intensive fattening conditions in western Mediterranean Sea. *Aquaculture Research*, *37*, 1254-1258.

Aguado-Gimenez, F., Marin, A., Montoya, S., Marin-Guirao, L., Piedecausa, A. & Garcia-Garcia, B. (2007). Comparison between some procedures for monitoring offshore cage culture in western Mediterranean Sea: Sampling methods and impact indicators in soft substrata. *Aquaculture*, *271*, 357-370.

Aguado-Gimenez, F., Marin, A., Montoya, S., Marin-Guirao, L., Piedecausa, A. & Garcia-Garcia, B, (2007). Comparison between some procedures for monitoring offshore cage culture in western Mediterranean Sea: Sampling methods and impact indicators in soft substrata. *Aquaculture*, *271*, 357-370.

Aguilar-Macías, O. L., Ojeda-Ramírez, J. J., Campa-Córdova, A. I. & Saucedo, P. E. (2010). Evaluation of natural and commercial probiotics for improving growth and survival of the pearl oyster, *Pinctada mazatlanica*, during late hatchery and early field culturing. *Journal of the World Aquaculture Society*, *41*, 447-454.

Ahlgren, M. O. (1998). Consumption and assimilation of salmon net pen fouling debris by the Red Sea cucumber *Parastichopus californicus*: Implications for polyculture. *Journal of the World Aquaculture Society*, *29*, 133-139.

Aksu, M. & Kocatas. A. (2007). Environmental effects of the three fish farms in Izmir Bay (Aegean Sea, Turkey) on water column and sediment. Rapport Commission International de la Mer Mediterranee, *38*, 414.

Aksu, M., Kaymakci-Basaran, A. & Egemen, O. (2010). Long-term monitoring of the impact of a capture-based bluefin tuna aquaculture on water column nutrient levels in the Eastern Aegean Sea, Turkey. *Environmental Monitoring and Assessment*, *171*, 681-688.

Aksu, M., Kaymakci-Basaran, A. & Egemen, O. (2010). Long-term monitoring of the impact of a capture-based bluefin tuna aquaculture on water column nutrient levels in the Eastern Aegean Sea, Turkey. *Environmental Monitoring and Assessment*, *171*, 681-688.

Aksu, M., Kaymakci-Basaran, A. & Egemen. O. (2010). Long-term monitoring of the impact of a capture-based bluefin tuna aquaculture on water column nutrient levels in the Eastern Aegean Sea, Turkey. *Environmental Monitoring and Assessment*, *171*, 681-688.

Aksu, M., Kaymakci-Basaran, A. & Egemen. O. (2010). Long-term monitoring of the impact of a capture-based bluefin tuna aquaculture on water column nutrient levels in the Eastern Aegean Sea, Turkey. *Environmental Monitoring and Assessment*, *171*, 681-688.

Akyol, O. & Ertosluk, O. (2010). Fishing near sea-cage farms along the coast of the Turkish Aegean Sea. *Journal of Applied Ichthyology*, *26*, 11-15.

Alday, V., Guichard, B., Smith, P. & Uhland, C. (2006). Toward a risk analysis of antimicrobial use in aquaculture. Joint FAO/WHO/OIE expert consultation on antimicrobial use in aquaculture and antimicrobial resistance, 13-16 June 2006, Seoul, South Korea.

Aleffi, I. F., Brizzi, G. & Zamboni, R. (2005). Effects of an intensive cage farm on the macrobenthos in the Gulf of Trieste (northern Adriatic Sea). *Annales Series Historia Naturalis*, *15*, 5-10.

Alongi, D. M. (2002). Present state and future of the world's mangrove forests. *Environmental Conservation*, *29*, 331-349.

Alongi, D. M., Chong, V. C., Dixon, P., Sasekumar, A. & Tirendi, F. (2003). The influence of fish cage aquaculture on pelagic carbon flow and water chemistry in tidally dominated mangrove estuaries of peninsular Malaysia. *Marine Environmental Research*, *55*, 313-333.

Alongi, D. M., Chong, V. C., Dixon, P., Sasekumar, A. & Tirendi. F. (2003). The influence of fish cage aquaculture on pelagic carbon flow and water chemistry in tidally dominated mangrove estuaries of peninsular Malaysia. *Marine Environmental Research*, *55*, 313-333.

Alston, D. E., Cabarcas, A., Capella, J., Benetti, D. D., Keene-Meltzoff, S., Bonilla, J. & Cortes. R. (2005). Report on the environmental and social impacts of sustainable offshore cage culture production in Puerto Rican waters. Final Report to the National Oceanic and Atmospheric Administration, Contract NA16RG1611. Available at: www.lib.noaa.gov/retiredsites/docaqua/reports_noaaresearch/finaloffshorepuertorico.pdf. Accessed: 27 September 2012.

Alston, D. E., Cabarcas, A., Capella, J., Benetti, D. D., Keene-Meltzoff, S. Bonilla, J. & Cortes. R. (2005). Report on the environmental and social impacts of sustainable offshore cage culture production in Puerto Rican waters. Final Report to the National Oceanic and Atmospheric Administration, Contract NA16RG1611. Available at: www.lib.noaa.gov/retiredsites/docaqua/reports_noaaresearch/finaloffshorepuertorico.pdf. Accessed: 27 September 2012.

Alston, D. E., Cabarcas, A., Capella, J., Benetti, D. D., Keene-Meltzoff, S., Bonilla, J. & Cortes, R. (2005). Report on the environmental and social impacts of sustainable offshore cage culture production in Puerto Rican waters. Final Report to the National Oceanic and

Atmospheric Administration, Contract NA16RG1611. Available at: www.lib.noaa.gov/retiredsites/docaqua/reports_noaaresearch/finaloffshorepuertorico.pdf. Accessed: 27 September 2012.

Alston, D. E., Cabarcas, A., Capella, J., Benetti, D. D., Keene-Meltzoff, S., Bonilla, J. & Cortes, R. (2005). Report on the environmental and social impacts of sustainable offshore cage culture production in Puerto Rican waters. Final Report to the National Oceanic and Atmospheric Administration, Contract NA16RG1611. Available at: www.lib.noaa.gov/ retiredsites/docaqua/reports_noaaresearch/finaloffshorepuertorico.pdf. Accessed: 27 September 2012.

Anderson, D. M., Burkholder, J. M., Cochlan, W. P., Glibert, P. M., Gobler, C. J., Heil, C. A., Kudela, R. M., Parsons, M. L., Rensel, J. E. J., Townsend, D. W., Trainer, V. L. & Vargo. G. A. (2008). Harmful algal blooms and eutrophication: Examining linkages from selected coastal regions of the United States. *Harmful Algae*, *8*, 39-53.

Anderson, D. M., Burkholder, J. M., Cochlan, W. P., Glibert, P. M., Gobler, C. J., Heil, C. A., Kudela, R. M., Parsons, M. L., Rensel, J. E. J., Townsend, D. W., Trainer, V. L. & Vargo, G. A. (2008). Harmful algal blooms and eutrophication: Examining linkages from selected coastal regions of the United States. *Harmful Algae*, *8*, 39-53.

Anderson, D. M., Glibert, P. M. & Burkholder, J. M. (2002). Harmful algal blooms and eutrophication: Nutrient sources, composition, and consequences. *Estuaries*, *25*, 704-726.

Anderson, D. M., Glibert, P. M. & Burkholder. J. M. (2002). Harmful algal blooms and eutrophication: Nutrient sources, composition, & consequences. *Estuaries*, *25*, 704-726.

Anderson, M. R., Tlusty, M. F. & Pepper. V. A. (2005). Organic enrichment at cold water aquaculture sites—the case of coastal Newfoundland. Pages 99-113 *in* B.T. Hargrave, editor. *Handbook of Environmental Chemistry*, Volume *5M*. Springer-Verlag, Berlin.

Angel, D, & Freeman, S. (2009). Integrated aquaculture (INTAQ) as a tool for an ecosystem approach to the marine farming sector in the Mediterranean Sea. Pages 133-183 *in* D. Soto, editor. Integrated mariculture: a global review. FAO Fisheries and Aquaculture Technical Paper. No. 529. FAO, Rome. Available at: ftp://ftp.fao.org/docrep/fao/012/ i1092e/i1092e04a.pdf. Accessed: 01 October 2012.

Angel, D. L., Verghese, S., Lee, J. J., Saleh, A. M., Zuber, D., Lindell, D. & Symons, A. (2000). Impact of a net cage fish farm on the distribution of benthic foraminifera in the Northern Gulf of Eilat (Aqaba, Red Sea). *Journal of Foraminiferal Research*, *30*, 54-65.

Apostolaki, E. T., Tsagaraki, T., Tsapaki, M. & Karakassis, I. (2007). Fish farming impact on sediments and macrofauna associated with seagrass meadows in the Mediterranean. *Estuarine*, *Coastal and Shelf Science*, *75*, 408-416.

Apostolaki, E. T., Tsagaraki, T., Tsapaki, M. & Karakassis. I. (2007). Fish farming impact on sediments and macrofauna associated with seagrass meadows in the Mediterranean. *Estuarine*, *Coastal and Shelf Science*, *75*, 408-416.

Armstrong, S. M., Hargrave, B. T. & Haya, K. (2005). Antibiotic use in finfish aquaculture: Modes of action, environmental fate, and microbial resistance. Pages 341-357 *in* B.T. Hargrave, editor. Environmental effects of marine finfish aquaculture. *Handbook of Environmental Chemistry*, Volume *5M*, Springer, Dordrecht, London.

Australian Government. (2009). White shark issues paper. Department of the Environment, Water, Heritage and the Arts. Commonwealth of Australia. Available at: www.environment. gov.au/biodiversity/threatened/publications/recovery/pubs/white-shark-issues-paper.pdf. Accessed: 01 October 2012.

Barbier, E. B. (2003). Habitat-fishery linkages and mangrove loss in Thailand. *Contemporary Economic Policy*, *21*, 59-77.

Barrington, K., Chopin, T. & Robinson, S. (2009). Integrated multi-trophic aquaculture (IMTA) in marine temperate waters. Pages 7-46 *in* D. Soto, editor. Integrated mariculture: a global review. FAO Fisheries and Aquaculture Technical Paper. No. 529. FAO, Rome. Available at: ftp://ftp.fao.org/docrep/fao/012/i1092e/i1092e02a.pdf. Accessed: 01 October 2012.

Barrington, K., Ridler, N., Chopin, T., Robinson, S. & Robinson, B. (2010). Social aspects of the sustainability of integrated multi-trophic aquaculture. *Aquaculture International*, *18*, 201-211.

Basaran, A. K., Aksu, M. & Egemen, O. (2010). Impacts of the fish farms on the water column nutrient concentrations and accumulation of heavy metals in the sediments in the eastern Aegean Sea (Turkey). *Environmental Monitoring and Assessment*, *162*, 439-451.

Basaran, A. K., Aksu, M. & Egemen. O. (2007). Monitoring the impacts of the offshore cage fish farm on water quality located in Ildir Bay (Izmir-Aegean Sea). *Journal of Agricultural Sciences*, *13*, 22-28.

Bazes, A., Silkina, A., Defer, D., Bernedebauduin, C., Quemener, E., Braud, J. & Bourgougnon, N. (2006). Active substances from *Ceramium botryocarpum* used as antifouling products in aquaculture. *Aquaculture*, *258*, 664-674.

Belant, J. L., Tyson, L. A. & A. Mastrangelo, P. (2000). Effects of lethal control at aquaculture facilities on populations of piscivorous birds. *Wildlife Society Bulletin*, *28*, 379-384.

Belias, C. V., Bikas, V. G., Dassenakis, M. J. & Scoullos, M. J. (2003). Environmental impacts of coastal aquaculture in eastern Mediterranean bays: The case of Astakos Gulf, Greece. *Environmental Science and Pollution Research*, *10*, 287-295.

Belias, C., Bikas, V., Dassenakis, M. & Scoullos, M. (2003). Environmental impacts of coastal aquaculture in eastern Mediterranean Bays. The case of Astakos Gulf, Greece. *Environmental Science and Pollution Research International*, *10*, 287-295.

Belias, C., Bikas, V., Dassenakis, M. & Scoullos. M. (2003). Environmental impacts of coastal aquaculture in eastern Mediterranean Bays. The case of Astakos Gulf, Greece. *Environmental Science and Pollution Research International*, *10*, 287-295.48

Belias, C., Ladakis, E., Papakonstantinou, K., Dessenakis, E. & Scoullos. M. (2005). The influence of fish farming in the addition of particulate nitrogen in coastal environments. The 9th International Conference on Environmental Science and Technology. 1-3 September 2005, Rhodes Island, Greece. Available at: http://www.srcosmos.gr/srcosmos/showpub.aspx?aa=6817. Accessed: 01 October 2012.

Belle, S. M, & Nash, C. E. (2008). Better management practices for net-pen aquaculture. Pages 261-330 *in* C.S. Tucker and J. Hargreaves, editors. Environmental Best Management Practices for Aquaculture. Blackwell Publishing, Ames, Iowa.

Belle, S. M. & Nash, C. E. (2008). Better management practices for net-pen aquaculture. Pages 261-330 *in* C.S. Tucker and J. Hargreaves, editors. Environmental Best Management Practices for Aquaculture. Blackwell Publishing, Ames, Iowa.

Belle, S. M. & Nash, C. E. (2008). Better management practices for net-pen aquaculture. Pages 261-330 *in* C.S. Tucker and J. Hargreaves, editors. Environmental Best Management Practices for Aquaculture. Blackwell Publishing, Ames, Iowa.

Belle, S. M. & Nash, C. E. (2008. Better management practices for net-pen aquaculture. Pages 261-330 *in* C.S. Tucker and J. Hargreaves, editors. Environmental Best Management Practices for Aquaculture. Blackwell Publishing, Ames, Iowa.

Belle, S. M. & Nash. C. E. 2008). Better management practices for net-pen aquaculture. Pages 261-330 *in* C.S. Tucker and J. Hargreaves, editors. Environmental Best Management Practices for Aquaculture. Blackwell Publishing, Ames, Iowa.

Benbrook, C. M. (2002). Antibiotic drug use in U.S. aquaculture. Report to the Institute for Agriclture and Trade Policy, Minneapolis, Minnesota. Available at: http://m.iatp.org/files/421_2_37397.pdf. Accessed: 01 October 2012.

Benetti, D. D., O'Hanlon, B., Rivera, J., Welch, A., Maxey, C. & Orhun. M. R. (2010). Growth rates of cobia (*Rachycentron canadum*) cultured in open ocean submerged cages in the Caribbean. *Aquaculture*, *302*, 195-201.

Benetti, D., Brand, L., Collins, J., Brooks, G., Orhun, R., Maxey, C., Danylchuk, A., Walton, G., Freeman, B., Kenworthy, J. & Scheidt. J. (2005). Final report on Cape Eleuthra offshore aquaculture project. Cape Eleuthra Institute and AquaSense LLC, Bahamas.

Benetti, D., Brand, L., Collins, J., Brooks, G., Orhun, R., Maxey, C., Danylchuk, A., Walton, G., Freeman, B., Kenworthy, J. & Scheidt, J. (2005). Final report on Cape Eleuthra offshore aquaculture project. Cape Eleuthra Institute and AquaSense LLC, Bahamas.

Beveridge, M. (2004). Cage aquaculture. Blackwell Publishing, Oxford, UK.

Beveridge, M. (2004). Cage aquaculture. Blackwell Publishing, Oxford, UK.

Beveridge, M. (2004). Cage aquaculture. Blackwell Publishing, Oxford, UK.

Beveridge, M. (2004). Cage aquaculture. Blackwell Publishing, Oxford, UK.

Black, K. D., Blackstock, J., Gillibrand, P. A., Moffat, C., Needham, H., Nickell, T. D., Pearson, T. H., Powell, H., Sammes, P. A., Somerfield, P. & Willis, K. J. (2005). Ecological effects of sea lice medicines in Scottish sea lochs. Final report of the Scottish Association of Marine Science, Oban, Scotland. Available at: www.sams.ac.uk/kenny-black/pamp-section-1.pdf. Accessed: 01 October 2012.

Black, K. D., Cook, E. J., Jones, K. J., Kelly, M. S., Leakey, R. J., Nickell, T. D., Sayer, M. D. J., Tett, P. & Willis. K. (2002). Review and synthesis of the environmental impacts of aquaculture. Scottish Association for Marine Science and Napier University. Scottish Executive Central Research Unit, Edinburgh, Scotland. Available at: www.scotland.gov.uk/ Publications/2002/08/15170/9405. Accessed: 01 October 2012.

Black, K. D., Cook, E. J., Jones, K. J., Kelly, M. S., Leakey, R. J., Nickell, T. D., Sayer, M. D. J., Tett, P. & Willis, K. (2002). Review and synthesis of the environmental impacts of aquaculture. Scottish Association for Marine Science and Napier University. Scottish Executive Central Research Unit, Edinburgh, Scotland. Available at: www.scotland.gov.uk/Publications/2002/08/15170/9405. Accessed: 01 October 2012.

Black, K. D., Cook, E. J., Jones, K. J., Kelly, M. S., Leakey, R. J., Nickell, T. D., Sayer, M. D. J., Tett, P. & Willis, K. (2002). Review and synthesis of the environmental impacts of aquaculture. Scottish Association for Marine Science and Napier University. Scottish Executive Central Research Unit, Edinburgh, Scotland. Available at: www.scotland.gov.uk/Publications/2002/08/15170/9405. Accessed: 01 October 2012.

Black, K. D., Cook, E. J., Jones, K. J., Kelly, M. S., Leakey, R. J., Nickell, T. D., J. Sayer, M. D., Tett, P. & Willis, K. (2002). Review and synthesis of the environmental impacts of aquaculture. Scottish Association for Marine Science and Napier University. Scottish

Executive Central Research Unit, Edinburgh, Scotland. Available at: www.scotland.gov.uk/Publications/2002/08/15170/9405. Accessed: 01 October 2012.

Black, K. D., Hansen, P. K. & Holmer, M. (2008). Salmon Aquaculture Dialogue: Working group report on benthic impacts and farm siting. World Wildlife Fund. Available at: www.fiskerifond.no/files/projects/attach/working_group_report_on_benthic_impacts_and_farm_siting.pdf. Accessed 28 September 2012.

Black, K. D., Hansen, P. K. & Holmer. M. (2008). Salmon Aquaculture Dialogue: Working group report on benthic impacts and farm siting. World Wildlife Fund. Available at: www.fiskerifond.no/files/projects/attach/working_group_report_on_ benthic_impacts_and_farm_siting.pdf. Accessed 28 September 2012.

Black, K., Cromey, C. & Nickell, T. (2012). SARF030 Final report: Benthic Recovery Project. Scottish Association for Marine Science, Oban, Scotland. Available at: www.sarf.org.uk/cms-assets/documents/43892-181648.sarf030.pdf. Accessed 28 September 2012.

Black, K., Cromey, C. & Nickell. T. (2012). SARF030 Final report: Benthic Recovery Project. Scottish Association for Marine Science, Oban, Scotland. Available at: www.sarf.org.uk/cms-assets/documents/43892-181648.sarf030.pdf. Accessed 28 September 2012.

Blouin, N., Xiugeng, F., Peng, J., Yarish, C. & Brawley, S. H. (2007). Seeding nets with neutral spores of the red alga Porphyra umbilicalis (L.) Kützing for use in integrated multi-trophic aquaculture (IMTA). *Aquaculture*, *270*, 77-91.

Bodennec, G., Arzul, G., Crassous, M. P. & Youenou, A. (2002). Influence of dead fish and uneaten fish feed elutriates on the toxic potential of certain microalgae. 0761-3962284433072X.

Bodennec, G., Arzul, G., Crassous, M. P. & Youenou. A. (2002). Influence of dead fish and uneaten fish feed elutriates on the toxic potential of certain microalgae. 0761-3962 284433072X.

Bongiorni, L., Shafir, S. & Rinkevich, B. (2003b). Effects of particulate matter released by a fish farm (Eilat, Red Sea) on survival and growth of *Stylophora pistillata* coral nubbins. *Marine Pollution Bulletin*, *46*, 1120-1124.

Bongiorni, L., Shafir, S., Angel, D. & Rinkevich, B. (2003a). Survival, growth and gonad development of two hermatypic corals subjected to in situ fish-farm nutrient enrichment. *Marine Ecology Progress Series*, *253*, 137-144.

Borg, J. A., Crosetti, D. & Massa, F. (2011). Site selection and carrying capacity in Mediterranean marine aquaculture: Key issues. Draft Report GFCM, XXXV/2011/ Dma.9. General Fisheries Commission for the Mediterranean, 35th Session, 9-14 May 2011, Rome, Italy. Available at: http://151.1.154.86/GfcmWebSite/GFCM/35/ GFCM_XXXV_2011_Dma.9.pdf. Accessed: 01 October 2012.

Borg, J. A., Crosetti, D. & Massa, F. (2011). Site selection and carrying capacity in Mediterranean marine aquaculture: Key issues. Draft Report GFCM, XXXV/2011/Dma.9. General Fisheries Commission for the Mediterranean, 35th Session, 9-14 May 2011, Rome, Italy. Available at: http://151.1.154.86/GfcmWebSite/GFCM/ 35/GFCM_XXXV_2011_Dma.9.pdf. Accessed: 01 October 2012.

Borg, J. A., Crosetti, D. & Massa. F. (2011). Site selection and carrying capacity in Mediterranean marine aquaculture: Key issues. Draft Report GFCM, XXXV/2011/Dma.9. General Fisheries Commission for the Mediterranean, 35th Session,

9-14 May 2011, Rome, Italy. Available at: http://151.1.154.86/GfcmWebSite/GFCM/35/GFCM_XXXV_2011_Dma.9.pdf. Accessed: 01 October 2012.

Borja, A., German Rodriguez, J., Black, K., Bodoy, A., Emblow, C., Fernandes, T. F., Forte, J., Karakassis, I., Muxika, I., Nickell, T. D., Papageorgiou, N., Pranovi, F., Sevastou, K., Tomassetti, P. & Angel, D. (2009). Assessing the suitability of a range of benthic indices in the evaluation of environmental impact of fin and shellfish aquaculture located in sites across Europe. *Aquaculture*, *293*, 231-240.

Boyd, C. E., McNevin, A. A., Clay, J. & Johnson, H. M. (2005). Certification issues for some common aquaculture species. *Reviews in Fisheries Science*, *13*, 231-279.125

Boyra, A., Nascimento, F. J. A., Tuya, F., Sanchez-Jerez, P. & Haroun, R. J. (2004). Impact of sea-cage fish farms on intertidal macrobenthic assemblages. *Journal of the Marine Biological Association of the United Kingdom*, *84*, 665-668.

Braaten, B, (2007). Cage culture and environmental impacts. Pages 49-91 *in* A. Bergheim, editor. Aquacultural Engineering and Environment. Research Signpost, Kerala, India.

Braaten, B. (2007). Cage culture and environmental impacts. Pages 49-91 *in* A. Bergheim, editor. Aquacultural Engineering and Environment. Research Signpost, Kerala, India.

Braaten, B. (2007). Cage culture and environmental impacts. Pages 49-91 *in* A. Bergheim, editor. Aquacultural Engineering and Environment. Research Signpost, Kerala, India.

Braaten, B. (2007). Cage culture and environmental impacts. Pages 49-91 *in* A. Bergheim, editor. Aquacultural Engineering and Environment. Research Signpost, Kerala, India.

Braithwaite, R. A., Carrascosa, M. C. & McEvoy, L. A. (2007). Biofouling of salmon cage netting and the efficacy of a typical copper-based antifoulant. *Aquaculture*, *262*, 219-226.

Brambilla, F., Pais, A., Serra, S., Terova, G. & Saroglia. M. (2007). A Meramod (R) model approach for the Environmental Impact Assessment (EIA) of the off-shore aquaculture improvement in the Alghero Bay (North western Sardinia, Italy). *Italian Journal of Animal Science*, *6*, 791-793.

Bridger, C. J. & Neal, B. (2004). Technical and economic considerations for exposed aquaculture site development in the Bay of Fundy. New Brunswick Salmon Growers Association, L'Etang, NB, Canada. Available at: http://s3.amazonaws.com/zanran_storage/nbsga.com/ContentPages/48254452.pdf. Accessed: 01 October 2012.

Brooks, K. M. & Mahnken, C. V. W. (2003). Interactions of Atlantic salmon in the Pacific Northwest environment. III. Accumulation of zinc and copper. *Fisheries Research*, *62*, 295-305.

Brooks, K. M., Stierns, A. R. & Backman, C. (2004). Seven year remediation study at the Carrie Bay Atlantic salmon (*Salmo salar*) farm in the Broughton Archipelago, British Columbia, Canada. *Aquaculture*, *239*, 81-123.

Brooks, K. M., Stierns, A. R. & Backman, C. (2004). Seven year remediation study at the Carrie Bay Atlantic salmon (*Salmo salar*) farm in the Broughton Archipelago, British Columbia, Canada. *Aquaculture*, *239*, 81-123.144

Brooks, K. M., Stierns, A. R. & Backman. C. (2004). Seven year remediation study at the Carrie Bay Atlantic salmon (*Salmo salar*) farm in the Broughton Archipelago, British Columbia, Canada. *Aquaculture*, *239*, 81-123.

Brooks, K. M., Stierns, A. R., Mahnken, C. V. W. & Blackburn, D. B. (2003). Chemical and biological remediation of the benthos near Atlantic salmon farms. *Aquaculture*, *219*, 355-377.

Brooks, K. M., Stierns, A. R., Mahnken, C. V. W. & Blackburn, D. B. (2003). Chemical and biological remediation of the benthos near Atlantic salmon farms. *Aquaculture, 219*, 355-377.
Brooks, K. M., Stierns, A. R., Mahnken, C. V. W. & Blackburn, D. B. (2003). Chemical and biological remediation of the benthos near Atlantic salmon farms. *Aquaculture, 219*, 355-377.
Brooks, K. M., Stierns, A. R., Mahnken, C. V. W. & Blackburn. D. B. (2003). Chemical and biological remediation of the benthos near Atlantic salmon farms. *Aquaculture, 219*, 355-377.
Bruun, M. S., Madsen, L. & Dalsgaard, I. (2003). Efficiency of oxytetracycline treatment in rainbow trout experimentally infected with *Flavobacterium psychrophilum* strains having different in vitro antibiotic susceptibilities. *Aquaculture and Fisheries Management, 215*, 11–20.
Buck, B. H, & Buchholz, C. M. (2004). The offshore-ring: A new system design for the open ocean aquaculture of macroalgae. *Journal of Applied Phycology, 16*, 355-368.
Burridge, L, (2003). Chemical use in marine finfish aquaculture in Canada: A review of current practices and possible environmental effects. Pages 97-131 *in* Canadian Technical Report of Fisheries and Aquatic Sciences 2450, Volume 1. Available at: http://mmc.gov/drakes_estero/pdfs/bivalve_aquaculture_03.pdf. Accessed: 27 September 2012.
Burridge, L., Weis, J. S., Cabello, F., Pizarro, J. & Bostick, K. (2010). Chemical use in salmon aquaculture: A review of current practices and possible environmental effects. *Aquaculture, 306*, 7-23.
Burridge, L., Weis, J., Cabello, F. & Pizarro, J. (2008). Chemical use in salmon aquaculture: A review of current practices and possible environmental effects. World Wildlife Fund. Available at: www.farmedanddangerous.org/wp-content/uploads/2011/01/SAD_chemicals_report.pdf. Accessed: 01 October 2012.
Buryniuk, M., Petrell, R. J., Baldwin, S. & Lo. K. V. (2006). Accumulation and natural disintegration of solid wastes caught on a screen suspended below a fish farm cage. *Aquacultural Engineering, 35*, 78-90.
Buschmann, A. H. Cabello, ., F, Young, K., Carvajal, J., Varela, D. A. & Henriquez, L. (2009). Salmon aquaculture and coastal ecosystem health in Chile: Analysis of regulations, environmental impacts and bioremediation systems. *Ocean and Coastal Management, 52*, 243-249.
Buschmann, A. H., Cabello, F., Young, K., Carvajal, J., Varela, D. A. & Henriquez. L. (2009. Salmon *Aquaculture*, and coastal ecosystem health in Chile: Analysis of regulations, environmental impacts and bioremediation systems. *Ocean & Coastal Management, 52*, 243-249.49
Buschmann, A. H., Cabello, F., Young, K., Carvajal, J., Varela, D. A. & Henriquez, L. (2009). Salmon aquaculture and coastal ecosystem health in Chile: Analysis of regulations, environmental impacts and bioremediation systems. *Ocean and Coastal Management, 52*, 243-249.
Buschmann, A. H., Cabello, F., Young, K., Carvajal, J., Varela, D. A. & Henriquez, L. (2009). Salmon aquaculture and coastal ecosystem health in Chile: Analysis of regulations, environmental impacts and bioremediation systems. *Ocean and Coastal Management, 52*, 243-249.

Bybee, D. R. (2001). Environmental impact of an open ocean aquaculture cage on the benthic invertebrate community off Oahu, Hawaii, USA (Abstract). Page 32 *in* C.J. Bridger and T.H. Reid, editors. Open Ocean Aquaculture IV, 17-20 Jun 2001, St. Andrews, NB, Canada. Available at: http://nsgl.gso.uri.edu/masgc/masgcw01001.pdf. Accessed: 01 October 2012.

Cañadas, A. & Hammond, P. S. (2008). Abundance and habitat preferences of the short-beaked common dolphin *Delphinus delphis* in the southwestern Mediterranean: Implications for conservation. *Endangered Species Research*, *4*, 309-331.

Cancemi, G., D. Falco, G. & Pergent, G. (2003). Effects of organic matter input from a fish farming facility on a *Posidonia oceanica* meadow. *Estuarine, Coastal and Shelf Science*, *56*, 961-968.

Cancemi, G., De Falco, G. & Pergent. G. (2000). Impact of a fish farming facility on a *Posidonia oceanica* meadow. *Biologia Marina Mediterranea*, *7*, 341-344.

Capone, D. G., Weston, D. P., Miller, V. & Shoemaker, C. (1996). Antibacterial residues in marine sediments and invertebrates following chemotherapy in aquaculture. *Aquaculture*, *145*, 55-75.

Cardia, F. & Lovatelli, A. (2007). A review of cage aquaculture: Mediterranean Sea. Pages 156-187 *in* D. Halwart, D. Soto, and J.R. Arthur, editors. Cage aquaculture: Regional reviews and global overview. Food and Agricultural Organization of the United Nations, Rome, Italy. Available at: ftp://ftp.fao.org/docrep/fao/010/a1290e/a1290e.pdf. Accessed: 01 October 2012.

Cardia, F. & Lovatelli. A. (2007). A review of cage aquaculture: Mediterranean Sea. Pages 156-187 *in* D. Halwart, D. Soto, & J.R. Arthur, editors. Cage aquaculture: Regional reviews and global overview. Food and Agricultural Organization of the United Nations, Rome, Italy. Available at: ftp://ftp.fao.org/docrep/fao/010/a1290e/a1290e.pdf. Accessed: 01 October 2012.

Carroll, M. L., Cochrane, S., Fieler, R., Velvin, R. & White, P. (2003). Organic enrichment of sediments from salmon farming in Norway: Environmental factors, management practices, and monitoring techniques. *Aquaculture*, *226*, 165-180.

Carroll, M. L., Cochrane, S., Fieler, R., Velvin, R. & White. P. (2003). Organic enrichment of sediments from salmon farming in Norway: Environmental factors, management practices, & monitoring techniques. *Aquaculture*, *226*, 165-180.

Carroll, M. L., S., Cochrane, Fieler, R., Velvin, R. & White, P. (2003). Organic enrichment of sediments from salmon farming in Norway: Environmental factors, management practices, and monitoring techniques. *Aquaculture*, *226*, 165-180.

Chamberlain, J. & Stucchi, D. (2007). Simulating the effects of parameter uncertainty on waste model predictions of marine finfish aquaculture. *Aquaculture*, *272*, 296-311.

Chavez-Crooker, P. & Obreque-Contreras, J., (2010). Bioremediation of aquaculture wastes. *Current Opinion in Biotechnology*, *21*, 313-317.

Chelossi, E., Vezzulli, L., Milano, A., Branzoni, M., Fabiano, M., Riccardi, G. & Banat, I. M. (2003). Antibiotic resistance of benthic bacteria in fish-farm and control sediments of the Western Mediterranean. *Aquaculture*, *219*, 83-97.

Chen, Y. S., Beveridge, M. C. M., Telfer, T. C. & Roy. W. J. (2003). Nutrient leaching and settling rate characteristics of the faeces of Atlantic salmon (*Salmo salar* L.) and the implications for modelling of solid waste dispersion. *Journal of Applied Ichthyology*, *19*, 114-117.

Cheshuk, B. W., Purser, G. J. & Quintana, R. (2003). Integrated open-water mussel (Mytilus planulatus) and Atlantic salmon (Salmo salar) culture in Tasmania, Australia. *Aquaculture*, *218*, 357-378.

Chopin, T. (2006). Integrated Multi-Trophic Aquaculture. Northern Aquaculture July/ August, 4.

Chopin, T., Yarish, C., Wilkes, R., Belyea, E., Lu, S. & Mathieson, A. (1999). Developing Porphyra/salmon integrated aquaculture for bioremediation and diversification of the aquaculture industry. *Journal of Applied Phycology*, *11*, 463-472.

Chou, C. L., Haya, K., Paon, L. A. & Moffatt, J. D. (2003). Metals in the green sea urchin (*Strongylocentrotus droebachiensis*) as an indicator for the near field effects of chemical wastes from salmon aquaculture sites in New Brunswick, Canada. *Bulletin of Environmental Contamination and Toxicology*, *70*, 948-956.

Chou, C. L., Haya, K., Paon, L. A. & Moffatt, J. D. (2004). A regression model using sediment chemistry for the evaluation of marine environmental impacts associated with salmon aquaculture cage wastes. *Marine Pollution Bulletin*, 49, 465-472.

Chou, C. L., Haya, K., Paon, L. A. & Moffatt. J. D. (2004). A regression model using sediment chemistry for the evaluation of marine environmental impacts associated with salmon *Aquaculture*, cage wastes. *Marine Pollution Bulletin*, *49*, 465-472.

Chou, C. L., Haya, K., Paon, L. A., Burridge, L. & Moffatt, J. D. (2002a). Inorganic chemicals in sediments, wild lobsters, and sea urchins near aquaculture sites as indicators of marine environmental quality. *Canadian Technical Report of Fisheries and Aquatic Sciences*, *2403*, 132-140.

Chou, C. L., Haya, K., Paon, L. A., Burridge, L. & Moffatt, J. D. (2002b). Aquaculture-related trace metals in sediments and lobsters and relevance to environmental monitoring program ratings for near-field effects. *Marine Pollution Bulletin*, *44*, 1259-1268.126

Christensen, P. B., Rysgaard, S., Sloth, N. P., Dalsgaard, T. & Schwaerter. S. (2000). Sediment mineralization, nutrient fluxes, denitrification and dissimilatory nitrate reduction to ammonium in an estuarine fjord with sea cage trout farms. *Aquatic Microbial Ecology*, *21*, 73-84.

Chung, I. K., Kang, Y. H., Yarish, C., Kraemer, G. P. & Lee, J. A. (2002). Application of seaweed cultivation to the bioremediation of nutrient-rich effluent. *Algae*, *17*, 187-194.

Clement, J. & Janowicz. M. (2003). *Aquaculture*, physical remediation: Workshop proceedings, 20-21 September 2001. Gulf of Maine Council on the Marine Environment. Available at: www.gulfofmaine.org/council/publications/aquacultureworkshopreport.pdf. Accessed: 02 October 2012.

Cloern, J. E. (2001). Our evolving conceptual model of the coastal eutrophication problem. *Marine Ecology Progress Series*, *210*, 223-253.

Cloern, J. E. (2001). Our evolving conceptual model of the coastal eutrophication problem. *Marine Ecology Progress Series*, *210*, 223-253.

Cloern, J. E. (2001). Our evolving conceptual model of the coastal eutrophication problem. *Marine Ecology Progress Series*, *210*, 223-253.

Cole, R. (2002). Impacts of marine farming on wild fish populations. Final Research Report for Ministry of Fisheries Research Project ENV2000/08 Objective One, National Institute of Water and Atmospheric Research, New Zealand. Available at: aquaculture.govt.nz/ files/pdfs/Impacts_of_marine_farming_on_wild_fish_stocks.pdf. Accessed: 27 September 2012.

Cole, R. (2002). Impacts of marine farming on wild fish populations. Final Research Report for Ministry of Fisheries Research Project ENV2000/08 Objective One, National Institute of Water and Atmospheric Research, New Zealand. Available at: aquaculture.govt.nz/ files/pdfs/Impacts_of_marine_farming_on_wild_fish_stocks.pdf. Accessed: 27 September 2012.

Cole, R. (2002). Impacts of marine farming on wild fish populations. Final Research Report for Ministry of Fisheries Research Project ENV2000/08 Objective One, National Institute of Water and Atmospheric Research, New Zealand. Available at: aquaculture.govt.nz/ files/pdfs/Impacts_of_marine_farming_on_wild_fish_stocks.pdf. Accessed: 27 September 2012.

Cole, R. 2002). Impacts of marine farming on wild fish populations. Final Research Report for Ministry of Fisheries Research Project ENV2000/08 Objective One, National Institute of Water and Atmospheric Research, New Zealand. Available at: aquaculture.govt.nz/ files/pdfs/Impacts_of_marine_farming_on_wild_fish_stocks.pdf. Accessed:27 September 2012.

Cook, E. J., Black, K. D., Sayer, M. D. J., Cromey, C. J., Angel, D. L., Spanier, E., Tsemel, A., Katz, T., Eden, N., Karakassis, I., Tsapakis, M., Apostolaki, E. T. & Malej, A. (2006). The influence of caged mariculture on the early development of sublittoral fouling commnities: a pan-European study. *ICES Journal of Marine Science*, *63*, 637-649.

Costa-Pierce, B. A., Buschmann, A., Cross, S., Iriarte, J. L., Olsen, Y. O. & Reid. G. (2007). Nutrient impacts of farmed Atlantic salmon (*Salmo salar*) on pelagic ecosystems and implications for carrying capacity. Report of the Technical Working Group on Nutrients and Carrying Capacity of the World Wildlife Fund Salmon Aquaculture Dialogue. World Wildlife Federation, Washington, D.C. Available at: www.fiskerifond.no/files/projects/ attach/final_report_____nutrient_impacts_of_farmed_atlantic_salmon_salmo_salar_on.p df. Accessed: 28 September 2012.

Costa-Pierce, B. A., Buschmann, A., Cross, S., Iriarte, J. L., Olsen, Y. O. & Reid, G. (2007). Nutrient impacts of farmed Atlantic salmon (*Salmo salar*) on pelagic ecosystems and implications for carrying capacity. Report of the Technical Working Group on Nutrients and Carrying Capacity of the World Wildlife Fund Salmon *Aquaculture*, Dialogue. World Wildlife Federation, Washington, D.C. Available at: www.fiskerifond. no/files/projects/attach/final_report_____nutrient_impacts_of_farmed_atlantic_salmon_s almo_salar_on.pdf. Accessed: 28 September 2012.

Costa-Pierce, B. A., Buschmann, A., Cross, S., Iriarte, J. L., Olsen, Y. O. & Reid, G. (2007). Nutrient impacts of farmed Atlantic salmon (*Salmo salar*) on pelagic ecosystems and implications for carrying capacity. Report of the Technical Working Group on Nutrients and Carrying Capacity of the World Wildlife Fund Salmon Aquaculture Dialogue. World Wildlife Federation, Washington, D.C. Available at: www.fiskerifond. no/files/projects/attach/final_report_____nutrient_impacts_of_farmed_atlantic_salmon_s almo_salar_on.pdf. Accessed: 28 September 2012.

Costello, M. J., Grant, A., Davies, I. M., Cecchini, S., Papoutsoglou, S., Quigley, D. & Saroglia, M. (2001). The control of chemicals used in aquaculture in Europe. *Journal of Applied Ichthyology*, *17*, 173-180.

Crawford, C, (2003). Environmental management of marine aquaculture in Tasmania, Australia. *Aquaculture*, *226*, 129-138.

Crawford, C. (2003). Environmental management of marine aquaculture in Tasmania, Australia. *Aquaculture*, *226*, 129-138.

Cromey, C. J., Nickell, T. D. & Black, K. D. (2002). DEPOMOD—modelling the deposition and biological effects of waste solids from marine cage farms. *Aquaculture*, *214*, 211-239.

Cromey, C. J., Nickell, T. D. & Black, K. D. (2002). DEPOMOD—modelling the deposition and biological effects of waste solids from marine cage farms. *Aquaculture*, *214*, 211-239.

Cromey, C. J., Thetmeyer, H., Lampadariou, N., Black, K. D., Kögeler, J. & Karakassis, I. (2012). MERAMOD: Predicting the deposition and benthic impact of aquaculture in the eastern Mediterranean Sea. *Aquaculture Environment Interactions*, *2*, 157-176.

Cromey, C. J., Thetmeyer, H., Lampadariou, N., Black, K. D., Kögeler, J. & Karakassis, I. (2012). MERAMOD: Predicting the deposition and benthic impact of aquaculture in the eastern Mediterranean Sea. *Aquaculture Environment Interactions*, *2*, 157-176.

Dalsgaard, T. & Krause-Jensen, D. (2006). Monitoring nutrient release from fish farms with macroalgal and phytoplankton bioassays. *Aquaculture*, *256*, 302-310.

Danovaro, R., Corinaldesi, C., La Rosa, T., Luna, G. M., Mazzola, A., Mirto, S., Vezzulli, L. & Fabiano, M., (2003). Aquaculture impact on benthic microbes and organic matter cycling in coastal Mediterranean sediments: A synthesis. *Chemistry and Ecology*, *19*, 59-65.

Das, S., Ward, L. R. & Burke, C. (2008). Prospects of using marine actinobacteria as probiotics in aquaculture. *Applied Microbiology and Biotechnology*, *81*, 419-429.

Davidson, K., Miller, P., Wilding, T. A., Shutler, J., Bresnan, E., Kennington, K. & Swan, S., (2009). A large and prolonged bloom of *Karenia mikimotoi* in Scottish waters in 2006. *Harmful Algae*, *8*, 349-361.

Davies, I. M., Drinkwater, J. & McKie, J. C. (1988). Effects of tributyltin compounds from antifoulants on Pacific oysters (*Crassostrea gigas*) in Scottish sea lochs. *Aquaculture*, *74*, 319-330.

Davis, M., O'Hanlon, B., Rivera, J., Corsaut, J., Wadley, T., Creswell, L., Ayvazian, J. & Benetti, D. (2006). Recruitment of spiny lobsters, *Panulirus argus*, to submerged sea cages off Puerto Rico, and its implication for the development of an aquaculture operation. *Proceedings of the Gulf and Caribbean Fisheries Institute*, *57*, 975-980.

Dean, R. J., Shimmield, T. M. & Black, K. D. (2007). Copper, zinc and cadmium in marine cage fish farm sediments: An extensive survey. *Environmental Pollutionm*, *145*, 84-95. Epub 2006 Jun 8.

Debruyn, A. M., Trudel, M., Eyding, N., Harding, J., McNally, H., Mountain, R., Orr, C., Urban, D., Verenitch, S. & Mazumder, A. (2006). Ecosystemic effects of salmon farming increase mercury contamination in wild fish. *Environmental Science and Technology*, *40*, 3489-3493.

Demirak, A., Balci, A. & Tufekci, M. (2006). Environmental impact of the marine aquaculture in Gulluk Bay, Turkey. *Environmental Monitoring and Assessment*, *123*, 1-12.

Demirak, A., Balci, A. & Tufekci. M. (2006). Environmental impact of the marine aquaculture in Gulluk Bay, Turkey. *Environmental Monitoring and Assessment*, *123*, 1-12.

Dempster, T., Fernandez-Jover, D., Sanchez-Jerez, P., Tuya, F., Bayle-Sempere, J., Boyra, A. & Haroun, R. J. (2005). Vertical variability of wild fish assemblages around sea-cage fish farms: Implications for management. *Marine Ecology Progress Series*, *304*, 15-29.

Dempster, T., Korsoen, O., Folkedal, O., Juell, J. E. & Oppedal. F. (2009). Submergence of Atlantic salmon (*Salmo salar* L.) in commercial scale sea-cages: A potential short-term solution to poor surface conditions. *Aquaculture*, *288*, 254-263.

Dempster, T., Sanchez-Jerez, P., Tuya, F., Fernandez-Jover, D., Bayle-Sempere, J., Boyra, A. & Haroun, R. (2006). Coastal aquaculture and conservation can work together. *Marine Ecology Progress Series*, *314*, 309-310.

Devlin, M., Painting, S. & Best, M. (2007). Setting nutrient thresholds to support an ecological assessment based on nutrient enrichment, potential primary production and undesirable disturbance. *Marine Pollution Bulletin*, *55*, 65-73.50

Dharmaraj, S. & Kandasamy, D. (2010). *Streptomyces* as probiotics for *X. helleri* growth. *Food Technology and Biotechnology*, *48*, 497-504.

Díaz López, B. & Bernal Shirai, J. A. (2007). Bottlenose dolphin (*Tursiops truncatus*) presence and incidental capture in a marine fish farm on the north-eastern coast of Sardinia (Italy). *Journal of the Marine Biological Association of the U.K.*, *87*, 113-117.

Díaz López, B. (2012). Bottlenose dolphins and aquaculture: interaction and site fidelity on the north-eastern coast of Sardinia (Italy). *Marine Biology*, *159*, 1-12, DOI: 10.1007/s00227-00012-02002-x.

Diaz Lopez, B., Marini, L. & Polo, F. (2005). The impact of a fish farm on a bottlenose dolphin population in the Mediterranean Sea. *Thalassas*, *21*, 65-70.

Diaz-Castaneda, V. & Valenzuela-Solano, S. (2009). Polychaete fauna in the vicinity of bluefin tuna sea-cages in Ensenada, Baja California, Mexico. Magnolia Press, *Zoosymposia*, 2, 505-526. Available at: www.mapress.com/zoosymposia/content/2009/v2/f/v002p505-526f.pdf. Accessed: 01 October 2012.

Diaz-Castaneda, V. & Valenzuela-Solano, S. (2009). Polychaete fauna in the vicinity of bluefin tuna sea-cages in Ensenada, Baja California, Mexico. Magnolia Press, Zoosymposia 2: 505-526. Available at: www.mapress.com/zoosymposia/content/2009/v2/f/v002p505-526f.pdf. Accessed: 01 October 2012.

Dimitriou, P., Castritsi-Catharios, J. & Miliou, H. (2003). Acute toxicity effects of tributyltin chloride and triphenyltin chloride on gilthead seabream, *Sparus aurata* L., embryos. *Ecotoxicology and Environmental Safety*, *54*, 30-35.

Doglioli, A. M., Magaldi, M. G., Vezzulli, L. & Tucci, S. (2004). Development of a numerical model to study the dispersion of wastes coming from a marine fish farm in the Ligurian Sea (western Mediterranean). *Aquaculture*, *231*, 215-235.

Doglioli, A. M., Magaldi, M. G., Vezzulli, L. & Tucci. S. (2004). Development of a numerical model to study the dispersion of wastes coming from a marine fish farm in the Ligurian Sea (western Mediterranean). *Aquaculture*, *231*, 215-235.

Doglioli, A. M., Magaldi, M. G., Vezzulli, L. & Tucci. S. (2004). Development of a numerical model to study the dispersion of wastes coming from a marine fish farm in the Ligurian Sea (western Mediterranean). *Aquaculture*, *231*, 215-235.

Dolenec, T., Lojen, S., Kniewald, G., Dolenee, M. & Rogan, N. (2007). Nitrogen stable isotope composition as a tracer of fish farming in invertebrates *Aplysina aerophoba*, *Balanus perforatus* and *Anemonia sulcata* in central Adriatic. *Aquaculture*, *262*, 237-249.

Dolenec, T., Lojen, S., Kniewald, G., Dolenee, M. & Rogan. N. (2007). Nitrogen stable isotope composition as a tracer of fish farming in invertebrates *Aplysina aerophoba*, *Balanus perforatus* and *Anemonia sulcata* in central Adriatic. *Aquaculture*, *262*, 237-249.

Dolenec, T., Lojen, S., Lambasa, S. & Dolenec, M. (2006). Effects of fish farm loading on sea grass *Posidonia oceanica* at Vrgada Island (Central Adriatic): A nitrogen stable isotope study. *Isotopes in Environmental and Health Studies*, *42*, 77-85.

Dominguez, L. M., Calero, G. L., Martin, J. M. V. & Robaina, L. R. (2001). A comparative study of sediments under a marine cage farm at Gran Canaria Island (Spain). Preliminary results. *Aquaculture*, *192*, 225-231.

Duarte, P., Meneses, R., Hawkins, A. J. S., Zhu, M., Fang, J. & Grant, J. (2003). Mathematical modelling to assess the carrying capacity for multi-species culture within coastal waters. *Ecological Modelling*, *168*, 109-143.

Early, G. (2001). The impact of aquaculture on marine mammals. Pages 211-214 *in* M. Tlusty, D. Bengtson, H.O. Halvorson, S. Oktay, J. Pearce, and R.B. Rheault, Jr., editors. Marine Aquaculture and the Environment: A meeting for stakeholders in the northeast, Cape Cod Press, Falmouth, Massachusetts. Available at: www.neaq.org/conservation_and_research/projects/publications_and_presentations/pdf/12__.pdf. Accessed: 01 October 2012.

Eden, N., Katz, T. & Angel, D. L. (2003). Dynamic response of a mud snail *Nassarius sinusigerus* to changes in sediment biogeochemistry. *Marine Ecology Progress Series*, *263*, 139-147.

Edgar, G. J., Davey, A. & Shepherd, C. (2010). Application of biotic and abiotic indicators for detecting benthic impacts of marine salmonid farming among coastal regions of Tasmania. *Aquaculture*, *307*, 212-218.

Edgar, G. J., Davey, A. & Shepherd, C. (2010). Application of biotic and abiotic indicators for detecting benthic impacts of marine salmonid farming among coastal regions of Tasmania. *Aquaculture*, *307*, 212-218.

Edgar, G. J., Macleod, C. K., Mawbey, R. B. & Shields, D. (2005). Broad-scale effects of marine salmonid *Aquaculture*, on macrobenthos and the sediment environment in southeastern Tasmania. *Journal of Experimental Marine Biology and Ecology*, *327*, 70-90.

Edgar, G. J., Macleod, C. K., Mawbey, R. B. & Shields, D. (2005). Broad-scale effects of marine salmonid aquaculture on macrobenthos and the sediment environment in southeastern Tasmania. *Journal of Experimental Marine Biology and Ecology*, *327*, 70-90.

Endo, A., Srithongouthai, S., Nashiki, H., Teshiba, I., Iwasaki, T., Hama, D. & Tsutsumi. H. (2008). DO-increasing effects of a microscopic bubble generating system in a fish farm. *Marine Pollution Bulletin*, *57*, 78-85.

Engstroem-Oest, J., Lehtiniemi, M., Jonasdottir, S. H. & Viitasalo, M. (2005). Growth of pike larvae (Esox lucius) under different conditions of food quality and salinity. *Ecology of Freshwater Fish*, *14*, 385-393.

Environment Canada. (2001). Environmental assessment of marine finfish aquaculture projects: Guidelines for consideration of environment Canada expertise. Environment Canada, Dartmouth, Nova Scotia.

Fairgrieve, W. T. & Rust, M. B. (2003). Interactions of Atlantic salmon in the Pacific northwest V. Human health and safety. *Fisheries Research*, *62*, 329-338.

Felsing, B., Glencross, B. & Telfer, T. (2005). Preliminary study on the effects of exclusion of wild fauna from aquaculture cages in a shallow marine environment. *Aquaculture*, *243*, 159-174.

Findlay, R. H., Watling, L. & Mayer, L. M. (1995). Environmental impact of salmon net-pen culture on marine benthic communities in Maine: a case study. *Estuaries*, *18*, 145-179.

Forrest, B. M., Hopkins, G. A., Dodgshun, T. J. & Gardner, J. P. A. (2007). Efficacy of acetic acid treatments in the management of marine biofouling. *Aquaculture*, *262*, 319-332.

Fortt, A. & Buschmann, A. R. (2007). Use and abuse of antibiotics in salmon farming. Document 23. Oceana, Santiago, Chile. Available at: http://oceana.org/sites/default/files/reports/Uso_antibioticos_en_la_salmonicultura__version_ingles_1.pdf. Accessed: 02 October 2012.

Fortt, A. & Buschmann, A. R. (2007). Use and abuse of antibiotics in salmon farming. Document 23. Oceana, Santiago, Chile. Available at: http://oceana.org/sites/default/files/reports/Uso_antibioticos_en_la_salmonicultura__version_ingles_1.pdf. Accessed: 02 October 2012.

Galloway, J. N., Dentener, F. J., Capone, D. G., Boyer, E. W., Howarth, R. W., Seitzinger, S. P., Asner, G. P., Cleveland, C. C., Green, P. A., Holland, E. A., Karl, D. M., Michaels, A. F., Porter, J. H., Townsend, A. R. & Vörösmarty. C. J. (2004). Nitrogen cycles: past, present, and future. *Biogeochemistry*, *70*, 153-226.

Gao, Q. F., Cheung, K. L., Cheung, S. G. & Shin, P. K. S. (2005). Effects of nutrient enrichment derived from fish farming activities on macroinvertebrate assemblages in a subtropical region of Hong Kong. *Marine Pollution Bulletin*, *51*, 994-1002.

Gao, Q. F., Cheung, K. L., Cheung, S. G. & Shin, P. K. S. (2005). Effects of nutrient enrichment derived from fish farming activities on macroinvertebrate assemblages in a subtropical region of Hong Kong. *Marine Pollution Bulletin*, *51*, 994-1002.

Garren, M., Raymundo, L., Guest, J., Harvell, C. D. & Azam, F. (2009). Resilience of coral-associated bacterial communities exposed to fish farm effluent. *PLoS ONE*, *4*, e731, DOI, 710.1371/journal.pone.0007319.

Giles, H. (2008). Using Bayesian networks to examine consistent trends in fish farm benthic impact studies. *Aquaculture*, *274*, 181-195.

Giles, H. (2008). Using Bayesian networks to examine consistent trends in fish farm benthic impact studies. *Aquaculture*, *274*, 181-195.

Gillibrand, P. A., Cromey, C. J., Black, K. D., Inall, M. E. & Gontarek. S. J. (2006). Identifying the risk of deoxygenation in Scottish sea lochs with isolated deep water. A report to the Scottish Aquaculture Research Forum, Scottish Association for Marine Science, Oban, Scotland. Available at: www.sarf.org.uk/cms-assets/documents/28798-260528.sarf017-final-report-rev---may2007.pdf. Accessed: 27 September 2012.

Gillibrand, P. A., Gubbins, M. J., Greathead, C. & Davies. I. M. (2002). Scottish Executive locational guidelines for fish farming: Predicted levels of nutrient enhancement and benthic impact. Fisheries Research Services, Marine Laboratory, Aberdeen, Scotland. Available at: www.scotland.gov.uk/Uploads/Documents/Report63.pdf. Accessed: 02 October 2012.

Global Aquaculture Alliance. (2011). Salmon farms: Best aquaculture practices certification standards, guidelines. Global Aquaculture Alliance, St. Louis, Missouri. Available at: www.gaalliance.org/cmsAdmin/uploads/BAP-SalmonF-611.pdf. Accessed: 01 October 2012.145

Goldburg, R. & Naylor. R. (2005). Future seascapes, fishing, and fish farming. *Frontiers in Ecology and the Environment*, *3*, 21-28.

Goldburg, R. & Triplett, T. (1997). Murky waters: Environmental effects of aquaculture in the United States. Environmental Defense Fund, Washington, D.C. Available at: apps.edf.org/documents/490_AQUA.pdf. Accessed: 27 September 2012.

Goldburg, R. & Triplett, T. (1997). Murky waters: Environmental effects of aquaculture in the United States. Environmental Defense Fund, Washington, D.C. Available at: apps.edf.org/documents/490_AQUA.pdf. Accessed: 27 September 2012.127

Goldburg, R. & Triplett. T. (1997). Murky waters: Environmental effects of aquaculture in the United States. Environmental Defense Fund, Washington, D.C. Available at: apps.edf.org/documents/490_AQUA.pdf. Accessed: 27 September 2012.

Goldburg, R. & Triplett. T. (1997). Murky waters: Environmental effects of *Aquaculture*, in the United States. Environmental Defense Fund, Washington, D.C. Available at: apps.edf.org/documents/490_AQUA.pdf. Accessed: 27 September 2012.

Goldburg, R. J., Elliott, M. S. & Naylor, R. L. (2001). Marine *Aquaculture*, in the United States: Environmental impacts and policy options. Pew Oceans Commission, Arlington, Virginia. Available at: www.pewtrusts.org/uploadedFiles/wwwpewtrustsorg/Reports/Protecting_ocean_life/env_pew_oceans_aquaculture.pdf. Accessed: 28 September 2012.

Goldburg, R. J., Elliott, M. S. & Naylor, R. L. (2001). Marine aquaculture in the United States: Environmental impacts and policy options. Pew Oceans Commission, Arlington, Virginia. Available at: www.pewtrusts.org/uploadedFiles/wwwpewtrustsorg/Reports/Protecting_ocean_life/env_pew_oceans_aquaculture.pdf. Accessed: 28 September 2012.

Goldburg, R. J., Elliott, M. S. & Naylor, R. L. (2001). Marine aquaculture in the United States: Environmental impacts and policy options. Pew Oceans Commission, Arlington, Virginia. Available at: www.pewtrusts.org/uploadedFiles/wwwpewtrustsorg/Reports/Protecting_ocean_life/env_pew_oceans_aquaculture.pdf. Accessed: 28 September 2012.

Goldburg, R. J., Elliott, M. S. & Naylor. R. L. (2001). Marine aquaculture in the United States: Environmental impacts and policy options. Pew Oceans Commission, Arlington, Virginia. Available at: www.pewtrusts.org/uploadedFiles/wwwpewtrustsorg/Reports/Protecting_ocean_life/env_pew_oceans_aquaculture.pdf. Accessed: 28 September 2012.

Gowland, B. T. G., Moffat, C. F., Stagg, R. M., Houlihan, D. F. & Davies, I. M. (2002). Cypermethrin induces glutathione S-transferase activity in the shore crab, *Carcinus maenas*. *Marine Environmental Research*, *54*, 169-177.

Grant, A. & Briggs, A. D. (1998). Toxicity of ivermectin to estuarine and marine invertebrates. *Marine Pollution Bulletin*, *36*, 540-541.

Graslund, S. & Bengtsson, B.-E. (2001). Chemicals and biological products used in south-east Asian shrimp farming, and their potential impact on the environment: A review. *Science of the Total Environment*, *280*, 93-131.

Grigorakis, K. & Rigos, G. (2011). Aquaculture effects on environmental and public welfare - the case of Mediterranean mariculture. *Chemosphere*, *855*, 899-919.

Grigorakis, K. & Rigos, G. (2011). *Aquaculture*, effects on environmental and public welfare - the case of Mediterranean mariculture. *Chemosphere*, *855*, 899-919.

Grigorakis, K. & Rigos. G. (2011). Aquaculture effects on environmental and public welfare - the case of Mediterranean mariculture. *Chemosphere*, *855*, 899-919.

Guecluesoy, H. & Savas, Y. (2003). Interaction between monk seals *Monachus monachus* (Hermann, 1779) and marine fish farms in the Turkish Aegean and management of the problem. *Aquaculture Research*, *34*, 777-783.

Guenther, J., Carl, C. & Sunde, L. M. (2009). The effects of colour and copper on the settlement of the hydroid *Ectopleura larynx* on aquaculture nets in Norway. *Aquaculture*, *292*, 252-255.

Gulf of Mexico Fishery Management Council. (2009). Final Fishery Management Plan for Regulating Offshore Marine Aquaculture in the Gulf of Mexico. Available at: www.gulfcouncil.org/Beta/GMFMCWeb/Aquaculture/Aquaculture%20FMP%20PEIS%20Final%202-24-09.pdf. Accessed: 27 September 2012.

Hall, S. J., Delaporte, A., Phillips, M. J., Beveridge, M. C. M. & O'Keefe, M. (2011). Blue frontiers: Managing the environmental costs of aquaculture. The World Fish Center, Penang, Malaysia. Available at: www.worldfishcenter.org/sites/default/files/report.pdf. Accessed: 02 October 2012.51

Hall, S. J., Delaporte, A., Phillips, M. J., Beveridge, M. C. M. & O'Keefe, M. (2011). Blue frontiers: Managing the environmental costs of aquaculture. The World Fish Center, Penang, Malaysia. Available at: www.worldfishcenter.org/sites/default/files/report.pdf. Accessed: 02 October 2012.

Halling-Sorensen, B., Nors Nielsen, S., Lanzky, P. F., Ingerslev, F., Holten Luetzhoft, H. C. & Jorgensen, S. E. (1998). Occurrence, fate and effects of pharmaceutical substances in the environment - A review. *Chemosphere*, *36*, 357-393.

Hall-Spencer, J., White, N., Gillespie, E., Gillham, K. & Foggo, A. (2006). Impact of fish farms on maerl beds in strongly tidal areas. *Marine Ecology Progress Series*, *326*, 1-9.

Halwart, M., Soto, D. & Arthur, J. R. (2007). Cage aquaculture: Regional reviews and global overview. FAO Fisheries Technical Paper No. 498, FAO, Rome, Italy. Available at: ftp.fao.org/docrep/fao/010/a1290e/a1290e.pdf. Accessed: 27 September 2012.

Halwart, M., Soto, D. & Arthur, J. R. (2007). Cage aquaculture: Regional reviews and global overview. FAO Fisheries Technical Paper No. 498, FAO, Rome, Italy. Available at: ftp.fao.org/docrep/fao/010/a1290e/a1290e.pdf. Accessed: 27 September 2012.

Halwart, M., Soto, D. & Arthur, J. R. (2007). Cage aquaculture: Regional reviews and global overview. FAO Fisheries Technical Paper No. 498, FAO, Rome, Italy. Available at: ftp.fao.org/docrep/fao/010/a1290e/a1290e.pdf. Accessed: 27 September 2012.

Halwart, M., Soto, D. & Arthur, J. R. (2007). Cage aquaculture: Regional reviews and global overview. FAO Fisheries Technical Paper No. 498, FAO, Rome, Italy. Available at: ftp.fao.org/docrep/fao/010/a1290e/a1290e.pdf. Accessed: 27 September 2012.

Halwart, M., Soto, D. & Arthur. J. R. (2007). Cage aquaculture: Regional reviews and global overview. FAO Fisheries Technical Paper No. 498, FAO, Rome, Italy. Available at: ftp.fao.org/docrep/fao/010/a1290e/a1290e.pdf. Accessed: 27 September 2012.

Halwart, M., Soto, D., Arthur. J. R. (2007). Cage aquaculture: Regional reviews and global overview. FAO Fisheries Technical Paper No. 498, FAO, Rome, Italy. Available at: ftp.fao.org/docrep/fao/010/a1290e/a1290e.pdf. Accessed: 27 September 2012.

Hansen, P. K., Ervik, A., Schaanning, M., Johannessen, P., Aure, J., Jahnsen, T. & Stigebrandt. A. (2001). Regulating the local environmental impact of intensive, marine fish farming - II. The monitoring programme of the MOM system (Modelling-Ongrowing fish farms-Monitoring). *Aquaculture*, 194, 75-92.

Hansen, P. K., Ervik, A., Schaanning, M., Johannessen, P., Aure, J., Jahnsen, T. & Stigebrandt, A. (2001. Regulating the local environmental impact of intensive, marine fish farming - II. The monitoring programme of the MOM system (Modelling-Ongrowing fish farms-Monitoring). *Aquaculture*, *194*, 75-92.

Hargrave, B. T. (2003). Far-field environmental effects of marine finfish aquaculture. Pages 1-49 *in* Canadian Technical Report of Fisheries and Aquatic Sciences 2450, Volume 1. Available at: http://mmc.gov/drakes_estero/pdfs/bivalve_aquaculture_03.pdf. Accessed: 27 September 2012.

Hargrave, B. T. (2003). Far-field environmental effects of marine finfish aquaculture. Pages 1-49 *in* Canadian Technical Report of Fisheries and Aquatic Sciences 2450, Volume 1. Available at: http://mmc.gov/drakes_estero/pdfs/bivalve_aquaculture_03.pdf. Accessed: 27 September 2012.

Hargrave, B. T. (2003). Far-field environmental effects of marine finfish aquaculture. Pages 1-49 *in* Canadian Technical Report of Fisheries and Aquatic Sciences 2450, Volume 1. Available at: http://mmc.gov/drakes_estero/pdfs/bivalve_aquaculture_03.pdf. Accessed: 27 September 2012.

Hargrave, B. T. (2005). Environmental effects of marine finfish aquaculture. *Handbook of Environmental Chemistry*, Volume *5M*, Springer-Verlag, Berlin.

Hargrave, B. T. (2005). Environmental effects of marine finfish aquaculture. *Handbook of Environmental Chemistry*, Volume *5M*, Springer-Verlag, Berlin.

Hargrave, B., Holmer, M. & Newcombe, C. (2008). Towards a classification of organic enrichment in marine sediments based on biogeochemical indicators. *Marine Pollution Bulletin*, *56*, 810-824.

Hargrave, B., Holmer, M. & Newcombe, C. (2008). Towards a classification of organic enrichment in marine sediments based on biogeochemical indicators. *Marine Pollution Bulletin*, *56*, 810-824.

Harrison, W. G., Perry, T. & Li, W. K. W. (2005). Ecosystem indicators of water quality, Part I. Plankton biomass,primary production and nutrient demand. Pages 59-82 *in* B.T. Hargrave, editor. Environmental Effects of Marine Finfish Aquaculture. *Handbook of Environmental Chemistry*, Volume *5M*, Springer-Verlag, Berlin.

Harrison, W. G., Perry, T. & Li, W. K. W. (2005). Ecosystem indicators of water quality, Part I. Plankton biomass, primary production and nutrient demand. Pages 59-82 *in* B.T. Hargrave, editor. Environmental Effects of Marine Finfish Aquaculture. *Handbook of Environmental Chemistry*, Volume *5M*, Springer-Verlag, Berlin.

Haya, K., Burridge, L. E. & Chang, B. D. (2001). Environmental impact of chemical wastes produced by the salmon aquaculture industry. *ICES Journal of Marine Science*, *58*, 492-496.

Haya, K., Burridge, L. E., Davies, I. M. & Ervik, A. (2005). A review and assessment of environmental risk of chemicals used for the treatment of sea lice infestations of cultured salmon. Pages 305-340 *in* B.T. Hargrave, editor. Environmental effects of marine finfish aquaculture. *Handbook of Environmental Chemistry*, Volume *5M*. Springer-Verlag, Berlin.

Heath, M. R., Edwards, A. C., Patsch, J. & R. Turrell, W. (2003). Application of the European Regional Seas Ecosystem Model (ERSEM) to assessing the eutrophication status in the OSPAR Maritime Area, with particular reference to nutrient discharges from Scottish salmonid aquaculture. OSPAR Environmental Assessment and Monitoring

Committee (ASMO), London, U.K. Available at: http://strathprints.strath.ac.uk/18590/. Accessed: 02 October 2012.

Heinig, C. (2001). The impacts of salmon aquaculture: The difficulties of establishing acceptability limits and standards. Pages 41-68 *in* M.F. Tlusty, D.A. Bengston, H.O. Halvorson, S.D. Oktay, J.B. Pearce, and R.B. Rheault, editors. Marine Aquaculture and the Environment: A Meeting for Stakeholders in the Northeast. Cape Cod Press, Falmouth, Massachusetts. Available at: http://www.neaq.org/conservation_and_research/projects/publications_and_presentations/pdf/12__.pdf. Accessed: 02 October 2012.100

Heinig, C. S., Sowles, J. W. & Gustafson, L. (2006). Evaluation of LiftUp® system in the mitigation of environmental impacts and fish health in net-pen aquaculture. Saltonstall-Kennedy Final Project Report (FNA03NMF427015), NOAA, National Marine Fisheries Service. Available at: www.nmfs.noaa.gov/mb/financial_services/skpdfs/Attachment10.pdf. Accessed: 01 October 2012.

Heip, C. H. R., Herman, P. M. J. & Soetaert, K. (1998). Indices of diversity and evenness. *Oceanis*, *24*, 61-87.

Hellou, J., Haya, K., Steller, S. & Burridge, L. (2005). Presence and distribution of PAHs, PCBs and DDE in feed and sediments under salmon aquaculture cages in the Bay of Fundy, New Brunswick, Canada. *Aquatic Conservation: Marine and Freshwater Ecosystems*, *15*, 349-365.

Helsley, C. E. (2007). Environmental observations around offshore cages in Hawaii. Pages 41-44 *in* C.S. Lee and P.J. O'Bryen, editors. Open Ocean Aquaculture - Moving Forward. Oceanic Institute, Waimanalo, Hawaii. Available at: nsgl.gso.uri.edu/ocei/oceiw06001.pdf. Accessed: 01 October 2012.

Helsley, C. E. (2007). Environmental observations around offshore cages in Hawaii. Pages 41-44 *in* C.S. Lee and P.J. O'Bryen, editors. Open Ocean Aquaculture - Moving Forward. Oceanic Institute, Waimanalo, Hawaii. Available at: nsgl.gso.uri.edu/ocei/oceiw06001.pdf. Accessed: 01 October 2012.

Hernandez, I., A. Fernandez-Engo, M., L. Perez-Llorens, J. & Vergara, J. J. (2005). Integrated outdoor culture of two estuarine macroalgae as biofilters for dissolved nutrients from *Sparus auratus* waste waters. *Journal of Applied Phycology*, *17*, 557-567.

Hernandez, I., Fernandez-Engo, M. A., Perez-Llorens, J. L. & Vergara, J. J. (2005). Integrated outdoor culture of two estuarine macroalgae as biofilters for dissolved nutrients from *Sparus auratus* waste waters. *Journal of Applied Phycology*, *17*, 557-567.

Holmer, M. (2010). Environmental issues of fish farming in offshore waters: Perspectives, concerns, and research needs. *Aquaculture Environment Interactions*, 1, 57-70.

Holmer, M. (2010). Environmental issues of fish farming in offshore waters: Perspectives, concerns, & research needs. *Aquaculture*, *Environment Interactions*, *1*, 57-70.

Holmer, M. (2010). Environmental issues of fish farming in offshore waters: Perspectives, concerns, and research needs. *Aquaculture Environment Interactions*, *1*, 57-70.

Holmer, M. (2010). Environmental issues of fish farming in offshore waters: Perspectives, concerns, and research needs. *Aquaculture*, *Environment Interactions*, *1*, 57-70.

Holmer, M., Black, K., Duarte, C. M., Marba, N. & Karakassis, I., editors. (2008a). Aquaculture, in the Ecosystem. Springer, Dordrecht, London.

Holmer, M., Hansen, P. K., Karakassis, I., Borg, J. A. & Schembri. P. (2008). Monitoring of environmental impacts of marine aquaculture. Pages 47-85 *in* M. Holmer, K. Black, C.M.

Duarte, N. Marba, and I. Karakassis, editors. Aquaculture in the Ecosystem. Springer, Dordrecht, London.

Holmer, M., Hansen, P. K., Karakassis, I., Borg, J. A. & Schembri, P. (2008b). Monitoring of environmental impacts of marine aquaculture. Pages 47-85 *in* M. Holmer, K. Black, C.M. Duarte, N. Marba, & I. Karakassis, editors. *Aquaculture*, in the Ecosystem. Springer, Dordrecht, London.

Holmer, M., Hansen, P. K., Karakassis, I., Borg, J. A. & Schembri, P. (2008). Monitoring of environmental impacts of marine aquaculture. Pages 47-85 *in* M. Holmer, K. Black, C.M. Duarte, N. Marba, and I. Karakassis, editors. Aquaculture in the Ecosystem. Springer, Dordrecht, London.

Holmer, M., Hansen, P. K., Karakassis, I., Borg, J. A. & Schembri, P. (2008). Monitoring of environmental impacts of marine aquaculture. Pages 47-85 *in* M. Holmer, K. Black, C.M. Duarte, N. Marba, and I. Karakassis, editors. *Aquaculture, in the Ecosystem*. Springer, Dordrecht, London.

Holmer, M., Marba, N., Terrados, J., Duarte, C. M. & Fortes, M. D. (2002). Impacts of milkfish (*Chanos chanos*) *Aquaculture*, on carbon and nutrient fluxes in the Bolinao area, Philippines. *Marine Pollution Bulletin*, *44*, 685-696.

Holmer, M., Perez, M. & Duarte, C. M. (2003). Benthic primary producers - a neglected environmental problem in Mediterranean maricultures? *Marine Pollution Bulletin*, *46*, 1372-1376.

Holmer, M., Wildish, D. & Hargrave, B. (2005). Organic enrichment from marine finfish *Aquaculture*, and effects on sediment biogeochemical processes. Pages 181-206 *in* B.T. Hargrave, editor. Environmental effects of marine finfish aquaculture. *Handbook of Environmental Chemistry*, Volume *5M*. Springer-Verlag, Berlin.

Holmer, M., Wildish, D. & Hargrave, B. (2005). Organic enrichment from marine finfish aquaculture and effects on sediment biogeochemical processes. Pages 181-206 *in* B.T. Hargrave, editor. Environmental effects of marine finfish aquaculture. *Handbook of Environmental Chemistry*, Volume *5M*. Springer-Verlag, Berlin.

Honkanen, T. & Helminen, H. (2000). Impacts of fish farming on eutrophication: Comparisons among different characteristics of ecosystem. *International Review of Hydrobiology*, *85*, 673-686.

Huang, Y. C. A., Hsieh, H. J., Huang, S. C., Meng, P. J., Chen, Y. S., Keshavmurthy, S., Nozawa, Y. & Chen, C. A. (2011). Nutrient enrichment caused by marine cage culture and its influence on subtropical coral communities in turbid waters. *Marine Ecology Progress Series*, *423*, 83-93.

Hung, J. J., Hung, C. S. & Su, H. M. (2008). Biogeochemical response to the removal of maricultural structures from an eutrophic lagoon (Tapong Bay) in Taiwan. *Marine Environmental Research*, *65*, 1-17.

Hung, J. J., Hung, C. S. & Su. H. M. (2008). Biogeochemical response to the removal of maricultural structures from an eutrophic lagoon (Tapong Bay) in Taiwan. *Marine Environmental Research*, *65*, 1-17.

Huntington, T. C., Roberts, H., Cousins, N., Pitta, V., Marchesi, N., Sanmamed, A., Hunter-Rowe, T., Fernandes, T. F., Tett, P., McCue, J. & Brockie. N. (2006). Some aspects of the environmental impact of aquaculture in sensitive areas. Final Report to the Directorate-General Fish and Maritime Affairs of the European Commission, Poseidon Aquatic Resource Management Ltd., U.K. Available at: ec.europa.eu/fisheries/

documentation/studies/aquaculture_environment_2006_en.pdf. Accessed: 27 September 2012.

Huntington, T. C., Roberts, H., Cousins, N., Pitta, V., Marchesi, N., Sanmamed, A., Hunter-Rowe, T., Fernandes, T. F., Tett, P., McCue, J. & Brockie, N. (2006). Some aspects of the environmental impact of *Aquaculture*, in sensitive areas. Final Report to the Directorate-General Fish and Maritime Affairs of the European Commission, Poseidon Aquatic Resource Management Ltd., U.K. Available at: ec.europa.eu/fisheries/documentation/studies/aquaculture_environment_2006_en.pdf. Accessed: 27 September 2012.52

Huntington, T. C., Roberts, H., Cousins, N., Pitta, V., Marchesi, N., Sanmamed, A., Hunter-Rowe, T., Fernandes, T. F., Tett, P., McCue, J. & Brockie, N. (2006). Some aspects of the environmental impact of aquaculture in sensitive areas. Final Report to the Directorate-General Fish and Maritime Affairs of the European Commission, Poseidon Aquatic Resource Management Ltd., U.K. Available at: ec.europa.eu/fisheries/documentation/studies/aquaculture_environment_2006_en.pdf. Accessed: 27 September 2012.

Huntington, T. C., Roberts, H., Cousins, N., Pitta, V., Marchesi, N., Sanmamed, A., Hunter-Rowe, T., Fernandes, T. F., Tett, P., McCue, J. & Brockie, N. (2006). Some aspects of the environmental impact of aquaculture in sensitive areas. Final Report to the Directorate-General Fish and Maritime Affairs of the European Commission, Poseidon Aquatic Resource Management Ltd., U.K. Available at: ec.europa.eu/fisheries/documentation/studies/aquaculture_environment_2006_en.pdf. Accessed: 27 September 2012.128

Huntington, T. C., Roberts, H., Cousins, N., Pitta, V., Marchesi, N., Sanmamed, A., Hunter-Rowe, T., Fernandes, T. F., Tett, P., McCue, J. & Brockie, N. (2006). Some aspects of the environmental impact of aquaculture in sensitive areas. Final Report to the Directorate-General Fish and Maritime Affairs of the European Commission, Poseidon Aquatic Resource Management Ltd., U.K. Available at: ec.europa.eu/fisheries/documentation/studies/aquaculture_environment_2006_en.pdf. Accessed: 27 September 2012.

Imelda-Joseph, Joseph, S., Ignatius, B., Rao, G. S., Sobhana, K. S., Prema, D. & Varghese, M. (2010). A pilot study on culture of Asian seabass *Lates calcarifer* (Bloch) in open sea cage at Munambam, Cochin coast, India. *Indian journal of fisheries*, *57*, 29-33.

International Council for the Exploration of the Seas. (2002). Report of the working group on environmental interactions of mariculture. ICES, Copenhagen, Denmark. Mariculture Committee. 8-12 April 2002. Available at: www.ices.dk/reports/MCC/2002/WGEIM02.pdf. Accessed: 28 September 2012.

International Council for the Exploration of the Seas. (2002). Report of the working group on environmental interactions of mariculture. ICES, Copenhagen, Denmark. Mariculture Committee. 8-12 April 2002. Available at: www.ices.dk/reports/MCC/2002/WGEIM02.pdf. Accessed: 28 September 2012.

International Council for the Exploration of the Seas. (2002). Report of the working group on environmental interactions of mariculture. ICES, Copenhagen, Denmark. Mariculture Committee. 8-12 April 2002. Available at: www.ices.dk/reports/MCC/2002/WGEIM02.pdf. Accessed: 28 September 2012.

International Council for the Exploration of the Seas. (2002). Report of the working group on environmental interactions of mariculture. ICES, Copenhagen, Denmark. Mariculture Committee. 8-12 April 2002. Available at: www.ices.dk/reports/MCC/2002/WGEIM02.pdf. Accessed: 28 September 2012.

International Union for Conservation of Nature. (2007). Guide for the sustainable development of Mediterranean aquaculture. Interaction between aquaculture and the environment. IUCN, Gland Switerland and Malaga, Spain. Available at: cmsdata.iucn.org/downloads/acua_en_final.pdf. Accessed: 27 September 2012.

International Union for Conservation of Nature. (2007). Guide for the sustainable development of Mediterranean aquaculture. Interaction between *Aquaculture*, and the environment. IUCN, Gland Switerland and Malaga, Spain. Available at: cmsdata.iucn.org/downloads/ acua_en_final.pdf. Accessed: 27 September 2012.

International Union for Conservation of Nature. (2007). Guide for the sustainable development of Mediterranean aquaculture. Interaction between aquaculture and the environment. IUCN, Gland Switerland and Malaga, Spain. Available at: cmsdata.iucn.org/downloads/ acua_en_final.pdf. Accessed: 27 September 2012.

International Union for Conservation of Nature. (2007). Guide for the sustainable development of Mediterranean aquaculture. Interaction between aquaculture and the environment. IUCN, Gland Switerland and Malaga, Spain. Available at: cmsdata.iucn.org/downloads/acua_en_final.pdf. Accessed: 27 September 2012.

International Union for Conservation of Nature. (2007). Guide for the sustainable development of Mediterranean aquaculture. Interaction between aquaculture and the environment. IUCN, Gland Switerland and Malaga, Spain. Available at: cmsdata.iucn.org/downloads/acua_en_final.pdf. Accessed: 27 September 2012.

Islam, M. (2005). Nitrogen and phosphorus budget in coastal and marine cage aquaculture and impacts of effluent loading on ecosystem: review and analysis towards model development. *Marine Pollution Bulletin, 50*, 48-61.

Islam, M. (2005). Nitrogen and phosphorus budget in coastal and marine cage *Aquaculture*, and impacts of effluent loading on ecosystem: review and analysis towards model development. *Marine Pollution Bulletin*, *50*, 48-61.

Jacobs, S. R. & Terhune, J. M. (2000). Harbor seal (*Phoca vitulina*) numbers along the New Brunswick coast of the Bay of Fundy in autumn in relation to aquaculture. *Northeastern Naturalist*, *7*, 289-296.

Jamieson, G. & Olesiuk, P. (2002). Salmon farm-pinniped interactions in British Columbia: An analysis of predator control, its justification and alternative approaches. Fisheries and Oceans Science, Research Document 2001/142. Canadian Science Advisory Secretariat, Ottawa, Canada. Available at: http://www.dfo-mpo.gc.ca/csas/Csas/DocREC/2001/RES2001_142e.pdf. Accessed: 02 October 2012.

Jiang, Z. J., Fang, J. G., Mao, Y. Z. & Wang, W. (2010). Eutrophication assessment and bioremediation strategy in a marine fish cage culture area in Nansha Bay, China. *Journal of Applied Phycology*, *22*, 421-426.

Johannessen, D. I., Macdonald, J. S., Harris, K. A. & Ross, P. S. (2007). Marine environmental quality in the Pacific North coast integrated management area (PNCIMA), British Columbia, Canada: A summary of contaminant sources, types, & risks. Pages 1-53 *in* Canadian Technical Report of Fisheries and Aquatic Sciences 2716. Available at: www.dfo-mpo.gc.ca/Library/328420.pdf. Accessed: 28 September 2012.

Johannessen, D. I., Macdonald, J. S., Harris, K. A. & Ross, P. S. (2007). Marine environmental quality in the Pacific North coast integrated management area (PNCIMA), British Columbia, Canada: A summary of contaminant sources, types, and risks. Pages 1-53 *in* Canadian Technical Report of Fisheries and Aquatic Sciences 2716. Available at: www.dfo-mpo.gc.ca/Library/328420.pdf. Accessed: 28 September 2012.

Johnson, M. R., Boelke, C., Chiarella, L. A., Colosi, P., Greene, K., Lellis-Dibble, K., Ludemann, H., Ludwig, M., McDermott, S., Ortiz, J., Rusanowsky, D., Scott, M. & Smith. J. (2008). Impacts to marine fisheries habitat from nonfishing activities in the northeastern United States. NOAA Technical Memorandum NMFS-NE-209, NOAA, Gloucester, Massachusetts. Available at: www.nefsc.noaa.gov/publications/tm/tm209/index.html. Accessed 28 September 2012.

Johnson, M. R., Boelke, C., Chiarella, L. A., Colosi, P., Greene, K., Lellis-Dibble, K., Ludemann, H., Ludwig, M., McDermott, S., Ortiz, J., Rusanowsky, D., Scott, M. & Smith, J. (2008). Impacts to marine fisheries habitat from nonfishing activities in the northeastern United States. NOAA Technical Memorandum NMFS-NE-209, NOAA, Gloucester, Massachusetts. Available at: www.nefsc.noaa.gov/publications/tm/tm209/index.html. Accessed 28 September 2012.

Johnson, M. R., Boelke, C., Chiarella, L. A., Colosi, P., Greene, K., Lellis-Dibble, K., Ludemann, H., Ludwig, M., McDermott, S., Ortiz, J., Rusanowsky, D., Scott, M. & Smith, J. (2008). Impacts to marine fisheries habitat from nonfishing activities in the northeastern United States. NOAA Technical Memorandum NMFS-NE-209, NOAA, Gloucester, Massachusetts. Available at: www.nefsc.noaa.gov/publications/tm/tm209/index.html. Accessed 28 September 2012.

Joint Group of Experts on the Scientific Aspects of Marine Environmental Protection. (1997). Towards safe and effective use of chemicals in coastal aquaculture. GESAMP Report and Studies No. 65. IMO/FAO/UNESCO-IOC/WMO/WHO/IAEA/UN/UNEP. FAO, Rome, Italy. Available at: www.fao.org/docrep/meeting/003/w6435e.htm. Accessed: 02 October 2012.

Jusup, M., Klanjscek, J., Petricioli, D. & Legovic, T. (2009). Predicting aquaculture-derived benthic organic enrichment: Model validation. *Ecological Modelling*, *220*, 2407-2414.

Jusup, M., Klanjscek, J., Petricioli, D. & Legovic, T. (2009). Predicting aquaculture-derived benthic organic enrichment: Model validation. *Ecological Modelling*, *220*, 2407-2414.

Kaczmarsky, L. T., Draud, M. & Williams, E. (2005). Is there a relationship between proximity to sewage effluent and the prevalence of coral disease? *Caribbean Journal of Science*, *41*, 124-137.

Kalantzi, L. & Karakassis, L. (2006). Benthic impacts of fish farming: Meta-analysis of community and geochemical data. *Marine Pollution Bulletin*, *52*, 484-493.

Kalantzi, L. & Karakassis, L. (2006). Benthic impacts of fish farming: Meta-analysis of community and geochemical data. *Marine Pollution Bulletin*, *52*, 484-493.

Kan-atireklap, S., Tanabe, S., Sanguansin, J., Tabucanon, M. S. & Hungspreugs, M. (1997). Contamination by butyltin compounds and organochlorine residues in green mussel (*Perna viridis*, L.) from Thailand coastal waters. *Environmental Pollution*, *97*, 79-89.

Kang, Y. H., Shin, J. A., Kim, M. S. & Chung, I. K. (2008). A preliminary study of the bioremediation potential of *Codium fragile* applied to seaweed integrated multi-trophic aquaculture (IMTA) during the summer. *Journal of Applied Phycology*, *20*, 183-190.

Karakassis, I., Pitta, P. & Krom. M. D. (2005). Contribution of fish farming to the nutrient loading of the Mediterranean. *Scientia Marina*, *69*, 313-321.

Karakassis, I., Tsapakis, M., Smith, C. J. & Rumohr, H. (2002). Fish farming impacts in the Mediterranean studied through sediment profiling imagery. *Marine Ecology Progress Series*, *227*, 125-133.

Karakassis, I., Tsapakis, M., Smith, C. J. & Rumohr, H. (2002). Fish farming impacts in the Mediterranean studied through sediment profiling imagery. *Marine Ecology Progress Series*, *227*, 125-133.

Karney, B. & Venayagamoorthy. S. K. *in press*. Sustainability implications. Pages xx-xx *in* H.J.S. Fernando, editor. Handbook of Environmental Fluid Dynamics. CRC Press, Boca Raton, Florida.

Karpouzi, V. S., Watson, R. & Pauly, D. (2007). Modelling and mapping resource overlap between seabirds and fisheries on a global scale: A preliminary assessment. *Marine Ecology Progress Series*, *343*, 87-99.

Katz, T., Herut, B., Genin, A. & Angel, D. L. (2002). Gray mullets ameliorate organically enriched sediments below a fish farm in the oligotrophic Gulf of Aqaba (Red Sea). *Marine Ecology Progress Series*, *234*, 205-214.

Katz, T., Herut, B., Genin, A. & Angel. D. L. (2002). Gray mullets ameliorate organically enriched sediments below a fish farm in the oligotrophic Gulf of Aqaba (Red Sea). *Marine Ecology Progress Series*, *234*, 205-214.

Keeley, N. B., Forrest, B. M., Crawford, C. & Macleod, C. K. (2012). Exploiting salomn farm benthic enrichment gradients to evaluate the regional performance of biotic indices and environmental indicators. *Ecological Indicators*, *23*, 453-466.

Keeley, N. B., Forrest, B. M., Crawford, C. & Macleod, C. K. (2012). Exploiting salmon farm benthic enrichment gradients to evaluate the regional performance of biotic indices and environmental indicators. *Ecological Indicators*, *23*, 453-466.

Kerry, J., Coyne, R., Gilroy, D., Hiney, M. & Smith, P. (1996). Spatial distribution of oxytetracycline and elevated frequencies of oxytetracycline resistance in sediments beneath a marine salmon farm following oxytetracycline therapy. *Aquaculture*, *145*, 31-39.

Kim, D. S., Choi, J. W. & Je, J. G. (1998). Community structure of meiobenthos for pollution monitoring in mariculture farms in Tongyong coastal area, southern Korea. *Journal of the Korean Fisheries Society*, *31*, 217-225.

Kitadai, Y. & Kadowaki, S. (2007). Growth, nitrogen and phosphorous uptake rates and O^2 production rate of seaweeds cultured on coastal fish farms. *Bulletin of the Fisheries Research Agency*, *19*, 149-154.

Klaoudatos, S. D. (2000). Environmental impacts of aquaculture in Greece. Practical experiences. Proceedings of the seminar of the CIHEAM network on technology of aquaculture in the Mediterranean. 1873692099, Zaragoza, Spain, 17-21 January 2000. Available at: http://ressources.ciheam.org/om/pdf/c55/01600226.pdf. Accessed: 01 October 2012.

Klaoudatos, S. D., Klaoudatos, D. S., Smith, J., Bogdanos, K. & Papageorgiou, E. (2006). Assessment of site specific benthic impact of floating cage farming in the eastern Hios island, Eastern Aegean Sea, Greece. *Journal of Experimental Marine Biology and Ecology*, *338*, 96-111.

Kovac, N., Cermelj, B., Vrišer, B. & Lojen, S. (2004). Case Study: The influence of fish farming on coastal marine sediment in Slovenia. Annex II *in* United Nations Environment Programme/Mediterranean Action Plan, Mariculture in the Mediterranean, MAP Technical Reports Series No.140, Athens, Greece. Available at: www.faosipam. org/ htm/Uploads/med%20unep.pdf. Accessed: 02 October 2012.

Kovac, N., Cermelj, B., Vrišer, B. & Lojen. S. (2004). Case Study: The influence of fish farming on coastal marine sediment in Slovenia. Annex II *in* United Nations Environment Programme/Mediterranean Action Plan, Mariculture in the Mediterranean, MAP Technical Reports Series No.140, Athens, Greece. Available at: www.faosipam.org/htm/Uploads/med%20unep.pdf. Accessed: 02 October 2012.53

Kovac, N., Vriser, B. & Cermelj, B. (2001. Impacts of net cage fish farm on sedimentary biogeochemical and meiofaunal properties of the Gulf of Trieste. *Annales Series Historia Naturalis*, *23*, 65-74.

Kraemer, G. P., Carmona, R., Neefus, C., Chopin, T., Miller, S., Tang, X. & Yarish, C. (2004). Preliminary examination of the bioremediation and mariculture potential of a Northeast U.S.A. and and Asian species of *Porphyra. Bulletin of Fisheries Research Agency Supplement*, *1*, 77-82.

Kramarsky-Winter, E., Downs, C. A., Downs, A. & Loya, Y. (2009). Cellular responses in the coral *Stylophora pistillata* exposed to eutrophication from fish mariculture. *Evolutionary Ecology Research*, *11*, 1-21.

Kraufvelin, P., Sinisalo, B., Leppäkoski, E., Mattila, J. & Bonsdorff, E. (2001). Changes in zoobenthic community structure after pollution abatement from fish farms in the Archipelago Sea (N. Baltic Sea). *Marine Environmental Research*, *51*, 229-245.

Kraufvelin, P., Sinisalo, B., Leppäkoski, E., Mattila, J. & Bonsdorff, E. (2001). Changes in zoobenthic community structure after pollution abatement from fish farms in the Archipelago Sea (N. Baltic Sea). *Marine Environmental Research*, *51*, 229-245.

Kraufvelin, P., Sinisalo, B., Leppäkoski, E., Mattila, J. & Bonsdorff, E. (2001). Changes in zoobenthic community structure after pollution abatement from fish farms in the Archipelago Sea (N. Baltic Sea). *Marine Environmental Research*, *51*, 229-245.

Kummerer, K. (2004). Resistance in the environment. *Journal of Antimicrobial Chemotherapy*, *54*, 311-320.

Kutti, T., Ervik, A. & Hansen, P. K. (2007a). Effects of organic effluents from a salmon farm on a fjord system. I. Vertical export and dispersal processes. *Aquaculture*, *262*, 367-381.

Kutti, T., Hansen, P. K., Ervik, A., Høisæter, T. & Johannessen, P. (2007b). Effects of organic effluents from a salmon farm on a fjord system. II. Temporal and spatial patterns in infauna community composition. *Aquaculture*, *262*, 355-366.

Kutti, T., Hansen, P. K., Ervik, A., Høisæter, T. & Johannessen, P. (2007). Effects of organic effluents from a salmon farm on a fjord system. II. Temporal and spatial patterns in infauna community composition. *Aquaculture*, *262*, 355-366.

La Rosa, T., Mirto, S., Mazzola, A. & Danovaro, R. (2001). Differential responses of benthic microbes and meiofauna to fish-farm disturbance in coastal sediments. *Environmental Pollution*, *112*, 427-434.

La Rosa, T., Mirto, S., Mazzola, A. & Danovaro, R. (2001). Differential responses of benthic microbes and meiofauna to fish-farm disturbance in coastal sediments. *Environmental Pollution*, *112*, 427-434.

La Rosa, T., Mirto, S., Mazzola, A. & Danovaro, R. (2001). Differential responses of benthic microbes and meiofauna to fish-farm disturbance in coastal sediments. *Environmental Pollution*, *112*, 427-434.146

La Rosa, T., Mirto, S., Mazzola, A. & Maugeri, T. L. (2004). Benthic microbial indicators of fish farm impact in a coastal area of the Tyrrhenian Sea. *Aquaculture*, *230*, 153-167.

La Rosa, T., Mirto, S., Mazzola, A. & Maugeri. T. L. (2004). Benthic microbial indicators of fish farm impact in a coastal area of the Tyrrhenian Sea. *Aquaculture*, 230, 153-167.

Lalumera, G. (2004). Preliminary investigation on the environmental occurrence and effects of antibiotics used in aquaculture in Italy. *Chemosphere*, *54*, 661-668.

Lampadariou, N., Akoumianaki, I. & Karakassis, I. (2008). Use of the size fractionation of the macrobenthic biomass for the rapid assessment of benthic organic enrichment. *Ecological Indicators*, *8*, 729-742.

Lampadariou, N., Akoumianaki, I. & Karakassis, I. (2008). Use of the size fractionation of the macrobenthic biomass for the rapid assessment of benthic organic enrichment. *Ecological Indicators*, *8*, 729-742.

Lampadariou, N., Akoumianaki, I. & Karakassis, I. (2008). Use of the size fractionation of the macrobenthic biomass for the rapid assessment of benthic organic enrichment. *Ecological Indicators*, *8*, 729-742.

Lampadariou, N., Karakassis, I., Teraschke, S. & Arlt, G. (2005). Changes in the benthic meiofaunal assemblages in the vicinity of fish farms in the eastern Mediterranean. *Vie et Milieu*, *55*, 61-69.

Lander, T. R., Shaw, J. R. K., Robinson, S. M. C. & Martin, J. D. (2009). Blue mussel (*Mytilus edulis*) settlement patterns on antifoulant treated salmon nets and its implications for recycling used salmon nets for mussel spat collection in Integrated Multi-trophic Aquaculture (IMTA). Canadian Technical Report of Fisheries and Aquatic Sciences 2849. Available at: www.dfo-mpo.gc.ca/Library/338828.pdf. Accessed: 02 October 2012.

Langan, R. (2004). Balancing marine aquaculture inputs and extraction: Combined culture of finfish and bivlave molluscs in the open ocean. *Bulletin of Fisheries Research Agency Supplement*, *1*, 51-58.

Langan, R. (2007). Results of environmental monitoring at an experimental offshore farm in the Gulf of Maine: Environmental conditions after seven years of multi-species farming. Pages 57-60 *in* C.-S. Lee and P.J. O'Bryan, editors. Open Ocean *Aquaculture*, - Moving Forward. Oceanic Institute, Waimanalo, Hawaii. Available at: nsgl.gso.uri.edu/ocei/oceiw06001.pdf. Accessed: 01 October 2012.

Lardicci, C., Como, S., Corti, S. & Rossi, F. (2001). Changes and recovery of macrozoobenthic communities after restoration measures of the Orbetello Lagoon (Tyrrhenian coast, italy). *Aquatic Conservation: Marine and Freshwater Ecosystems*, *11*, 281-287.

Lauer, P. R., Fernandes, M., Fairweather, P. G., Tanner, J. & Cheshire. A. (2009). Benthic fluxes of nitrogen and phosphorus at southern bluefin tuna *Thunnus maccoyii* sea-cages. *Marine Ecology Progress Series*, *390*, 251-263.

Lauer, P. R., Fernandes, M., Fairweather, P. G., Tanner, J. & Cheshire, A. (2009). Benthic fluxes of nitrogen and phosphorus at southern bluefin tuna *Thunnus maccoyii* sea-cages. *Marine Ecology Progress Series*, *390*, 251-263.

Lee, C.-S. & O'Bryen, P. J. (2007). Discussion summary: Open ocean aquaculture—Moving forward. Pages 65-76 *in* C.-S. Lee and P.J. O'Bryen, editors. Open Ocean Aquaculture -

Moving Forward. Oceanic Institute, Waimanalo, Hawaii. Available at: nsgl.gso.uri.edu/ocei/oceiw06001.pdf. Accessed: 01 October 2012.

Lee, C.-S. & O'Bryen, P. J. (2007). Open ocean aquaculture - Moving forward. Oceanic Institute, Waimanalo, Hawaii. Available at: http://nsgl.gso.uri.edu/ocei/oceiw06001.pdf. Accessed: 01 October 2012.

Lee, H. W., Bailey-Brock, J. H. & McGurr, M. M. (2006). Temporal changes in the polychaete infaunal community surrounding a Hawaiian mariculture operation. *Marine Ecology Progress Series*, *307*, 175-185.

Lee, H. W., Bailey-Brock, J. H. & McGurr, M. M. (2006). Temporal changes in the polychaete infaunal community surrounding a Hawaiian mariculture operation. *Marine Ecology Progress Series*, *307*, 175-185.

Lee, H. W., Bailey-Brock, J. H. & McGurr. M. M. (2006). Temporal changes in the polychaete infaunal community surrounding a Hawaiian mariculture operation. *Marine Ecology Progress Series*, *307*, 175-185.

Levy, O., Rosenfeld, M., Loya, Y., Yam, R., Mizrachi, I. & Shemesh, A. (2010). Anthropogenic stressors and eutrophication processes as recorded by stable isotopes compositions in coral skeletons. *Biogeosciences Discussions*, *7*, 7657-7672.

Lewitus, A. J., Horner, R. A., Caron, D. A., Garcia-Mendoza, E., Hickey, B. M., Hunter, M., Huppert, D. D., Kudela, R. M., Langlois, G. W., Largier, J. L., Lessard, E. J., RaLonde, R., Jack Rensel, J. E., Strutton, P. G., Trainer, V. L. & Tweddle, J. F. (2012). Harmful algal blooms along the North American west coast region: History, trends, causes, and impacts. *Harmful Algae*, *19*, 133-159.

Lin, D. T. & Bailey-Brock, J. H. (2008). Partial recovery of infaunal communities during a fallow period at an open-ocean aquaculture. *Marine Ecology Progress Series*, *371*, 65-72.

Lin, D. T. & Bailey-Brock, J. H. (2008). Partial recovery of infaunal communities during a fallow period at an open-ocean aquaculture. *Marine Ecology Progress Series*, *371*, 65-72.

Love, D. C., Rodman, S., Neff, R. A. & Nachman, K. E. (2011). Veterinary drug residues in seafood inspected by the European Union, United States, Canada, and Japan from 2000 to 2009. *Environmental Science and Technology*, *45*, 7232-7240.

Loya, Y. & Kramarsky-Winter, E. (2003). *In situ* eutrophication caused by fish farms in the northern Gulf of Eilat (Aqaba) is beneficial for its coral reefs: A critique. *Marine Ecology Progress Series*, *261*, 299-303.

Lu, L. & Wu, R. S. S. (1998). Recolonization and succession of marine macrobenthos in organic-enriched sediment deposited from fish farms. *Environmental Pollution*, *101*, 241-251.

MacDonald, B. A., Robinson, S. M. C. & Barrington, K. A. (2011). Feeding activity of mussels (Mytilus edulis) held in the field at an integrated multi-trophic aquaculture (IMTA) site (Salmo salar) and exposed to fish food in the laboratory. *Aquaculture*, *314*, 244-251.

Machias, A., Giannoulaki, M., Somarakis, S., Maravelias, C. D., Neofitou, C., Koutsoubas, D., Papadopoulou, K. N. & Karakassis, I. (2006). Fish farming effects on local fisheries landings in oligotrophic seas. *Aquaculture 261*, 809-816.

Machias, A., Karakassis, I., Giannoulaki, M., Papadopoulou, K. N., Smith, C. J. & Somarakis, S. (2005). Response of demersal fish communities to the presence of fish farms. *Marine Ecology Progress Series*, *288*, 241-250.

Machias, A., Karakassis, I., Labropoulou, M., Somarakis, S., Papadopoulou, K. N. & Papaconstantinou, C. (2004). Changes in wild fish assemblages after the establishment of a fish farming zone in an oligotrophic marine ecosystem. *Estuarine, Coastal and Shelf Science*, *60*, 771-779.

Macleod, C. K., Crawford, C. M. & Moltschaniwskyj, N. A. (2004). Assessment of long term change in sediment condition after organic enrichment: Defining recovery. *Marine Pollution Bulletin*, *49*, 79-88.

Macleod, C. K., Crawford, C. M. & Moltschaniwskyj, N. A. (2004). Assessment of long term change in sediment condition after organic enrichment: Defining recovery. *Marine Pollution Bulletin*, *49*, 79-88.

Macleod, C. K., Crawford, C. M. & Moltschaniwskyj. N. A. (2004). Assessment of long term change in sediment condition after organic enrichment: Defining recovery. *Marine Pollution Bulletin*, *49*, 79-88.

Maldonado, M., Carmona, M. C., Echeverria, Y. & Riesgo, A. (2005). The environmental impact of Mediterranean cage fish farms at semi-exposed locations: Does it need a re-assessment? *Helgoland Marine Research*, *59*, 121-135.

Maldonado, M., Carmona, M. C., Echeverria, Y. & Riesgo, A. (2005). The environmental impact of Mediterranean cage fish farms at semi-exposed locations: Does it need a re-assessment? *Helgoland Marine Research*, *59*, 121-135.

Mansell, B., Powell, M. D., Ernst, I. & Nowak, B. F. (2005). Effects of the gill monogenean *Zeuxapta seriolae* (Meserve, 1938) and treatment with hydrogen peroxide on pathophysiology of kingfish, Seriola lalandi Valenciennes, 1833. *Journal of Fish Diseases*, *28*, 253-262.129

Mantzavrakos, E., Kornaros, M., Lyberatos, G., Kaspiris, P. & Lekkas. T. D. (2005). Impacts of a marine fish farm in Argolikos Gulf on the water column and the sediment.*in* Proceedings of the 9th International Conference on Environmental Science and Technology, Rhodes Island, Greece, 1-3 September 2005. Available at: www.srcosmos.gr/srcosmos/showpub.aspx?aa=6625. Accessed: 01 October 2012.

Mantzavrakos, E., Kornaros, M., Lyberatos, G., Kaspiris, P. & Lekkas. T. D. (2005). Impacts of a marine fish farm in Argolikos Gulf on the water column and the sediment.*in* Proceedings of the 9th International Conference on Environmental Science and Technology, Rhodes Island, Greece, 1-3 September 2005. Available at: www.srcosmos.gr/srcosmos/showpub.aspx?aa=6625. Accessed: 01 October 2012.

Markowitz, T. M., Harlin, A. D., Wursig, B. & McFadden, C. J. (2004). Dusky dolphin foraging habitat: Overlap with aquaculture in New Zealand. *Aquatic Conservation: Marine and Freshwater Ecosystems*, *14*, 133-149.

Matijevic, S., Kuspilic, G., Kljakovic-Gaspic, Z. & Bogner, D. (2008). Impact of fish farming on the distribution of phosphorus in sediments in the middle Adriatic area. *Marine Pollution Bulletin*, *56*, 535-548.54

Matijevic, S., Kuspilic, G., Morovic, M., Grbec, B., Bogner, D., Skejic, S. & Veza. J. (2009). Physical and chemical properties of the water column and sediments at sea bass/sea bream farm in the middle Adriatic (Maslinova Bay). *Acta Adriatica*, *50*, 59-76.

Matijevic, S., Kuspilic, G., Morovic, M., Grbec, B., Bogner, D., Skejic, S. & Veza, J. (2009). Physical and chemical properties of the water column and sediments at sea bass/sea bream farm in the middle Adriatic (Maslinova Bay). *Acta Adriatica*, *50*, 59-76.

Maule, A. G., Joergensen, E. H., Vijayan, M. M. & Killie, J. E. (2005). Aroclor 1254 exposure reduces disease resistance and innate immune responses in fasted Arctic charr. *Environmental Toxicology and Chemistry*, *24*, 117-124.

Mazzola, A., Mirto, S. & Danovaro, R. (1999). Initial fish-farm impact on meiofaunal assemblages in coastal sediments of the western Mediterranean. *Marine Pollution Bulletin*, *38*, 1126-1133.

McKinnon, D., Trott, L., Duggan, S., Brinkman, R., Alongi, D., Castine, S. & Patel. F. (2008). The environmental impacts of sea cage aquaculture in a Queensland context — Hinchinbrook Channel case study (SD57/06) Final Report. Australian Institute of Marine Science, Townsville, Queensland, Australia. Available at: www.aims.gov.au/c/document_library/get_file?uuid=965f17c9-b42b-4e41-a5a5-e568a37a5459&groupId=30301. Accessed: 27 September 2012.

McKinnon, D., Trott, L., Duggan, S., Brinkman, R., Alongi, D., Castine, S. & Patel, F. (2008). The environmental impacts of sea cage *Aquaculture*, in a Queensland context — Hinchinbrook Channel case study (SD57/06) Final Report. Australian Institute of Marine Science, Townsville, Queensland, Australia. Available at: www.aims.gov.au/c/document_library/get_file?uuid=965f17c9-b42b-4e41-a5a5-e568a37a5459&groupId=30301. Accessed: 27 September 2012.

McKinnon, D., Trott, L., Duggan, S., Brinkman, R., Alongi, D., Castine, S. & Patel, F. (2008). The environmental impacts of sea cage aquaculture in a Queensland context — Hinchinbrook Channel case study (SD57/06) Final Report. Australian Institute of Marine Science, Townsville, Queensland, Australia. Available at: www.aims.gov.au/c/document_library/get_file?uuid=965f17c9-b42b-4e41-a5a5-e568a37a5459&groupId=30301. Accessed: 27 September 2012.

Mendiguchia, C., Moreno, C., Manuel-Vez, M. P. & Garcia-Vargas, M. (2006). Preliminary investigation on the enrichment of heavy metals in marine sediments originated from intensive aquaculture effluents. *Aquaculture*, *254*, 317-325.

Meng, P., Wang, J., Liu, L., Chen, M. & Hung, T. (2005). Toxicity and bioaccumulation of tributyltin and triphenyltin on oysters and rock shells collected from Taiwan maricuture area. *Science of the Total Environment*, *349*, 140-149.

Michler-Cieluch, T., Krause, G. & Buck, B. H. (2009). Reflections on integrating operation and maintenance activities of offshore wind farms and mariculture. *Ocean & Coastal Management*, *52*, 57-68.

Migliore, L., Civitareale, C., Brambilla, G. & Di Delupis, G. D. (1997). Toxicity of several important agricultural antibiotics to Artemia. *Water Research*, *31*, 1801-1806.

Milligan, T. G. & Law, B. A. (2005). The effect of marine *Aquaculture*, on fine sediment dynamics in coastal inlets. Pages 239-251 *in* B.T. Hargrave, editor. Environmental effects of marine finfish aquaculture. *Handbook of Environmental Chemistry*, Volume *5M*. Springer-Verlag, Berlin.

Minchin, D., Bauer, B., Oehlmann, J., Schulte-Oehlmann, U. & Duggan, C. B. (1997). Biological indicators used to map organotin contamination from a fishing port, Killybegs, Ireland. *Marine Pollution Bulletin*, *34*, 235-243.

Minchin, D., Stroben, E., Oehlmann, J., Bauer, B., Duggan, C. B. & Keatinge, M. (1996). Biological indicators used to map organotin contamination in Cork Harbour, Ireland. *Marine Pollution Bulletin*, *32*, 188-195.

Mirto, S., La Rosa, T., Gambi, C., Danovaro, R. & Mazzola, A. (2002). Nematode community response to fish-farm impact in the western Mediterranean. *Environmental Pollution, 116*, 203-214.

Mirto, S., La Rosa, T., Gambi, C., Danovaro, R. & Mazzola, A. (2002). Nematode community response to fish-farm impact in the western Mediterranean. Environmental Pollution 116, 203-214.

Modica, A., Scilipoti, D., La Torre, R., Manganaro, A. & Sara, G. (2006). The effect of mariculture facilities on biochemical features of suspended organic matter (southern Tyrrhenian, Mediterranean). *Estuarine, Coastal and Shelf Science, 66*, 177-184.

Molina Dominguez, L., Lopez Calero, G., Vergara Martin, J. M. & Robaina Robaina, L. (2001). A comparative study of sediments under a marine cage farm at Gran Canaria Island (Spain). Preliminary results. *Aquaculture, 192*, 225-231.

Morrisey, D. J., Gibbs, M. M., Pickmere, S. E. & Cole, R. G. (2000). Predicting impacts and recovery of marine-farm sites in Stewart Island, New Zealand, from the Findlay-Watling Model. *Aquaculture, 185*, 257-271.

Morrisey, D. J., Gibbs, M. M., Pickmere, S. E. & Cole, R. G. (2000). Predicting impacts and recovery of marine-farm sites in Stewart Island, New Zealand, from the Findlay-Watling Model. *Aquaculture, 185*, 257-271.

Morrisey, D. J., Gibbs, M. M., Pickmere, S. E. & Cole, R. G. (2000). Predicting impacts and recovery of marine-farm sites in Stewart Island, New Zealand, from the Findlay-Watling Model. *Aquaculture, 185*, 257-271.

Morton, A. (2002). Displacement of *Orcinus orca* (L.) by high amplitude sound in British Columbia, Canada. *ICES Journal of Marine Science, 59*, 71-80.

Mulsow, S., Krieger, Y. & Kennedy, R. (2006). Sediment profile imaging (SPI) and micro-electrode technologies in impact asssessment studies: Examples from two fjords in southern Chile used for fish farming. *Journal of Marine Systems, 62*, 152-163.

Mulsow, S., Krieger, Y. & Kennedy, R. (2006). Sediment profile imaging (SPI) and micro-electrode technologies in impact assessment studies: Example from two fjords in southern Chile used for fish farming. *Journal of Marine Systems, 62*, 152-163.

Muxika, I., Borja, A. & Bonne, W. (2005). The suitability of the marine biotic index (AMBI) to new impact sources along European coasts. *Ecological Indicators, 5*, 19-31.

Nagelkerken, I. (2006). Relationship between anthropogenic impacts and bleaching-associated tissue mortality of corals in Curaçao (Netherlands Antilles). *Revista de Biologia Tropical, 54* (Suppt. 3), 31-44.

Najdek, M., Travizi, A., Bogner, D. & Blazina, M. (2007). Low impact of marine fish farming on sediment and meiofauna in Limski channel (northern Adriatic, Croatia). *Fresenius Environmental Bulletin, 16*, 784-791.

Nash, C. E, & Waknitz, F. W. (2003). Interactions of Atlantic salmon in the Pacific Northwest. I. Salmon enhancement and the net-pen farming industry. *Fisheries Research, 62*, 237-254.

Nash, C. E. & Waknitz, F. W. (2003). Interactions of Atlantic salmon in the Pacific Northwest. I. Salmon enhancement and the net-pen farming industry. *Fisheries Research, 62*, 237-254.

Nash, C. E. (2001). The net-pen salmon farming industry in the Pacific Northwest. U.S. Department of Commerce. NOAA Technical Memorandum NMFS-NWFSC-49.

Available at: http://www.nwfsc.noaa.gov/publications/techmemos/tm49/tm49.htm. Accessed: 27 September 2012.

Nash, C. E. (2001). The net-pen salmon farming industry in the Pacific Northwest. U.S. Department of Commerce. NOAA Technical Memorandum NMFS-NWFSC-49. Available at: http://www.nwfsc.noaa.gov/publications/techmemos/tm49/tm49.htm. Accessed: 27 September 2012.

Nash, C. E. (2001). The net-pen salmon farming industry in the Pacific Northwest. NOAA Technical Memorandum NMFS-NWFSC-49, National Oceanic and Atmospheric Administration, Silver Springs, Maryland, 125 pp.

Nash, C. E. (2001). The net-pen salmon farming industry in the Pacific Northwest. U.S. Department of Commerce. NOAA Technical Memorandum NMFS-NWFSC-49. Available at: http://www.nwfsc.noaa.gov/publications/techmemos/tm49/tm49.htm. Accessed: 27 September 2012.

Nash, C. E. (2001). The net-pen salmon farming industry in the Pacific Northwest. U.S. Department of Commerce. NOAA Technical Memorandum NMFS-NWFSC-49. Available at: http://www.nwfsc.noaa.gov/publications/techmemos/tm49/tm49.htm. Accessed: 27 September 2012.

Nash, C. E. (2003). Interactions of Atlantic salmon in the Pacific Northwest. VI. A synopsis of the risk and uncertainty. *Fisheries Research*, *62*, 339-347.

Nash, C. E. (2003). Interactions of Atlantic salmon in the Pacific Northwest. VI. A synopsis of the risk and uncertainty. *Fisheries Research*, *62*, 339-347.

Nash, C. E. (2003). Interactions of Atlantic salmon in the Pacific Northwest. VI. A synopsis of the risk and uncertainty. *Fisheries Research*, *62*, 339-347.

Nash, C. E., Burbridge, P. R. & Volkman, J. K. (2005). Guidelines for ecological risk assessment of marine fish aquaculture. U.S. Department of Commerce. NOAA Technical Memorandum NMFS-NWFSC-71. Available at: www.nwfsc.noaa.gov/assets/25/6450_01302006_155445_NashFAOFinalTM71.pdf. Accessed: 27 September 2012.

Nash, C. E., Burbridge, P. R. & Volkman, J. K. (2005). Guidelines for ecological risk assessment of marine fish aquaculture. U.S. Department of Commerce. NOAA Technical Memorandum NMFS-NWFSC-71. Available at: www.nwfsc.noaa.gov/assets/25/6450_01302006_155445_NashFAOFinalTM71.pdf. Accessed: 27 September 2012.

Nash, C. E., Burbridge, P. R. & Volkman, J. K. (2005). Guidelines for ecological risk assessment of marine fish aquaculture. U.S. Department of Commerce. NOAA Technical Memorandum NMFS-NWFSC-71. Available at: www.nwfsc.noaa.gov/assets/25/6450_01302006_155445_NashFAOFinalTM71.pdf. Accessed: 27 September 2012.

Nash, C. E., Burbridge, P. R. & Volkman. J. K. (2005). Guidelines for ecological risk assessment of marine fish aquaculture. U.S. Department of Commerce. NOAA Technical Memorandum NMFS-NWFSC-71. Available at: www.nwfsc.noaa.gov/assets/25/6450_01302006_155445_NashFAOFinalTM71.pdf. Accessed: 27 September 2012.

Nash, C. E., Burbridge, P. R. & Volkman. J. K. (2005). Guidelines for ecological risk assessment of marine fish aquaculture. U.S. Department of Commerce. NOAA Technical Memorandum NMFS-NWFSC-71. Available at: www.nwfsc.noaa.gov/assets/25/6450_01302006_155445_NashFAOFinalTM71.pdf. Accessed: 27 September 2012.

Nash, C. E., Iwamoto, R. N. & Mahnken, C. V. W. (2000). Aquaculture risk management and marine mammal interactions in the Pacific Northwest. *Aquaculture*, *183*, 307-323.

National Oceanic and Atmospheric Administration. (2007). NOAA 10-year Plan for Marine Aquaculture. October 2007. Available at: www.nmfs.noaa.gov/aquaculture/docs/policy/final_noaa_10_yr_plan.pdf; Accessed: 27 September 2012.
National Oceanic and Atmospheric Administration. (2011). National Oceanic and Atmospheric Administration Marine Aquaculture Policy. June 9, 2011. Available at: http://www.nmfs.noaa.gov/aquaculture/docs/policy/noaa_aquaculture_policy_2011.pdf. Accessed: 27 September 2011.
Navarrete-Mier, F., Sanz-Lazaro, C. & Marin, A. (2010). Does bivalve mollusc polyculture reduce marine fin fish farming environmental impact? *Aquaculture*, *306*, 101-107.
Navarro, N., Leakey, R. J. G. & Black, K. D. (2008). Effect of salmon cage aquaculture on the pelagic environment of temperate coastal waters: Seasonal changes in nutrients and microbial community. *Marine Ecology Progress Series*, *361*, 47-58.
Neofitou, N. & Klaoudatos. S. (2008). Effect of fish farming on the water column nutrient concentration in a semi-enclosed gulf of the Eastern Mediterranean. *Aquaculture Research*, *39*, 482-490.
Neofitou, N., Vafidis, D. & Klaoudatos, S. (2010). Spatial and temporal effects of fish farming on benthic community structure in a semi-enclosed gulf of the Eastern Mediterranean. *Aquaculture, Environment Interactions*, *1*, 95-105.
Neofitou, N., Vafidis, D. & Klaoudatos, S. (2010). Spatial and temporal effects of fish farming on benthic community structure in a semi-enclosed gulf of the Eastern Mediterranean. *Aquaculture Environment Interactions*, *1*, 95-105.
Neori, A., Chopin, T., Troell, M., Buschmann, A. H., Kraemer, G. P., Halling, C., Shpigel, M. & Yarish, C. (2004). Integrated aquaculture: Rationale, evolution and state of the art emphasizing seaweed biofiltration in modem mariculture. *Aquaculture*, *231*, 361-391.147
Nilsson, H. C. & Rosenburg, R. (2000). Succession in marine benthic habitats and fauna in response to oxygen deficiency: Analyzed by sediment profile-imaging and by grab samples. *Marine Ecology Progress Series*, *197*, 139-149.
Norði, G., Glud, R. N., Gaard, E. & Simonse, K. (2011). Environmental impacts of coastal fish farming: Carbon and nitrogen budgets for trout farming in Kaldbaksfjørður (Faroe Islands). *Marine Ecology Progress Series*, *431*, 223-241.
Norði, G., Glud, R. N., Gaard, E. & Simonse. K. (2011). Environmental impacts of coastal fish farming: Carbon and nitrogen budgets for trout farming in Kaldbaksfjørður (Faroe Islands). *Marine Ecology Progress Series*, *431*, 223-241.
Nordvarg, L. & Hakanson, L. (2002). Predicting the environmental response of fish farming in coastal areas of the Aland Archipelago (Baltic Sea) using management models for coastal water planning. *Aquaculture*, *206*, 217-243.
Nordvarg, L. & Hakanson. L. (2002). Predicting the environmental response of fish farming in coastal areas of the Aland Archipelago (Baltic Sea) using management models for coastal water planning. *Aquaculture*, *206*, 217-243.
Nordvarg, L. & Johansson, T. (2002). The effects of fish farm effluents on the water quality in the Aland Archipelago, Baltic Sea. *Aquacultural Engineering*, *25*, 253-279.
Nordvarg, L. & Johansson. T. (2002). The effects of fish farm effluents on the water quality in the Aland Achipelago, Baltic Sea. *Aquacultural Engineering*, *25* (4), 253-279.
Northridge, S. P., Gordon, J. G., Booth, C., Calderan, S., Cargill, A., Coram, A., Gillespie, D., Lonergan, M. & Webb, A. (2010). Assessment of the impacts and utility of acoustic deterrent devices. Final report to the Scottish Aquaculture Research Forum, Project Code

SARF044. Avaialble at: www.sarf.org.uk/cms-assets/documents/28820-18834.sarf044---final-report.pdf. Accessed: 02 October 2012.

Oakes, C. T. & Pondella, D. J. II. (2009). The value of a net-cage as a fish aggregating device in southern California. *Journal of the World Aquaculture Society*, *40*, 1-21.

Ocean Conservancy, (2011). Right from the start: Open ocean aquaculture in the United States. Ocean Conservancy, Washington, D.C. Available at: http://www.oceanconservancy.org/our-work/aquaculture/assets/pdf/oc_rfts_v11_single.pdf. Accessed: 01 October 2012.

Ocean Conservancy. (2011). Right from the start: Open ocean aquaculture in the United States. Ocean Conservancy, Washington, D.C. Available at: http://www.oceanconservancy.org/our-work/aquaculture/assets/pdf/oc_rfts_v11_single.pdf. Accessed: 01 October 2012.

Oh, M. H., Kang, Y. H., Lee, C. H. &. Chung, I. K. (2005). Effects of six antibiotics on the activity of the photosynthetic apparatus and ammonium uptake of thallus of *Porphyra yezoensis*. *Algae*, *20*, 121-125.

Olesiuk, P. F., Nichol, L. M., Sowden, M. J. & Ford, J. K. B. (2002). Effect of the sound generated by an acoustic harassment device on the relative abundance and distribution of harbor porpoises (*Phocoena phocoena*) in Retreat Passage, British Columbia. *Marine Mammal Science*, *18*, 843-862.

Olsen, L., Holmer, M. & Olsen, Y. (2008). Perspectives of nutrient emission from fish *Aquaculture*, in coastal waters: Literature review with evaluated state of knowledge. Final Report FHF project no. 542014. The Fishery and Aquaculture, Industry Research Fund, Oslo, Norway.

Olsen, L., Holmer, M. & Olsen, Y. (2008). Perspectives of nutrient emission from fish aquaculture in coastal waters: Literature review with evaluated state of knowledge. Final Report FHF project no. 542014. The Fishery and Aquaculture Industry Research Fund, Oslo, Norway.

Olsen, L., Holmer, M. & Olsen. Y. (2008). Perspectives of nutrient emission from fish aquaculture in coastal waters: Literature review with evaluated state of knowledge. Final Report FHF project no. 542014. The Fishery and Aquaculture Industry Research Fund, Oslo, Norway.

Orth, R. J., Carruthers, T. J. B., Dennison, W. C., Duarte, C. M., Fourqurean, J. W., Kenneth, J., Heck, L., Hughes, A. R., Kendrick, G. A., Kenworthy, W. J., Olyarnik, S., Short, F. T., Waycott, M. & Williams, S. (2006). A global crisis for seagrass ecosystems. *BioScience*, *56*, 987-996.

Otero, E. (2009. Spatial and temporal patterns of water quality indicators in reef systems of southwestern Puerto Rico. *Caribbean Journal of Science*, *45*, 168-180.

Page, F. H., Losier, R., McCurdy, P., Greenberg, D., Chaffey, J. & Chang. B. (2005). Dissolved oxygen and salmon cage culture in the southwestern New Brunswick portion of the Bay of Fundy. Pages 1-28 *in* B.T. Hargrave, editor. *Handbook of Environmental Chemistry*, Volume *5M*. Springer-Verlag, Berlin.

Pan, C. (2005). China's aquaculture industry: developments and challenges. Rabobank International, Utrecht, Netherlands. Available at: http://dc.icsf.net/en/documention-centre/article/EN/9632-China%5C's-aquacul.html. Accessed: 02 October 2012.

Pang, S. J., Xiao, T. & Bao, Y. (2006). Dyanmic changes of total bacteria and *Vibrio* in an integrated seaweed-abalone culture system. *Aquaculture and Fisheries Management, 252*, 289-297.

Papastimatiou, Y. P., Itano, D. G., Dale, J. J., Meyer, C. G. & Hollan, K. N. (2010). Site fidelity and movements of sharks associated with ocean-farming cages in Hawaii. *Marine and Freshwater Research, 61*, 1366-1375.

Park, H. S., Choi, J. W. & Lee, H. G. (2000). Community structure of macrobenthic fauna under marine fish culture cages near Tong-yong, Southern Coast of Korea. *Journal of the Korean Fisheries Society, 33*, 1-8.

Parsons, G. J., Shumway, S. E., Kuenstner, S. & Gryska, A. (2002). Polyculture of sea scallops (*Placopecten magellanicus*) suspended from salmon cages. *Aquaculture International, 10*, 65-77.

Pawar, V., Matsuda, O. & Fujisaki, N. (2002). Relationship between feed input and sediment quality of the fish cage farms. *Fisheries Science, 68*, 894-903.55

Pearson, T. H. & Black, K. D. (2001). The environmental impacts of marine fish cage culture. Pages 1-31 *in* K.D. Black, editor. Environmental Impacts of Aquaculture. CRC Press, Boca Raton, Florida.

Pearson, T. H. & Black, K. D. (2001). The environmental impacts of marine fish cage culture. Pages 1-31 *in* K.D. Black, editor. Environmental Impacts of Aquaculture. CRC Press, Boca Raton, Florida.

Pearson, T. H. & Black. K. D. (2001). The environmental impacts of marine fish cage culture. Pages 1-31 *in* K.D. Black, editor. Environmental Impacts of Aquaculture. CRC Press, Boca Raton, Florida.

Pearson, T. H. & Rosenberg, R. (1978). Macrobenthic succession in relation to organic enrichment and pollution of the marine environment. Oceanography and Marine Biology. *An Annual Review, 16*, 229-311.

Pearson, T. H., & Black, K. D. (2001). The environmental impacts of marine fish cage culture. Pages 1-31 *in* K.D. Black, editor. Environmental Impacts of Aquaculture. CRC Press, Boca Raton, Florida.

Peel, M.C., Finlayson, B. L. & McMahon. T. A. (2007). Updated world map of the Köppen-Geiger climate classification. *Hydrology and Earth System Sciences, 11*, 1633-1644.

Pereira, P. M., Black, K. D., McLusky, D. S. & Nickell, T. D. (2004). Recovery of sediments after cessation of marine fish farm production. *Aquaculture, 235*, 315-330.

Pergent, G., Mendez, S., Pergent-Martini, C. & Pasqualini. V. (1999). Preliminary data on the impact of fish farming facilities on *Posidonia oceanica* meadows in the Mediterranean. *Oceanologica acta*, 22, 95-107.

Pergent-Martini, C., Boudouresque, C.-F., Pasqualini, V. & Pergent, G. (2006). Impact of fish farming facilities on *Posidonia oceanica* meadows: a review. *Marine Ecology, 27*, 310-319.

Phillips, S. (2005). Environmental impacts of marine aquaculture issue paper. Pacific States Marine Fisheries Commission, Portland, Oregon. Available at: www.aquaticnuisance.org/wordpress/wp-content/uploads/2009/01/Issue%20--%20Aquaculture%20Environmental%20Impacts,%20Atlantic%20Salmon,.pdf. Accessed: 28 September 2012.

Phillips, S. (2005). Environmental impacts of marine aquaculture issue paper. Pacific States Marine Fisheries Commission, Portland, Oregon. Available at: www.aquaticnuisance.org/

wordpress/wp-content/uploads/2009/01/Issue%20--%20Aquaculture%20Environmental %20Impacts,%20Atlantic%20Salmon,.pdf. Accessed: 28 September 2012.

Phillips, S. (2005). Environmental impacts of marine *Aquaculture*, issue paper. Pacific States Marine Fisheries Commission, Portland, Oregon. Available at: www.aquaticnuisance.org/ wordpress/wp-content/uploads/2009/01/Issue%20--%20Aquaculture%20Environmental %20Impacts,%20Atlantic%20Salmon,.pdf. Accessed: 28 September 2012.

Piedecausa, M. A., Aguado-Gimenez, F., Cerezo-Valverde, J., Hernandez-Llorente, M. D. & Garcia-Garcia, B. (2010). Simulating the temporal pattern of waste production in farmed gilthead seabream (*Sparus aurata*), European seabass (*Dicentrarchus labrax*) and Atlantic bluefin tuna (*Thunnus thynnus*). *Ecological Modelling*, *221*, 634-640.

Piedecausa, M. A., Aguado-Gimenez, F., Cerezo-Valverde, J., Hernandez-Llorente, M. D. & Garcia-Garcia, B. (2010). Simulating the temporal pattern of waste production in farmed gilthead seabream (*Sparus aurata*), European seabass (*Dicentrarchus labrax*) and Atlantic bluefin tuna (*Thunnus thynnus*). *Ecological Modelling*, *221*, 634-640.

Piniak, W. E. D., Mann, D. A., Eckert, S. A. & Harms, C. A. (2012). Amphibious Hearing in Sea Turtles. Pages 83-87 *in* A.N. Popper and A. Hawkins, editors. The Effects of Noise on Aquatic Life: Advances in Experimental Medicine and Biology 730. Springer Science+Business Media. DOI 10.1007/978-1-4419-7311-5_18.

Pitta, P., Apostolaki, E. T., Giannoulaki, M. & Karakassis, I. (2005). Mesoscale changes in the water column in response to fish farming zones in three coastal areas in the Eastern Mediterranean Sea. *Estuarine, Coastal and Shelf Science*, *65*, 501-512.

Pitta, P., Apostolaki, E. T., Giannoulaki, M. & Karakassis. I. (2005). Mesoscale changes in the water column in response to fish farming zones in three coastal areas in the Eastern Mediterranean Sea. *Estuarine, Coastal and Shelf Science*, *65*, 501-512.

Pitta, P., Tsapakis, M., Apostolaki, E. T., Tsagaraki, T., Holmer, M. & Karakassis, I. (2009). 'Ghost nutrients' from fish farms are transferred up the food web by phytoplankton grazers. *Marine Ecology Progress Series*, *374*, 1-6.

Pittenger, R., Anderson, B., Benetti, D. D., Dayton, P., Dewey, B., Goldburg, R., Rieser, A., Sher, B. & Sturgulewski. A. (2007). Sustainable marine aquaculture: Fulfilling the promise; managing the risks. Marine Aquaculture Task Force. Available at: www.pewtrusts.org/uploadedFiles/wwwpewtrustsorg/Reports/Protecting_ocean_life/Sustainable_Marine_Aquaculture_final_1_07.pdf. Accessed: 27 September 2012.

Pittenger, R., Anderson, B., Benetti, D. D., Dayton, P., Dewey, B., Goldburg, R., Rieser, A., Sher, B. & Sturgulewski, A. (2007). Sustainable marine aquaculture: Fulfilling the promise; managing the risks. Marine *Aquaculture*, Task Force. Available at: www.pewtrusts.org/uploadedFiles/wwwpewtrustsorg/Reports/Protecting_ocean_life/Sustainable_Marine_Aquaculture_final_1_07.pdf. Accessed: 27 September 2012.

Pittenger, R., Anderson, B., Benetti, D. D., Dayton, P., Dewey, B., Goldburg, R., Rieser, A., Sher, B. & Sturgulewski, A. (2007). Sustainable marine aquaculture: Fulfilling the promise; managing the risks. Marine Aquaculture Task Force. Available at: www.pewtrusts.org/uploadedFiles/wwwpewtrustsorg/Reports/Protecting_ocean_life/Sustainable_Marine_Aquaculture_final_1_07.pdf. Accessed: 27 September 2012.

Pittenger, R., Anderson, B., Benetti, D. D., Dayton, P., Dewey, B., Goldburg, R., Rieser, A., Sher, B. & Sturgulewski, A. (2007). Sustainable marine aquaculture: Fulfilling the promise; managing the risks. Marine Aquaculture Task Force. Available at:

www.pewtrusts.org/uploadedFiles/wwwpewtrustsorg/Reports/Protecting_ocean_life/Sustainable_Marine_Aquaculture_final_1_07.pdf. Accessed: 27 September 2012.130

Pohle, G., Frost, B. & Findlay, R. (2001). Assessment of regional benthic impact of salmon mariculture within the Letang Inlet, Bay of Fundy. *ICES Journal of Marine Science*, *58*, 417-426.

Pohle, G., Frost, B. & Findlay, R. (2001). Assessment of regional benthic impact of salmon mariculture within the Letang Inlet, Bay of Fundy. *ICES Journal of Marine Science*, *58*, 417-426.

Porrello, S., Tomassetti, P., Manzueto, L., Finoia, M. G., Persia, E., Mercatali, I. & Stipa, P. (2005). The influence of marine cages on the sediment chemistry in the western Mediterranean Sea. *Aquaculture*, *249*, 145-158.

Porrello, S., Tomassetti, P., Manzueto, L., Finoia, M. G., Persia, E., Mercatali, I. & Stipa, P. (2005). The influence of marine cages on the sediment chemistry in the western Mediterranean Sea. *Aquaculture*, *249*, 145-158.

Qi, Z., Zhang, X.-H., Boon, N. & Bossier, P. (2009). Probiotics in aquaculture of China — Current state, problems and prospect. *Aquaculture*, *290*, 15-21.

Qian, P. Y., Lau, S. C. K., Dahms, H. U., Dobretsov, S. & Harder, T. (2007). Marine biofilms as mediators of colonization by marine macroorganisms: Implications for antifouling and aquaculture. *Marine Biotechnology*, *9*, 399-410.

QRobinson, S.-M.-C., Auffrey, L.-M. & Barbeau, M.-A. (2005). Far-field impacts of eutrophication on the intertidal zone in the Bay of Fundy, Canada with emphasis on the soft-shell clam, *Mya arenaria*. Pages 253-274 *in* B.T. Hargrave, editor. Environmental effects of marine finfish aquaculture. *Handbook of Environmental Chemistry*, Volume *5M*. Springer-Verlag, Berlin.

Ramaswamy, B. R., Tao, H. & Hojo, M. (2004). Contamination and biomethylation of organotin compounds in pear/fish culture areas in Japan. *Analytical Sciences*, *20*, 45-53.

Raut, D., Ganesh, T., Murty, N. V. S. S. & Raman, A. V. (2005). Macrobenthos of Kakinada Bay in the Godavari delta, East coast of India: comparing decadal changes. *Estuarine, Coastal and Shelf Science*, *62*, 609-620.

Redmond, K. J., Magnesen, T., Hansen, P. K., Strand, Ø. & Meier, S. (2010). Stable isotopes and fatty acids as tracers of the assimilation of salmon fish feed in blue mussels (Mytilus edulis). *Aquaculture*, *298*, 202-210.

Reid, G. K., Cranford, P. J., Robinson, S. M. C., Filgueira, R. & Guyondet, T. (2011). Open-water integrated multi-trophic aquaculture (IMTA): Modelling the shellfish component. *Bulletin of the Aquaculture Association of Canada*, *109*, 3-12.

Reid, G. K., Liutkus, M., Bennett, A., Robinson, S. M. C., MacDonald, B. & Page, F. (2010). Absorption efficiency of blue mussels (*Mytilus edulis* and *M. trossulus*) feeding on Atlantic salmon (*Salmo salar*) feed and fecal particulates: Implications for integrated multi-trophic aquaculture. *Aquaculture*, *299*, 165-169.

Reid, G. K., Liutkus, M., Robinson, S. M. C., Chopin, T. R., Blair, T., Lander, T., Mullen, J., Page, F. & Moccia, R. D. (2009). A review of the biophysical properties of salmonid faeces: implications for aquaculture waste dispersal models and integrated multi-trophic aquaculture. *Aquaculture Research*, *40*, 257-273.

Rensel, J. E. & Forster, J. R. M. (2007). Beneficial environmental effects of marine finfish mariculture. Final Report to the National Oceanic and Atmospheric Administration

Award # NA040AR4170130, Washington, D.C. Available at: www.wfga.net/documents/ marine_finfish_finalreport.pdf. Accessed: 02 October 2012.

Rensel, J. E. & Forster, J. R. M. (2007). Beneficial environmental effects of marine finfish mariculture. Final Report to the National Oceanic and Atmospheric Administration Award # NA040AR4170130, Washington, D.C. Available at: www.wfga.net/documents/ marine_finfish_finalreport.pdf. Accessed: 02 October 2012.

Rensel, J. E. J., Haigh, N. & Tynan, T. J. (2010). Fraser River sockeye salmon marine survival decline and harmful blooms of *Heterosigma akashiwo*. *Harmful Algae*, *10*, 98-115.

Rensel, J. E. J., Kiefer, D. A., Forster, J. R. M., Woodruff, D. L. & Evans. N. R. (2007). Offshore finfish mariculture in the Strait of Juan de Fuca. *Bulletin of the Fisheries Research Agency*, *19*, 113-129.

Rensel, J. E. J., Kiefer, D. A., Forster, J. R. M., Woodruff, D. L. & Evans. N. R. (2007). Offshore finfish mariculture in the Strait of Juan de Fuca. *Bulletin of the Fisheries Research Agency*, *19*, 113-129.

Rensel, J. E. J., Kiefer, D. A., Forster, J. R. M., Woodruff, D. L. & Evans, N. R. (2007). Offshore finfish mariculture in the Strait of Juan de Fuca. *Bulletin of the Fisheries Research Agency*, *19*, 113-129.

Ridler, N., Wowchuk, M., Robinson, B., Barrington, K., Chopin, T., Robinson, S., Page, F., Reid, G., Szemerda, M., Sewuster, J. & Boyne-Travis, S. (2007). Integrated multi−trophic aquaculture (IMTA): A potential strategic choice for farmers. *Aquaculture Economics and Management*, *11*, 99-110.

Riera, R., Monterroso, Ó., Rodríguez, M., Ramos, E. & Sacramento, A. (2011). Six-year study of meiofaunal dynamics in fish farms in Tenerife (Canary Islands, NE Atlantic Ocean). *Aquatic Ecology*, *45*, 221-229.

Rigos, G. & Troisi, G. M. (2005). Antibacterial agents in mediterranean finfish farming: A synopsis of drug pharmacokinetics in important euryhaline fish species and possible environmental implications. *Reviews in Fish Biology and Fisheries*, *15*, 53-73.

Rigos, G., Nengas, I., Alexis, M. & Troisi, G. M. (2004). Potential drug (oxytetracycline and oxolinic acid) pollution from Mediterranean sparid fish farms. *Aquat Toxicol*, *69*, 281-288.

Rinkevich, B., Angel, D., Shafir, S. & Bongiorni, L. (2003). 'Fair is foul and foul is fair': Response to a critique. *Marine Ecology Progress Series*, *261*, 305-309.

Robinson, S. M. C., Martin, J. D., Cooper, J. A., Lander, T. R., Reid, G. K., Powell, F. & Griffin, R. (2011). The role of three dimensional habitats in the establishment of integrated multi-trophic aquculture (IMTA) systems. *Bulletin of the Aquaculture Association of Canada*, *190*, 23-29.

Rogers, S. I., Somerfield, P. J., Schratzberger, M., Warwick, R., Maxwell, T. A. & Ellis, J. R. (2008). Sampling strategies to evaluate the status of offshore soft sediment assemblages. *Marine Pollution Bulletin*, *56*, 880-894.

Rojas, A. & Wadsworth, S. (2007). A review of cage culture: Latin America and the Caribbean. Pages 70-100 *in* M. Halwart, D. Soto, and J.R. Arthur, editors. Cage aquaculture: Regional reviews and the global overview. FAO Fisheries Technical Paper. No. 498, Rome, FAO. Available at: ftp://ftp.fao.org/docrep/fao/010/a1290e/a1290e.pdf. Accessed: 01 October 2012.

Rudolph, A., Medina, P., Urrutia, C. & Ahumada, R. (2009). Ecotoxicological sediment evaluations in marine aquaculture areas of Chile. *Environmental Monitoring and Assessment*, *155*, 419-429.

Ruiz, J. M., Marco-Mendez, C. & Sanchez-Lizaso, J. L. (2010). Remote influence of off-shore fish farm waste on Mediterranean seagrass (*Posidonia oceanica*) meadows. *Marine Environmental Research*, *69*, 118-126.

Ruiz, J. M., Perez, M. & Romero. J. (2001). Effects of fish farm loadings on seagrass (*Posidonia oceanica*) distribution, growth and photosynthesis. *Marine Pollution Bulletin*, *42*, 749-760.

Rust, M. B., Barrows, F. T., Hardy, R. W., Lazur, A., Naughten, K. & Silverstein. J. (2010). The future of aquafeeds. Draft report to the NOAA/USDA Alternative Feeds Initiative. Available at: www.nmfs.noaa.gov/aquaculture/docs/feeds/the_future_of_aquafeeds_final.pdf. Accessed: 31 October 2012.

Rust, M. B., Barrows, F. T., Hardy, R. W., Lazur, A., Naughten, K. & Silverstein, J. (2010). The future of aquafeeds. Draft report to the NOAA/USDA Alternative Feeds Initiative. Available at: www.nmfs.noaa.gov/aquaculture/docs/feeds/the_future_of_aquafeeds_final.pdf. Accessed: 31 October 2012.

Sala, A. & Lucchetti, A. (2008). Low-cost tool to reduce biofouling in oyster longline culture. *Aquacultural Engineering*, *39*, 53-58.

Salazar, F. J. & Saldana, R. C. (2007). Characterization of manures from fish cage farming in Chile. *Bioresource Technology*, *98*, 3322-3327.

Sanchez-Jerez, P., Fernandez-Jover, D., Bayle-Sempere, J., Valle, C., Dempster, T., Tuya, F. & Juanes, F. (2008). Interactions between bluefish *Pomatomus saltatrix* (L.) and coastal sea-cage farms in the Mediterranean Sea. *Aquaculture*, *282*, 61-67.

Sarà, G. (2007). Aquaculture effects on some physical and chemical properties of the water column: A meta-analysis. *Chemistry and Ecology*, *23*, 251-262.

Sarà, G. (2007). *Aquaculture*, effects on some physical and chemical properties of the water column: A meta-analysis. *Chemistry and Ecology*, *23*, 251-262.

Sarà, G., Reid, G. K., Rinaldi, A., Palmeri, V., Troell, M. & Kooijman, S. A. L. M. (2012). Growth and reproductive simulation of candidate shellfish species at fish cages in the Southern Mediterranean: Dynamic Energy Budget (DEB) modelling for integrated multi-trophic aquaculture. *Aquaculture*, *324/325*, 259-266.148

Sara, G., Scilipoti, D., Mazzola, A. & Modica, A. (2004). Effects of fish farming waste to sedimentary and particulate organic matter in a southern Mediterranean area (Gulf of Castellammare, Sicily): A multiple stable isotope study δ13 C and δ15 N. *Aquaculture*, *234*, 199-213.

Sarà, G., Zenone, A. & Tomasello, A. (2009). Growth of *Mytilus galloprovincialis* (Mollusca, Bivalvia) close to fish farms: a case of integrated multi-trophic aquaculture within the Tyrrhenian Sea. *Hydrobiologia*, *636*, 129-136.

Sarver, D. (2009). Benthic sampling report for Kona Bluewater Farms samples taken at the offshore farm site on March 31, 2009. Deep Blue Research LLC, Kailua Kona, Hawaii. Available at: www.bofish.com/wp-content/files_mf/bmr2009.pdf. Accessed: 02 October 2012.

Sarver, D. (2009). Benthic sampling report for Kona Bluewater Farms samples taken at the offshore farm site on March 31, 2009. Deep Blue Research LLC, Kailua Kona, Hawaii.

Available at: www.bofish.com/wp-content/files_mf/bmr2009.pdf. Accessed: 02 October 2012.

Ščančar, J., Zuliani, T., Turk, T. & Milačič, R. (2006). Organotin compounds and selected metals in the marine environment of northern Adriatic Sea. *Environmental Monitoring and Assessment*, *127*, 271-282.

Schaaning, M.-T, & Hansen, P.-K. (2005). The suitability of electrode measurements for assessment of benthic organic impact and their use in a management system for marine fish farms. Pages 381-408 *in* B.T. Hargrave, editor. Environmental effects of marine finfish aquaculture. *Handbook of Environmental Chemistry*, Volume *5M*. Springer-Verlag, Berlin.

Schaaning, M.-T. & Hansen, P.-K. (2005). The suitability of electrode measurements for assessment of benthic organic impact and their use in a management system for marine fish farms. Pages 381-408 *in* B.T. Hargrave, editor. Environmental effects of marine finfish aquaculture. *Handbook of Environmental Chemistry*, Volume *5M*. Springer-Verlag, Berlin.

Schembri, P. J., Baldacchino, A. E., Mallia, A., Schembri, T., Sant, M. J., Stevens, D. T. & Vella. S. J. (2002). Living resources, fisheries and agriculture. Pages 162-346 *in* V. Axiak, V. Gauci, A. Mallia, E. Mallia, P.J. Schembri, A.J. Vella, and L. Vella, editors. State of the environment report for Malta, 2002. A report to the Ministry for Home Affairs and the Environment, Santa Venera, Malta. Available at: http://www.mepa.org.mt/soer2002. Accessed: 27 September 2012.

Schendel, E. K., Nordstroem, S. E. & Lavkulich, L. M. (2004). Floc and sediment properties and their environmental distribution from a marine fish farm. *Aquaculture Research*, *35*, 483-493.56

Schendel, E. K., Nordstroem, S. E. & Lavkulich, L. M. (2004). Floc and sediment properties and their environmental distribution from a marine fish farm. *Aquaculture Research*, *35*, 483-493.

Schmidt, A. S., Bruun, M. S., Dalsgaard, I., Pedersen, K. & Larsen, J. L. (2000). Occurrence of antimicrobial resistance in fish-pathogenic and environmental bacteria associated with four Danish rainbow trout farms. *Applied and Environmental Microbiology*, *66*, 4908-4915.

Schmidt, L. J., Gaikowski, M. P. & Gingerich, W. H. (2006). Environmental assessment for the use of hydrogen peroxide in aquaculture for treating external fungal and bacterial diseases of cultured fish and fish eggs. U.S. Geological Survey, Biological Resources Division, La Crosse, Wisconsin. Available at: www.fda.gov/ucm/groups/fdagov-public/@fdagov-av-gen/documents/document/ucm072399.pdf. Accessed: 02 October 2012.

Scholl, M. C. & Pade, N. (2005). Salmon farming in Gansbaai: An ecological disaster. White Shark Trust. Gansbaai, Western Cape, South Africa. Available at: www.whitesharktrust.org/media/salmonfarm/documents/salmonfarming.pdf. Accessed: 02 October 2012.

Schulz, H. D. (2000). Redox measurements in marine sediments. Pages 235-246 *in* J. Schuring, H.D. Schulz, W.R. Fisher, J. Bottcher, & W.H.M. Duijnisveld, editors. Redox: Fundamentals, processes and applications. Springer, New York. Available at: http://epic.awi. de/19778/ 1/Sch1999e.pdf#page=257. Accessed: 01 October 2012.

Scott, R. J. (2004). Environmental fate and effect of chemicals associated with Canadian freshwater aquaculture. *Canadian Technical Report of Fisheries and Aquatic Sciences, 2450*, 67-117.131

Shakouri, M. (2003). Impact of cage culture on sediment chemistry: A case study in Mjoifjordur. The United Nations University Fisheries Training Programme. Reykjavík, Iceland. Available at: www.unuftp.is/static/fellows/document/mehdiprf03.pdf. Accessed: 01 October 2012.

Silvert, W, (2001). Impact on habitats: Determing what is acceptable. Pages 16-40 *in* M.F. Tlusty, D.A. Bengston, H.O. Halvorson, S.D. Oktay, J.B. Pearce, and R.B. Rheault, editors. Marine Aquaculture and the Environment: A meeting for stakeholders in the northeast. Marine Aquaculture and the Environment: A meeting for stakeholders in the northeast, Cape Cod Press, Falmouth, Massachusetts. Available at: www.neaq.org/ conservation_and_research/projects/publications_and_presentations/pdf/12__.pdf. Accessed: 02 October 2012.

Smith, J.-N., Yeats, P.-A. & Milligan, T.-G. (2005). Sediment geochronologies for fish farm contaminants in Lime Kiln Bay, Bay of Fundy. Pages 221-238 *in* B.T. Hargrave, editor. Environmental effects of marine finfish aquaculture. *Handbook of Environmental Chemistry*, Volume *5M*. Springer-Verlag, Berlin.

Smith, T. B., Nemeth, R. S., Blondeau, J., Calnan, J. M., Kadison, E. & Herzlieb, S. (2008). Assessing coral reef health across onshore to offshore stress gradients in the US Virgin Islands. *Marine Pollution Bulletin*, *56*, 1983-1991.

Snow, M., King, J. A., Garden, A., Shanks, A. M. & Raynard, R. S. (2005). Comparative susceptibility of turbot *Scophthalmus maximus* to different genotypes of viral haemorrhagic septicaemia virus. *Diseases of Aquatic Organisms*, *67*, 31-38.

Solberg, C. B., Sæthre, L. & Julshamn, K. (2002). The effect of copper-treated net pens on farmed salmon (*Salmo salar*) and other marine organisms and sediments. *Marine Pollution Bulletin*, *45*, 126-132.

Song, X., Huang, L., Zhang, J., Huang, X., Zhang, J., Yin, J., Tan, Y. & Liu, S. (2004). Variation of phytoplankton biomass and primary production in Daya Bay during spring and summer. *Marine Pollution Bulletin*, *49*, 1036-1044.

Soto, D. & Norambuena, F. (2004). Evaluation of salmon farming effects on marine systems in the inner seas of southern Chile: A large-scale mensurative experiment. *Journal of Applied Ichthyology*, *20*, 493-501.

Soto, D. & Norambuena, F. (2004). Evaluation of salmon farming effects on marine systems in the inner seas of southern Chile: A large-scale mensurative experiment. *Journal of Applied Ichthyology*, *20*, 493-501.

Soto, D. & Norambuena. F. (2004). Evaluation of salmon farming effects on marine systems in the inner seas of southern Chile: A large-scale mensurative experiment. *Journal of Applied Ichthyology*, *20*, 493-501.

Soto, D., editor. (2009). Integrated mariculture: A global review. FAO Fisheries and Aquaculture Technical Paper. No. 529. Rome, FAO. Available at: ftp://ftp.fao.org/ docrep/fao/012/i1092e/i1092e04a.pdf. Accessed: 01 October 2012.

Sowles, J. W. (2005). Assessing nitrogen carrying capacity for Blue Hill Bay, Maine: A management case history. Pages 359-380 *in* B.T. Hargrave, editor. Environmental effects of marine finfish aquaculture. *Handbook of Environmental Chemistry*, Volume *5M*. Springer-Verlag, Berlin.

Sowles, J. W. (2005). Assessing nitrogen carrying capacity for Blue Hill Bay, Maine: A management case history. Pages 359-380 *in* B.T. Hargrave, editor. Environmental effects of marine finfish aquaculture. *Handbook of Environmental Chemistry*, Volume *5M.* Springer-Verlag, Berlin.

Sowles, J. W. (2005). Assessing nitrogen carrying capacity for Blue Hill Bay, Maine: A management case history. Pages 359-380 *in* B.T. Hargrave, editor. Environmental effects of marine finfish aquaculture. *Handbook of Environmental Chemistry*, Volume *5M.* Springer-Verlag, Berlin.

Srithongouthai, S., Endo, A., Inoue, A., Kinoshita, K., Yoshioka, M., Sato, A., Iwasaki, T., Teshiba, I., Nashiki, H., Hama, D. & Tsutsumi. H. (2006). Control of dissolved oxygen levels of water in net pens for fish farming by a microscopic bubble generating system. *Fisheries Science*, *72*, 485-493.

Stewart, J. E. (1998). Sharing the waters: An evaluation of site fallowing, year class separation and distances between sites for fish health purposes on Atlantic salmon farms. *Canadian Technical Report of Fisheries and Aquatic Sciences*, *2218*, 1-56.

Stickney, R. R. (2002). Impacts of cage and net-pen culture on water quality and benthic communities. Pages 105 -118 *in* J.R. Tomasso, editor. Aquaculture and the Environment in the United States. U.S. Aquaculture Society, World Aquaculture Society, Baton Rouge, Louisiana.

Stickney, R. R. (2002). Impacts of cage and net-pen culture on water quality and benthic communities. Pages 105 -118 *in* J.R. Tomasso, editor. *Aquaculture*, and the Environment in the United States. U.S. *Aquaculture*, Society, World *Aquaculture*, Society, Baton Rouge, Louisiana.

Stickney, R. R. (2002). Impacts of cage and net-pen culture on water quality and benthic communities. Pages 105 -118 *in* J.R. Tomasso, editor. Aquaculture and the Environment in the United States. U.S. Aquaculture Society, World Aquaculture Society, Baton Rouge, Louisiana.

Stickney, R. R. (2002). Impacts of cage and net-pen culture on water quality and benthic communities. Pages 105 -118 *in* J.R. Tomasso, editor. Aquaculture and the Environment in the United States. U.S. Aquaculture Society, World Aquaculture Society, Baton Rouge, Louisiana.

Stickney, R. R. (2002). Impacts of cage and net-pen culture on water quality and benthic communities. Pages 105 -118 *in* J.R. Tomasso, editor. Aquaculture and the Environment in the United States. U.S. Aquaculture Society, World Aquaculture Society, Baton Rouge, Louisiana.

Stirling, H. P. & Okumus, I. (1995). Growth and production of mussels (*Mytilus edulis* L.) suspended at salmon cages and shellfish farms in two Scottish sea lochs. *Aquaculture*, *134*, 193-210.

Stone, G., Cavagnaro, L., Hutt, A., Kraus, S., Baldwin, K. & Brown, J. (2000). Reactions of Hector's dolphins to acoustic gillnet pingers. Contract 3071, Conservation Services Levy, Department of Conservation, Wellington, Australia. Available at: www.cetaceanbycatch.org/Papers/stone.pdf. Accessed: 02 October 2012.

Strain, P. & Hargrave, B. (2005). Salmon aquaculture, nutrient fluxes and ecosystem processes in southwestern New Brunswick. Pages 29-57 *in* B.T. Hargrave, editor. Environmental effects of marine finfish aquaculture. *Handbook of Environmental Chemistry*, Volume 5M. Springer-Verlag, Berlin.

Strain, P. & Hargrave, B. (2005). Salmon aquaculture, nutrient fluxes and ecosystem processes in southwestern New Brunswick. Pages 29-57 *in* B.T. Hargrave, editor. Environmental effects of marine finfish aquaculture. Handbook of Environmental Chemistry, Volume 5M. Springer-Verlag, Berlin.

Strain, P. & Hargrave. B. (2005). Salmon aquaculture, nutrient fluxes and ecosystem processes in southwestern New Brunswick. Pages 29-57 *in* B.T. Hargrave, editor. Environmental effects of marine finfish aquaculture. *Handbook of Environmental Chemistry*, Volume *5M*. Springer-Verlag, Berlin.

Sudaryanto, A., Takahashi, S., Monirith, I., Ismail, A., Muchtar, M., Zheng, J., Richardson, B. J., Subramanian, A., Prudente, M., Hue, N. D. & Tanabe, S. (2002). Asia-Pacific mussel watch: Monitoring of butyltin contamination in coastal waters of Asian developing countries. *Environmental Toxicology and Chemistry*, *21*, 2119-2130.

Sudirman, H. Halide, Jompa, J., Zulfikar, Iswahyudin, & McKinnon, A. D. (2009. Wild fish associated with tropical sea cage aquaculture in South Sulawesi, Indonesia. *Aquaculture*, *286*, 233-239.

Sutherland, K. P., Shaban, S., Joyner, J. L., Porter, J. W. & Lipp, E. K. (2011). Human pathogen shown to cause disease in the threatened elkhorn coral *Acropora palmata*. *PLoS ONE*, *6*(8), e23468. doi, 23410.21371/journal.pone.0023468.

Sutherland, T. F. (2004). Framework for a benthic aquaculture monitoring program in the Pacific region. Fisheries and Oceans Canada, Vancouver, British Columbia, Canada. Available at: www.dfo-mpo.gc.ca/csas-sccs/Publications/ResDocs-DocRech/2004/2004_056-eng.htm. Accessed: 02 October 2012.

Sutherland, T. F., Martin, A. J. & Levings. C. D. (2001). Characterization of suspended particulate matter surrounding a salmonid net-pen in the Broughton Archipelago, British Columbia. *ICES Journal of Marine Science*, *58*, 404-410.

Sutherland, T. F., Petersen, S. A., Levings, C. D. & Martin, A. J. (2007). Distinguishing between natural and aquaculture-derived sediment concentrations of heavy metals in the Broughton Archipelago, British Columbia. *Marine Pollution Bulletin*, *54*, 1451-1460.

Swift, M., Fredriksson, D., Unrein, A., Fullerton, B., Patursson, O. & Baldwin, K. (2006). Drag force acting on biofouled net panels. *Aquacultural Engineering*, *35*, 292-299.

Tanaka, K. & Kodama. M. (2007). Effects of resuspended sediments on the environmental changes in the inner part of Ariake Bay, Japan. *Bulletin of the Fisheries Research Agency*, *19*, 9-15.

Telfer, T. C., Baird, D. J., McHenery, J. G., Stone, J., Sutherland, I. & Wislocki, P. (2006). Environmental effects of the anti-sea lice (Copepoda: Caligidae) therapeutant emamectin benzoate under commercial use conditions in the marine environment. *Aquaculture*, *260*, 163-180.

Terhune, J. M., Hoover, C. L. & Jacobs, S. R. (2002). Potential detection and deterrence ranges by harbor seals of underwater acoustic harassment devices (AHD) in the Bay of Fundy, Canada. *Journal of the World Aquaculture Society*, *33*, 176-183.

Terlizzi, A., De Falco, G., Felline, S., Fiorentino, D., Gambi, M. C. & Cancemi, G. (2010). Effects of marine cage aquaculture on macrofauna assemblages associated with *Posidonia oceanica* meadows. *Italian Journal of Zoology*, *77*, 362-371.

Tett, P. (2008). Fish farm waste in the ecosystem. Pages 1-46 *in* M. Holmer, K. Black, C.M. Duarte, N. Marba, and I. Karakassis, editors. Aquaculture in the Ecosystem. Springer, Dordrecht, London.

Tett, P. (2008). Fish farm waste in the ecosystem. Pages 1-46 *in* M. Holmer, K. Black, C.M. Duarte, N. Marba, & I. Karakassis, editors. *Aquaculture*, in the Ecosystem. Springer, Dordrecht, London.

Tett, P. (2008). Fish farm waste in the ecosystem. Pages 1-46 *in* M. Holmer, K. Black, C.M. Duarte, N. Marba, and I. Karakassis, editors. Aquaculture in the Ecosystem. Springer, Dordrecht, London.

Tett, P. (2008). Fish farm waste in the ecosystem. Pages 1-46 *in* M. Holmer, K. Black, C.M. Duarte, N. Marba, and I. Karakassis, editors. Aquaculture in the Ecosystem. Springer, Dordrecht, London.

Tett, P., Portilla, E., Gillibrand, P. A. & Inall, M. (2011). Carrying and assimilative capacities: The ACExR-LESV model for sea-loch aquaculture. *Aquaculture Research*, *42*, 51-67.

Tett, P., Portilla, E., Gillibrand, P. A. & Inall, M. (2011). Carrying and assimilative capacities: The ACExR-LESV model for sea-loch aquaculture. *Aquaculture Research*, *42*, 51-67.

The Mediterranean Science Commission. (2007). Impact of mariculture on coastal ecosystems. CIESM Workshop Monographs No. 32, Monaco. Available at: www.ciesm.org/online/monographs/lisboa07.pdf. Accessed: 27 September 2012.

The Mediterranean Science Commission. (2007). Impact of mariculture on coastal ecosystems. CIESM Workshop Monographs No. 32, Monaco. Available at: www.ciesm.org/ online/monographs/lisboa07.pdf. Accessed: 27 September 2012.

The Mediterranean Science Commission. (2007). Impact of mariculture on coastal ecosystems. CIESM Workshop Monographs No. 32, Monaco. Available at: www.ciesm.org/ online/monographs/lisboa07.pdf. Accessed: 27 September 2012.

Tlusty, M. F., Bengston, D. A., Halvorson, H. O., Oktay, S. D., Pearce, J. B. & Rheault, R. B. Jr. eds. (2001). Marine aquaculture and the environment: A meeting for stakeholders in the northeast. Cape Cod Press, Falmouth, Massachusetts. Available at: www.neaq.org/conservation_and_research/projects/publications_and_presentations/pdf/12__.pdf. Accessed: 02 October 2012.

Tlusty, M. F., Pepper, V. A. & Anderson, M. R. (2005). Reconciling aquaculture's influence on the water column and benthos of an estuarine fjord – A case study from Bay d'Espoir, Newfoundland. Pages 115-128 *in* B.T. Hargrave, editor. Environmental effects of marine finfish aquaculture. *Handbook of Environmental Chemistry*, Volume *5M.* Springer-Verlag, Berlin.

Tlusty, M. F., Pepper, V. A. & Anderson. M. R. (2005). Reconciling aquaculture's influence on the water column and benthos of an estuarine fjord – A case study from Bay d'Espoir, Newfoundland. Pages 115-128 *in* B.T. Hargrave, editor. Environmental effects of marine finfish aquaculture. *Handbook of Environmental Chemistry*, Volume *5M.* Springer-Verlag, Berlin.

Tlusty, M., Snook, K., Pepper, V. & Anderson, M. (2000). The potential for soluble and transport loss of particulate *Aquaculture*, wastes. *Aquaculture*, *Research*, *31*, 745-755.

Torres, J. L. (2001). Impacts of sedimentation on the growth rates of *Montastraea annularis* in southwest Puerto Rico. *Bulletin of Marine Science*, *69*, 631-637.

Tort, M. J., Wooster, G. A. & Bowser, P. R. (2003). Effects of hydrogen peroxide on hematology and blood chemistry parameters of walleye *Stizostedion vitreum*. *Journal of the World Aquaculture Society*, *34*, 236-242.

Troell, M., Joyce, A., Chopin, T., Neori, A., Buschmann, A. H. & Fang, J.-G. (2009). Ecological engineering in aquaculture - Potential for integrated multi-trophic aquaculture (IMTA) in marine offshore systems. *Aquaculture*, 297, 1-9.

Trushenski, J. T., Kasper, C. S. & Kohler. C. C. (2006). Challenges and opportunities in finfish nutrition. *North American Journal of Aquaculture*, *68*, 122-140.

Tsemel, A., Spanier, E. & Angel, D. L. (2006). Benthic communities of artificial structures: Effects of mariculture in the Gulf of Aqaba (Eilat) on development and bioaccumulation. *Bulletin of Marine Science*, *78*, 103-113.

Tsutsumi, H. (2007). Environmental deterioration of fish farms in Japanese enclosed bays and measures for their environmental management. Pages 45-49 *in* C.-S. Lee and P.J. O'Bryan, editors. Open Ocean *Aquaculture*, - Moving forward. Oceanic Institute, Waimanalo, Hawaii. Available at: http://nsgl.gso.uri.edu/ocei/oceiw06001.pdf. Accessed: 01 October 2012.

Tsutsumi, H., Srithongouthai, S., Inoue, A., Sato, A. & Hama, D. (2006). Seasonal fluctuations in the flux of particulate organic matter discharged from net pens for fish farming. *Fisheries Science*, *72*, 119-127.

Tucker, C. S. & Hargreaves, J. A. editors. (2008). *Environmental best management practices for aquaculture*. Wiley-Blackwell, Ames, Iowa.

Tucker, C. S. & Hargreaves, J. A. editors. 2008). Environmental best management practices for aquaculture. Wiley-Blackwell, Ames, Iowa.

Tucker, C. S. & Hargreaves, J. A., editors. (2008). Environmental best management practices for aquaculture. Wiley-Blackwell, Ames, Iowa.57

Tucker, C. S. & Hargreaves, J. A., editors. (2008). Environmental best management practices for aquaculture. Wiley-Blackwell, Ames, Iowa.

Tuya, F., Sanchez-Jerez, P., Dempster, T., Boyra, A. & Haroun, R. J. (2006). Changes in demersal wild fish aggregations beneath a sea-cage fish farm after the cessation of farming. *Journal of Fish Biology*, *69*, 682-697.

Tuya, F., Sanchez-Jerez, P., Dempster, T., Boyra, A. & Haroun, R. J. (2006). Changes in demersal wild fish aggregations beneath a sea-cage fish farm after the cessation of farming. *Journal of Fish Biology*, *69*, 682-697.

Tveterås, S. (2002). Norwegian salmon aquaculture and sustainability: The relationship between environmental quality and industry growth. *Marine Resource Economics*, *17*, 121-132.

Ueno, S., Susa, N., Furukawa, Y., Komatsu, Y., Koyama, S. & Suzuki, T. (1999). Butyltin and phenyltin compounds in some marine fishery products on the Japanese market. *Archives of Environmental Health*, *54*, 20-25.

Valdemarsen, T., Kristensen, E. & Holmer, M. (2009). Metabolic threshold and sulfide-buffering in diffusion controlled marine sediments impacted by continuous organic enrichment. *Biochemistry*, *95*, 335–353.

Valdemarsen, T., Kristensen, E. & Holmer, M. (2010). Sulfur, carbon, & nitrogen cycling in faunated sediments impacted by repeated organic enrichment. *Marine Ecology Progress Series*, *400*, 37-53.

Valdemarsen, T., Kristensen, E. & Holmer, M. (2010). Sulfur, carbon, and nitrogen cycling in faunated sediments impacted by repeated organic enrichment. *Marine Ecology Progress Series*, *400*, 37-53.

Vargas-Machuca, S. C., Ponce-Palafox, J. T., Arredondo-Figueroa, J. L., Chavez-Ortiz, E. A. & Vernon-Carter, E. J. (2008). Physico-chemical water parameters variation in the floating cages of snappers (*Lutjanus peru* and *L. guttatus*) farmed in tropical sea. *Revista Mexicana De Ingenieria Quimica, 7*, 237-242.

Vargo, G. A. (2009). A brief summary of the physiology and ecology of *Karenia brevis* Davis (G. Hansen and Moestrup comb. nov.) red tides on the West Florida Shelf and of hypotheses posed for their initiation, growth, maintenance, and termination. *Harmful Algae, 8*, 573-584.

Vezzuli, L., Marrale, D., Moreno, M. & Fabiano, M. (2003). Sediment organic matter and meiofauna community response to long-term fish-farm impact in the Ligurian Sea (Western Mediterranean). *Chemistry and Ecology. 19*, 431-440.

Vezzuli, L., Marrale, D., Moreno, M. & Fabiano, M. (2003). Sediment organic matter and meiofauna community response to long-term fish-farm impact in the Ligurian Sea (Western Mediterranean). *Chemistry and Ecology, 19*, 431-440.

Vezzulli, L., C. Pruzzo, & M. Fabiano. (2004. Response of the bacterial community to in situ bioremediation of organic-rich sediments. *Marine Pollution Bulletin*, 49, 740-751.

Vezzulli, L., Chelossi, E., Riccardi, G. & Fabiano, M. (2002). Bacterial community structure and activity in fish farm sediments of the Ligurian sea (Western Mediterranean). *Aquaculture International, 10*, 123-141.

Vezzulli, L., E. Chelossi, G. Riccardi, & M. Fabiano. (2002. Bacterial community structure and activity in fish farm sediments of the Ligurian sea (Western Mediterranean). *Aquaculture*, International 10, 123-141.

Vezzulli, L., M. Moreno, V. Marin, E. Pezzati, M. Bartoli, & Fabiano, M. (2008). Organic waste impact of capture-based Atlantic bluefin tuna *Aquaculture*, at an exposed site in the Mediterranean Sea. *Estuarine, Coastal and Shelf Science, 78*, 369-384.

Vezzulli, L., Moreno, M., Marin, V., Pezzati, E., Bartoli, M. & Fabiano, M. (2008). Organic waste impact of capture-based Atlantic bluefin tuna aquaculture at an exposed site in the Mediterranean Sea. *Estuarine, Coastal and Shelf Science, 78*, 369-384.

Vezzulli, L., Pruzzo, C. & Fabiano, M. (2004). Response of the bacterial community to in situ bioremediation of organic-rich sediments. *Marine Pollution Bulletin, 49*, 740-751.

Vezzulli, L., Pruzzo, C. & Fabiano, M. (2004). Response of the bacterial community to in situ bioremediation of organic-rich sediments. *Marine Pollution Bulletin, 49*, 740-751.

Villanueva, R. D., Yap, H. T. & Montaño, M. N. E. (2006). Intensive fish farming in the Philippines is detrimental to the reef-building coral *Pocillopora damicornis. Marine Ecology Progress Series, 316*, 165-174.

Vita, R., Marin, A., Madrid, J. A., Jimenez-Brinquis, B., Cesar, A. & Marin-Guirao, L. (2004). Effects of wild fishes on waste exportation from a Mediterranean fish farm. *Marine Ecology Progress Series, 277*, 253-261.

Vita, R., Marin, A., Madrid, J. A., Jimenez-Brinquis, B., Cesar, A. & Marin-Guirao, L. (2004). Effects of wild fishes on waste exportation from a Mediterranean fish farm. Marine Ecology Progress Series 277, 253-261.

Vizzini, S. & Mazzola, A. (2006). The effects of anthropogenic organic matter inputs on stable carbon and nitrogen isotopes in organisms from different trophic levels in a southern Mediterranean coastal area. *Science of the Total Environment, 368*, 723-731.

Vizzini, S. & Mazzola, A. (2006). The effects of anthropogenic organic matter inputs on stable carbon and nitrogen isotopes in organisms from different trophic levels in a southern Mediterranean coastal area. *Science of the Total Environment*, *368*, 723-731.

Vizzini, S., Savona, B., Caruso, M., Savona, A. & Mazzola, A. (2005). Analysis of stable carbon and nitrogen isotopes as a tool for assessing the environmental impact of aquaculture: a case study from the western Mediterranean. *Aquaculture International*, *13*, 157-165.

Whitmarsh, D. J., Cook, E. J. & Black, K. D. (2006). Searching for sustainability in aquaculture: An investigation into the economic prospects for an integrated salmon-mussel production system. *Marine Policy*, *30*, 293-298.

Wilding, T. A. (2011). A characterization and sensitivity analysis of the benthic biotopes around Scottish salmon farms with a focus on the sea pen *Pennatula phosphorea* L. Aquaculture Research 42, 35-40.

Wildish, D. J. & Pohle, G. (2005). Benthic macrofaunal changes resulting from finfish mariculture. Pages 239-251 *in* B.T. Hargrave, editor. Environmental effects of marine finfish aquaculture. *Handbook of Environmental Chemistry*, Volume *5M*. Springer-Verlag, Berlin.

Wildish, D. J. & Pohle, G. (2005). Benthic macrofaunal changes resulting from finfish mariculture. Pages 239-251 *in* B.T. Hargrave, editor. Environmental effects of marine finfish aquaculture. *Handbook of Environmental Chemistry*, Volume *5M*. Springer-Verlag, Berlin.

Wildish, D. J., Dowd, M., Sutherland, T. F. & Levings, C. D. (2004). Near-field organic enrichment from marine finfish aquaculture. Pages 1-51 *in* Canadian Technical Report of Fisheries and Aquatic Sciences 2450, Volume *3*. Available at: www.dfo-mpo.gc.ca/Library/285141.pdf. Accessed: 27 September 2012.

Wildish, D. J., Dowd, M., Sutherland, T. F. & Levings, C. D. (2004a). Near-field organic enrichment from marine finfish aquaculture. Pages 1-51 *in* Canadian Technical Report of Fisheries and Aquatic Sciences 2450, Volume *3*. Available at: www.dfo-mpo.gc.ca/Library/285141.pdf. Accessed: 27 September 2012.

Wildish, D. J., Dowd, M., Sutherland, T. F. & Levings. C. D. (2004). Near-field organic enrichment from marine finfish aquaculture. Pages 1-51 *in* Canadian Technical Report of Fisheries and Aquatic Sciences 2450, Volume *3*. Available at: www.dfo-mpo.gc.ca/Library/285141.pdf. Accessed: 27 September 2012.

Wildish, D. J., Hargrave, B. T. & Pohle, G. (2001). Cost-effective monitoring of organic enrichment resulting from salmon mariculture. *ICES Journal of Marine Science*, *58*, 469-476.

Wildish, D. J., Hargrave, B. T. & Pohle, G. (2001). Cost-effective monitoring of organic enrichment resulting from salmon mariculture. *ICES Journal of Marine Science*, *58*, 469-476.

Wildish, D. J., Hargrave, B. T., MacLeod, C. & Crawford, C. (2003). Detection of organic enrichment near finfish net-pens by sediment profile imaging at SCUBA-accessible depths. *Journal of Experimental Marine Biology and Ecology*, 285-286, 403-413.

Wildish, D. J., Hargrave, B. T., MacLeod, C. & Crawford, C. (2003). Detection of organic enrichment near finfish net-pens by sediment profile imaging at SCUBA-accessible depths. *Journal of Experimental Marine Biology and Ecology*, 285-286, 403-413.

Wildish, D. J., Hughes-Clarke, J. E., Pohle, G. W., Hargrave, B. T. & Mayer, L. M. (2004b). Acoustic detection of organic enrichment in sediments at a salmon farm is confirmed by independent groundtruthing methods. *Marine Ecology Progress Series*, *267*, 99–105.

Willemsen, P. R. (2005). Biofouling in European aquaculture: Is there an easy solution? European Aquaculture Society Special Publication No. 35. Available at: www.crabproject. com/client/files/Paper_Willemsen.pdf. Accessed 02 October 2012.132

Willis, K. J., Gillibrand, P. A., Cromey, C. J. & Black, K. D. (2005). Sea lice treatments on salmon farms have no adverse effects on zooplankton communities: A case study. *Marine Pollution Bulletin*, *50*, 806-816.

Wong, C. K., Wong, P. P. & Chu, L. M. (2001). Heavy metal concentrations in marine fishes collected from fish culture sites in Hong Kong. *Archives of Environmental Contamination and Toxicology*, *40*, 60-69.

Wu, R. S. S. (1995). The environmental impact of marine fish culture: Towards a sustainable future. *Marine Pollution Bulletin*, *31*, 159-166.

Wu, R. S. S. (1995). The environmental impact of marine fish culture: Towards a sustainable future. *Marine Pollution Bulletin*, *31*, 159-166.

Wu, R. S. S. (1995). The environmental impact of marine fish culture: Towards a sustainable future. *Marine Pollution Bulletin*, *31*, 159-166.

Wu, R. S. S. (1995). The environmental impact of marine fish culture: Towards a sustainable future. *Marine Pollution Bulletin*, *31*, 159-166.

Wu, R. S. S. (1995). The environmental impact of marine fish culture: Towards a sustainable future. *Marine Pollution Bulletin*, *31*, 159-166.

Würsig, B. & Gailey, G. A. (2002). Marine mammals and aquaculture: Conflicts and potential resolutions. Pages 45-59 *in* R.R. Stickney and J.P. McVey, editors. Responsible Marine Aquaculture. CAB International, New York.

Xue, X., Hong, H. & Charles, A. T. (2004). Cumulative environmental impacts and integrated coastal management: The case of Xiamen, China. *Journal of Environmental Management*, *71*, 271-283.

Yabanli, M. & Egemen. O. (2009). Monitoring the environmental impacts of marine aquaculture activities on the water column and sediment in vicinity of the Karaburun Peninsula (Turkey-Eastern Aegean Sea). *Journal of FisheriesSciences.com*, *3*, 207-213. DOI:210.3153/jfscom.2009025.

Yanong, R. P. E. (2008). Use of hydrogen peroxide in finfish aquaculture. Institute of Food and Agricultureal Sciences, Florida Cooperative Extension Service, University of Florida, Gainesville, Florida. Available at: http://edis.ifas.ufl.edu/pdffiles/FA/FA15700.pdf. Accessed: 02 October 2012.

Yeats, P. A., Milligan, T. G., Sutherland, T. F., Robinson, S. M. C., Smith, J. A., Lawton, P. & Levings, C. D. (2005). Lithium-normalized zinc and copper concentrations in sediments as measures of trace metal enrichment due to salmon aquaculture. Pages 207-220 *in* B.T. Hargrave, editor. Environmental effects of marine finfish aquaculture. *Handbook of Environmental Chemistry*, Volume *5M*. Springer-Verlag, Berlin.

Yokoyama, H, (2003). Environmental quality criteria for fish farms in Japan. *Aquaculture*, *226*, 45-56.

Yokoyama, H. (2003). Environmental quality criteria for fish farms in Japan. *Aquaculture*, *226*, 45-56.

Yokoyama, H., Inoue, M. & Abo, K. (2007). Macrobenthos, current velocity and topographic factors as indicators to assess the assimilative capacity of fish farms: Proposal of two indices. *Bulletin of Fisheries Research Agency*, *19*, 89-96.

Yokoyama, H., Takashi, T., Ishihi, Y. & Abo, K. (2009). Effects of restricted feeding on growth of red sea bream and sedimentation of *Aquaculture*, wastes. *Aquaculture*, 286, 80-88.

Yu, J., Tang, D. L., Oh, I. S. & Yao, L. J. (2007). Response of harmful algal blooms to environmental changes in Daya Bay, China. *Terrestrial Atmospheric and Oceanic Sciences*, *18*, 1011-1027.

Zorrilla, I. (2003). Bacteria recovered from diseased cultured gilthead sea bream (*Sparus aurata* L.) in southwestern Spain. *Aquaculture*, *218*, 11-20.

In: Marine Finfish Aquaculture and the Environment
Editor: Stewart White
ISBN: 978-1-63463-537-0

Chapter 2

GUIDELINES FOR ECOLOGICAL RISK ASSESSMENT OF MARINE FISH AQUACULTURE*

Colin E. Nash, Peter R. Burbridge and John K. Volkman

EXECUTIVE SUMMARY

Risk assessment is frequently defined as a tool for making decisions under conditions of uncertainty. Therefore it is far from being an exact science and yet it thrives on exact information. At this period in the emergence of the field of marine fish aquaculture, the scientific and technical knowledge of aquaculture's effects on the environment are a growing compendium of information, but at times the information appears to be ambiguous, conflicting, and incomplete. This is because it is being gleaned continuously from a diverse range of marine ecosystems where there are different fish and shellfish species being produced under a variety of systems and practices which, in some cases, are themselves still being developed.

But the time when knowledge is apparently deficient is arguably the right time for seriously addressing risks, and the constraint of incomplete scientific and technical knowledge has to be offset by the practical knowledge and professional experience of individuals in order to make balanced judgments. In the near future it will be possible to model the interactive ecological effects of marine aquaculture activities, and decisions will be made based on risks quantified numerically for different ecosystems. However, for the moment, informed decisions will have to be made by risk managers with the help of responsible individuals guided mostly by their past experiences and research.

The purpose of this document is to provide a basic set of guidelines for risk managers and other decision makers to use all information available to assess the different ecological risks of marine fish aquaculture in a variety of marine ecosystems. Ten areas of substantive risk in the interaction between marine fish aquaculture are perceived by the public and public administrators to be of most concern. In no order of priority they are: increased organic loading, increased inorganic loading, residual heavy metals, transmission of disease

* This is an edited, reformatted and augmented version of NOAA Technical Memorandum NMFS-NWFSC-71, issued by the National Oceanic and Atmospheric Administration, December 2005.

organisms, residual therapeutants, biological interaction of escapes with wild populations, physical interaction with marine wildlife, physical impact on marine habitat, using wild juveniles for grow-out, and harvesting industrial fisheries for aqua-feeds.

In this technical memorandum each of these 10 areas of risk is assessed for its degree of potential adversity, together with its mitigation, in an identical step-by-step process. This common analytical framework, which is first described and illustrated in detail, was developed by the World Health Organization many years ago, and its generic nature and international acceptance make its use for the risk assessment of marine fish aquaculture an ideal application.

Appendices A through J are templates that outline the approach for conducting a risk assessment for all 10 perceived issues. With the help of a flowchart, each template identifies the biological end points or entities and their attributes, both locally and far field, which might be affected for that respective area of risk. It also identifies appropriate methodologies that can be used for measuring or monitoring the effects of exposure to each specific risk.

The chances of any risk occurring can differ greatly in accordance with the natural characteristics of the local ecosystem and its geographic location. Therefore, each template contains a biological overview of its respective risk and briefly discusses factors that may enhance or mitigate the risk's occurrence. In addition, for the benefit of risk managers and risk assessors in all parts of the world, the 10 risks are framed in a matrix to suggest different orders of relevance for their application in different climatic zones. Each template contains a list of documents where further pertinent information can be found.

Finally, this technical memorandum contains a glossary of the risk assessment terms and marine terms used throughout, a list of the international participants of the workshop, and a list of references specific to marine aquaculture and risk assessment, but broader in nature.

INTRODUCTION

The Environment and the Intervention of Marine Aquaculture

Few, if any, human interventions in the environment fail to have impact. In some cases interventions are potentially so damaging that they must be eliminated. On the other hand, the majority of human interventions are purposeful and designed to be of benefit to humans, so it is necessary that they proceed responsibly, sharing equitably in the use of nature's vital resources. It is thus important that these interventions are carefully managed with good stewardship to ensure that benefits can be achieved over time frames of many decades.

Aquaculture, together with fisheries and agriculture, has long been a provider of food for human consumption. For over three millennia it has been a necessary and often the only source of animal protein for pastoral communities living at subsistence levels. But within the last century its history has dramatically changed, and science and technology have propelled modern aquaculture into semi-intensive and intensive farming systems. These systems have greatly increased its degree of exposure to the environment. Consequently, although aquaculture remains a crucial cornerstone of rural life in many countries, its modern practices and array of commercial end products are, to the rest of the world, dependent more on human lifestyle decisions governed by social choice.

Fortunately, an important factor in social choice as aquaculture emerges in the twenty-first century is not only to minimize the impact of all human interventions on the environment but also to sustain the existing integrity of its many ecosystems in perpetuity. This has become a challenge to all resource-based industries, not only marine aquaculture. There are innumerable aquatic ecosystems in which aquaculture intervention is feasible. Each and every ecosystem has its own very specific and desired values, and therefore for the stewards of these resources to set specific goals around these values it is necessary for them to know in advance 1) what integrity means for each ecosystem and what specifically needs to be protected; and 2) which ecological resources and processes have to be sustained and for what reason. Compared with that of terrestrial ecosystems, comprehensive knowledge of aquatic ecosystems is severely constrained. Partly this is because much of the ecosystem lies below water and is thus not readily observable, but also the need for extensive environmental research of marine ecosystems is only now becoming recognized in many countries.

Many aquatic and terrestrial ecosystems can be said to be equally fragile, but the factors differ as do the mechanisms available for remediation. Most human interventions in aquatic ecosystems, such as mineral extraction, fishing, and now aquaculture, may induce more lasting far-field effects unless properly managed. Nonetheless, these and any other industries that integrate with open waters, such as tourism and recreational boating, all have a right to exist equitably as stakeholders; the effects on the aquatic ecosystem by one should not eliminate the existence of another.

In enabling aquaculture to share aquatic resources responsibly, the stewards of these resources are faced with many options. Invariably these options cannot be quantified adequately, and thus managers must estimate their potential ecological risks through individual risk assessments. Nonetheless, although ecological risks are a paramount concern, the final decision is frequently decided by other factors brought to bear by social choice, such as economic benefits to a local community, or issues of public health.

Purpose of the Guidelines Document

The purpose of the document is to provide guidelines for risk managers, risk assessors, and anyone involved in the risk assessment process (Table 1) to address risks to the environment. The specific focus is on the possible effects or impacts of finfish aquaculture, but with several caveats:

1. The guidelines are limited only to the assessment of ecological risks. Although, as noted, final decisions are invariably made by risk managers using a broader range of factors, such assessments of economic risks and human health risks by any intervention of aquaculture are not part of this work.
2. The guidelines are applicable only to the risk assessment of marine fish aquaculture. The diversity of aquaculture, with its many systems and practices producing more than 200 species of aquatic animals and plants in a variety of fresh and saline waters, is too much to consider in a single document. However, it is anticipated that these guidelines will greatly simplify risk assessments in most other aquaculture fields.
3. The guidelines are confined to the risk assessment of marine fish aquaculture based on its effects on and not from other elements of the environment. Although marine

aquaculture is vulnerable to the degradation of water quality as a consequence of poorly managed development in the coastal zone, most countries have regulatory structures and guidelines in place to protect aquaculture, and in time these standards will be improved by combining the risks to the environment from all sources.

Using the Guidelines Document

Before any decisions can be made with regard to the siting or operation of a marine aquaculture facility, the first responsibility of risk managers, and that includes both managers of resources as well as managers of aquaculture operations, is to draw their conclusions from all information provided by the risk assessors that a perceived risk to a particular ecosystem has validity or not, and if so to estimate its degree of adverse effect. This may or may not be a straightforward task. In some cases the information reported to them by the risk assessors may be an excellent combination of field and laboratory data to compare with recognized benchmarks of stress, while in others it may be no more than the long-time experience of practitioners.

Irrespective of the final detail, it is important that the information is considered, collected, analyzed, characterized, and reported in a structured fashion. This ensures that the risk assessment report is not only complete as far as it can be (Table 2), but also that it can be compared directly with similar risk assessments made by other individuals elsewhere.

These guidelines for the risk assessment of marine fish aquaculture attempt to facilitate the work of risk assessors and risk managers to achieve these objectives. In brief, the guidelines:

- identify the 10 areas of substantive risk in the interaction between marine fish aquaculture operations and the environment;
- identify the biological end points or entities and their attributes, both locally and far field, that might be affected in those areas of risk;
- identify methodologies for measuring or monitoring the effects of exposure to each area of risk;
- provide a common framework, or step-by-step process, to estimate the degree of potential adversity of each area of risk, together with its mitigation; and
- provide a concept of the physical and environmental demands of marine fish aquaculture sites, and a matrix to suggest different orders of relevance for the application of each area of risk in different global ecosystems.

In planning a risk assessment, it is recommended that the risk managers and risk assessors, together with others with experience in marine fish aquaculture, first review the areas of risk identified as priorities in the guidelines, and establish their relevance in their own geographic region and to the particular local ecosystem where marine aquaculture facilities are to be sited. It is very probable that not all areas of risk will be applicable to every development site, and therefore a matrix has been developed as part of the guidelines to suggest some of the more common differences (see "Near-field and Far-field Effects" subsection on page 11). For those that are important, the respective templates (as described in Appendices A–J) can be used.

ECOLOGICAL RISK ASSESSMENT OF MARINE FISH AQUACULTURE

Framework

For more than 20 years, countries have been developing national guidelines for environmental risk assessment. At first their focus was predominantly on environmental risks to a single species (humans) and one end point (human health), but later nonhuman-oriented environmental risk assessments were included. These not only considered the risk to entire communities and addressed any number of selected end points, but they also included the possible effects of nonchemical stressors.

In order to accommodate the sudden burst of different views and approaches to environmental risk assessment by its member countries, the United Nations (UN) World Health Organization (WHO) developed a common analytical framework. The WHO Framework is adopted here for developing Guidelines for Ecological Risk Assessment of Marine Fish Aquaculture (this technical memorandum) because it provides a generic analytical framework that has been widely reviewed and accepted by international experts in UN-sponsored workshops.

The WHO Framework (Figure 1) represents the scope of the guidelines for undertaking ecological risk assessments. It represents a three-dimensional figure, with planes surrounding the actual risk assessment to depict the total process. These planes represent the continuum for all those who are involved in the decision-making process, and includes not only the interactions between risk managers and risk assessors (the scientific and technical experts), but also their interaction with stakeholders who may be affected by any decision. For marine aquaculture, participating stakeholders are typically the fish farmers and their trade associations, waterfront property owners, recreational users of waters, other fishing and aquaculture bodies, and environmental advocacy groups. The extent of stakeholder interaction, and at what point it is considered in the decision-making process, is the prerogative of the decision maker, and varies from one country to another in accordance to the regulatory, legal, and decision-making climate. Furthermore, stakeholders might perform their own risk assessments with or without the help of technical consultants, with differences arguable in court.

The risk assessment process is itself divided into three segments. These segments represent three distinct phases of work, but once again there is a continuum of interplay between the persons involved.

The following sections describe in broad terms a generic risk assessment process but without direct application to any specific category of risk. Detailed processes can be found for all the principal categories of risk from marine fish aquaculture in Appendices A–J.

Table 1. Definition of participants in the risk assessment process

Participant	Definition
Risk manager	Any individual and organization having the responsibility or the authority to take action or require action to mitigate an identified risk. Typically the term describes a decision maker in a government organization who has legal authority to protect or manage a resource. However, a risk manager may be any interested party who has the ability to take action to reduce or mitigate a risk; for example, the owner or manager of an aquaculture facility.
Risk assessor	A professional who brings a needed expertise to a risk assessment team from any number of relevant fields, including, for example, risk assessment, marine ecosystems, coastal zone management, marine engineering, marine biology, oceanography, aquaculture, fish nutrition, fish disease, etc.
Stakeholders	Any individual, company, or organization that has a direct or indirect interest in, or could be affected by, an aquaculture operation.

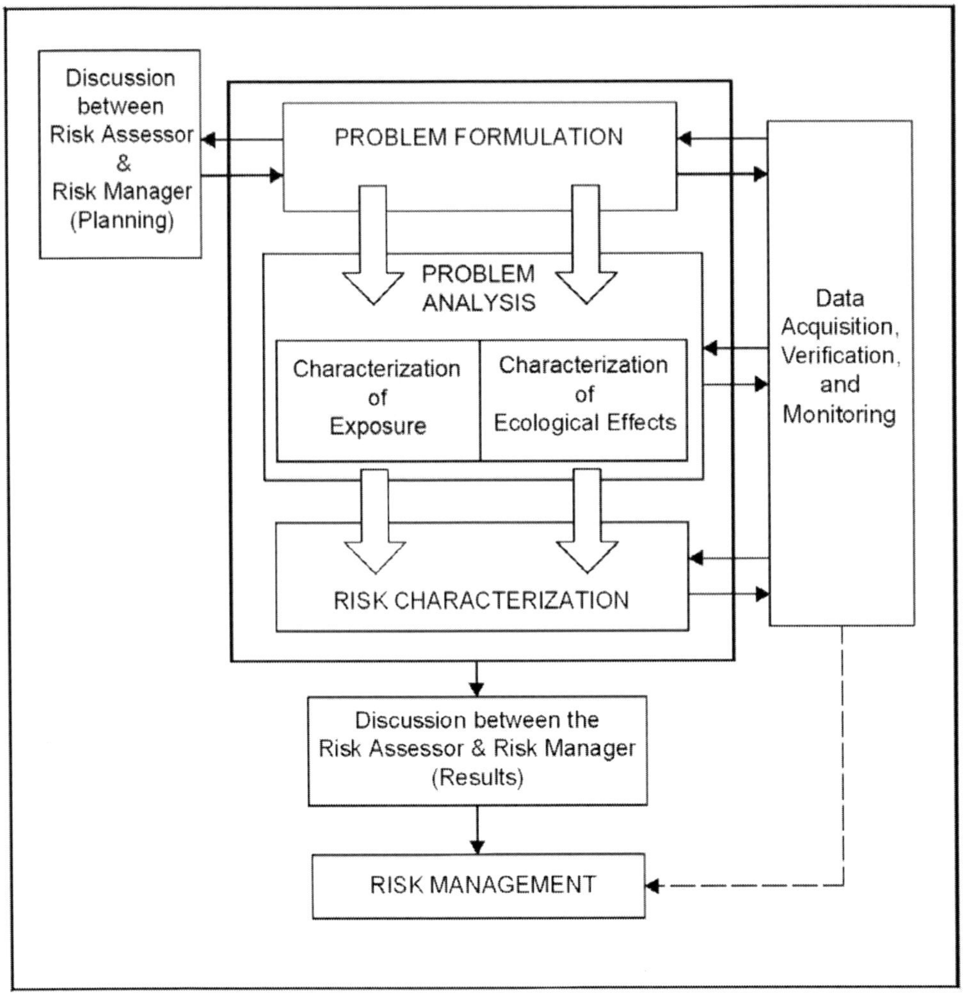

Figure 1. The WHO framework for ecological risk assessment.

Table 2. Possible contents of a risk assessment report

- Description of the preliminary objectives and plans
- Description of the environmental setting for the planned development
- Description of the proposed aquaculture practice and species to be cultured
- Review of the conceptual model and the assessment end points
- Discussion of the major data sources and analytical procedures used
- Review of the stressor response and exposure profiles
- Description of the risk to the assessment end points, including risk estimates and adversity evaluations
- Review and summary of major areas of uncertainty, and their direction, and approaches used to address them, such as:
  - Discussion of the degree of scientific consensus in key areas of uncertainty
  - Identification of major gaps and, where appropriate, indicate whether gathering additional data would add significantly to the overall confidence in the assessment results
  - Estimation of the risk probability by combining numerical data
  - Discussion of science policy judgments or default assumptions used to bridge information gaps and the basis for the assumptions
  - Discussion of how elements of quantitative uncertainty analysis are embedded in the estimate of risk

Problem Formulation for Marine Fish Aquaculture (Phase 1)

The first phase is problem formulation, or the identification of key factors to be considered in the risk assessment. Here all the necessary plans are made by the risk managers and risk assessors to determine how the analysis will be performed. These include, for example:

- the scope, focus, and sources to be considered (such as the type of marine aquaculture, and species);
- the biological or ecological end points and their attributes that are the concern for protection (such as sea grass preservation, maintenance of water quality, avoidance of low dissolved oxygen, avoidance of eutrophication, etc.);
- a conceptual model or diagram of how the system being assessed is thought to be organized; and finally,
- the plan for analyzing the information and conducting the rest of the assessment.

Problem formulation can be a long and difficult process. It depends on the degree of familiarity with the particular field of aquaculture, how contentious are any issues, and finally who is involved. Unfamiliar problems, such as the location of marine fish cages in the migratory routes or breeding grounds of cetaceans, unquestionably take longer to formulate compared with, say, the location of a land-based marine fish hatchery adjacent to an existing recreational marina or fish processing plant.

Modern marine fish aquaculture has been evolving for almost 50 years. Consequently, considerable experience has been building with regard to any real or perceived impact on marine ecosystems all over the world. Most of the practical knowledge and experience by fish farmers themselves has never been recorded, although some has been documented in gray literature, but a considerable volume of scientific and technical research can now be found in peer-reviewed journals. With this growing background information to draw on, it is possible for risk managers and risk assessors to undertake a very comprehensive problem formulation.

For the purpose of these guidelines the possible observed or perceived effects of marine aquaculture have been summarized in 10 categories (Table 3). Within these broad designations it is not possible to include all the possible effects which might be identifiable globally, and consequently the guidelines concentrate on the sources of effects, and the end points or entities of concern together with their attributes, of known importance to the majority of marine ecosystems. A risk assessment can include any number of other effects, but practical experience suggests that the 10 categories and their contents illustrated here provide a strong starting point. The biological end points of these possible effects are generalized in the following paragraph.

Biological end points of marine fish aquaculture and their attributes can be described in collective terms (such as the species abundance of the infauna), or very specifically by location (such as the discovery of giant tubeworms at hydrothermal vents). They may also be assessed generally (such as by the presence of certain species in the epifauna), or by specific measurements (such as by n, µg/g, or µg/L).

The end points identified in these guidelines for protection from marine fish aquaculture activities may include:

- the species richness and abundance of the seston, nekton, or infauna,
- the abundance of a specific species in the seston, nekton, or infauna,
- the species richness and abundance of the epifauna,
- the abundance of a specific species in the epifauna,
- the abundance of a specific species of marine mammal, reptile, or bird,
- the immune resistance of demersal and pelagic fishes,
- the number and fitness of the natural (conspecific) population,
- the fitness of another fish population, and
- the abundance of the industrial fisheries.

The choice of species may be guided by whether one is looking for a surrogate for system stressors, system response, or protection of some desirable biological attribute. Thus one might measure a toxic phytoplankton species because of the desire to avoid blooms of harmful or nuisance species, or one might choose a species that is indicative of degraded environmental condition (e.g., capitellid worms or the presence of *Beggiatoa* spp. in sediments), or one might measure sea grass distribution because of its high protection status.

Problem Analysis for Marine Fish Aquaculture (Phase 2)

Problem analysis is the second phase of risk assessment when all available scientific information relevant to the issue is collected and applied. For the most part it is carried out by technical experts. Problem analysis is divided into two parts. The first is the analysis of exposure, which predicts or measures the spatial and temporal distribution of a stressor and a point of concern; the second is the analysis of effects (sometimes called the exposure response), which identifies and quantifies any adverse effects caused by a stressor.

Characterizing the Background of an Aquaculture Site

It is important to know the characterization of the marine site(s) where the stressor originates and where it may have its adverse effects. Therefore the first step is a baseline survey, or stock-taking, of information about the near field, and in some cases the far field. The survey is in two parts, namely, collecting information through a literature search followed by assembling current information and data by field work.

Table 3. Categorization of observed or perceived effects associated with marine fish aquaculture, and the identifiable sources of the stressor

Effects	Sources

1. Increased organic loading	• Particulate organic loading
	  • Fish fecal material
	  • Uneaten fish feed
	  • Debris from biofouling organisms
	  • Decomposed fish mortalities on the farm
	• Soluble organic loading
	  • Dissolved components of uneaten feed
	  • Harvest wastes (blood)

2. Increased inorganic loading	• Nitrogen and phosphorus from fish excretory products • Trace elements and micronutrients (e.g., vitamins) in fish fecal matter and uneaten feed
3. Residual heavy metals	• Zinc compounds in fish fecal material • Zinc compounds in uneaten feed • Copper compounds in antifouling treatments
4. The transmission of disease organisms	• Indigenous parasites and pathogens • Exotic parasites and pathogens
5. Residual therapeutants	• Treatment by inoculation • Treatment in feed • Treatment in baths
6. Biological interaction of escapes with wild populations	• Unplanned release of farmed fish • Unplanned release of gametes and fertile eggs • Cross infection of parasites and pathogens • Planned release of cultured fish for enhancement or ranching
7. Physical interaction with marine wildlife	• Entanglement with lost nets and other jetsam • Entanglement with nets in place, structures, and moorings, etc. • Attraction of wildlife species (fish, birds, marine mammals, reptiles) • Predator control
8. Physical impact on marine habitat	• Buoyant fish containment structures and mooring lines • Anchors and moorings
9. Using wild juveniles for grow-out	• Harvest of target and nontarget species as larvae, juveniles, and subadults
10. Harvesting industrial fisheries for fish feed	• Increased fishing pressure on the shoaling small pelagic fish populations

Historical Information

A valuable part of the baseline survey is a search of existing literature of water and sediment quality parameters. These include, for example, data on water temperatures, salinity, dissolved oxygen, stratification, bottom currents, water depth, background nutrient concentrations, phytoplankton species and chlorophyll, sediment grain size, and organic matter content. In those cases where information is not available, then a program of data collection should be initiated to fill the gaps. It is hard to be prescriptive about spatial and temporal scales of measurement, but measurement of some water quality parameters may need to be taken on a weekly basis during seasons of high phytoplankton productivity.

Some additional information might be available on the background levels of contaminants in both the water and in the sediments. These include, for example, metals, and organics such as hydrocarbons, pesticides, polychlorinated biphenyls (PCBs), and polycyclic aromatic hydrocarbons (PAHs), etc. This information is particularly important (and more likely to be available) in near-shore coastal areas where there are significant anthropogenic inputs from agricultural and urban areas. In open waters there is little potential for the accumulation or discharge of these types of contaminants, and the need is reduced.

Finally, any documentation providing a broad description of the natural history of the area, together with any reports or local knowledge of the potential for noxious phytoplankton blooms or the prevalence and intensity of known parasites are potentially useful. Information on the incidence of blooms and parasites is more likely if there are commercial shellfish resources in the area.

Current Information

A typical baseline survey of current information for the lease area will include most of the items from the following checklist:

1. Identification of sensitive habitats. These may include, for example, beds of macroalgae and eelgrass, coral reefs, commercially valuable shellfish beds, spawning grounds and breeding areas, migratory pathways of aquatic species, rocky reef communities, and all other structures valuable as nurseries. Such habitats within 500 m of a proposed intensive farm site should be mapped, with the intention of avoiding them whenever possible.
2. The background physico-chemistry of the sediments. This may include, for example, total volatile solids (TVS) or organic matter content, redox potential (Eh), sediment grain size (SGS), free sulfide ($S^{=}$), and the two inorganic metals copper and zinc.
3. An inventory of the species and abundance of the macrobenthic communities. This may be carried out by stratification, or by the type of habitat.
4. The hydrographic variables, such as currents, tides and residence times, including acoustic doppler current profiler (ADCP) data collected over at least one lunar cycle, and bathymetry within 500 m of the proposed site.
5. A profile of water quality, including temperature, salinity, and the potential for stratification as a function of season (pycnoclines and haloclines).
6. A profile of primary productivity, including major species (including any toxic species), chlorophyll (Chl_a), phaeophytin, and dissolved oxygen (DO).
7. If possible, underwater surveys recorded on a video or a series of photographs to provide an overall, semiquantitative assessment of the benthic environment of the site, especially in deep water.
8. Finally, identification of activities by other resource users, such as marine sanctuaries, marine protected areas, fishing grounds, recreational areas, navigational

channels, oil and mineral extraction, military training areas, approved dumping grounds, etc.

The grid on which this information for the baseline survey is to be collected depends on the homogeneity of the system. A regression approach is recommended with single samples collected at intervals on four orthogonal transects beginning at the center of the proposed farm location. Samples should extend at least 500 m from the center. If video surveys are conducted first, the grab collections can be focused in areas where samples are possible, namely soft to mixed substrates. About 24 samples are adequate.

The profile of the macrobenthic community can be reduced in cost by using the smaller petite ponar grab (with a 0.0225 m^2 footprint) rather than the more standard van Veen grab (0.1 m^2).

Near-Field and Far-field Effects

Effects of aquaculture interventions on the ecosystem are spatial and temporal. They can be localized and immediate, or distant and sometime in the future. However, both near-field and far-field effects have to be considered in the risk assessment process.

Near-Field Effects

The near field can be defined as that area encompassing the limit of directly measurable effects. In the marine environment, the majority of human interventions, such as sand mining, dredging, drilling, waste disposal, fish processing, and recreational boating, etc., all have instant near-field effects, particularly on the sediments and their benthic communities in the immediate vicinity of the source. Consequently, because of the long history of these activities in marine waters, the extent and diversity of their effects are well known. They can be measured with accuracy, and the particulate data and benthic biological data linked in a number of empirical or mechanistic models to assess potential risk.

With regard to the relatively recent intervention of aquaculture in the marine environment, and its most localized and instant impact of wastes and contaminants accumulating on the bottom sediment beneath fish enclosures or in solution, there is a wealth of comparative information about the measurement of near-field effects on which to draw. For example: 1) in terms of sedimented organic waste, the near field describes that area in which statistically significant differences (t-tests, ANOVA, etc.) or significant clines (statistically significant coefficients on dependent variables in linear or nonlinear regression analysis) in either physico-chemical or biological end points associated with aquaculture-related effects can be demonstrated at the peak of farm production; and 2) in terms of reduced concentrations of dissolved contaminants or effects of metabolic waste, the near field describes that area in which statistically significant increases or decreases in the end point of interest can be measured in comparison with local reference conditions.

Because of the extent of good data, near-field effects are generally assessed using local computer models to predict the deposition of organic material released by the producer. The DEPOMOD computer modeling tool, for example, models benthic enrichment effects by combining particle tracking with empirical relationships between the spatial distribution of solids and changes in the structure of the benthic community.

Near-field effects are usually limited or managed by regulatory authorities setting performance standards, which are appropriate for the location or the region as a whole.

Typically, under the terms of a permit or license, the producer is responsible for conducting the necessary monitoring and complying with the management practices adopted to enable the performance standards to be met.

Far-Field Effects

Far-field effects are those effects that occur outside that area where statistically significant clines in relationship with the source cannot be measured. These are cumulative effects that normally can only be detected by long-term monitoring programs at locations not directly influenced by local effects. Assessment of far-field effects associated with aquaculture becomes increasingly important as the industry expands.

The maximum spatial extent of far-field effects is a hydrologic unit that includes all inputs potentially affecting the unit. It may include, for example, a single bay, several bays, or an entire estuary or delta. Far-field effects become increasingly difficult to measure in open bodies of water, such as those offshore where aquaculture may occur. However, even in large open bodies of water the same definitions could be applied.

Because of the vast scope of far-field effects, their potential is normally best assessed through computer models. These are monitored by consortiums of contributors to the cumulative effects in coordination with some level of government. Management of far-field effects is normally a public function in cooperation with all the contributors. With regard to organic loading, for example, from a number of marine fish farms into a bay 10 km distant, the regulatory authority may set Total Maximum Daily Loads (TDML) for the far field of interest (the bay), and apportion the TMDL to individual producers or farm complexes. The authority then manages the far-field effects by manipulating the respective TMDLs to meet one stated objective.

Risk Characterization for Marine Fish Aquaculture (Phase 3)

Risk characterization is the final phase when the two analyses of exposure and effects are brought together. It is best performed using models developed to estimate effects from hypothetical risks.

In a number of fields, such as the pharmaceutical industry or chemical engineering, risk characterization can be straightforward. The point estimate of exposure is compared with the point estimate of the threshold of effects, and if the ratio is greater than one then an effect is assumed. It can be taken further with an exposure-response model, when the distribution of the exposure and effects can be shown to accumulate over a period of time. However, in the marine aquaculture industry the process of risk characterization is complicated by the fact that most of the effects are interactive. Such complexity could be dealt with by modeling, but quantifiable information for many aspects of marine aquaculture is extremely scarce. Consequently, for risk characterization the only recourse at present is either to make use of a mechanistic model for a particular site, providing the assumptions are reasonable and that the model can be adequately calibrated and validated, or to rely on all existing information and especially the classical "dose and response" laboratory information.

In assessing a risk it is important both to qualify and quantify, where possible, the associated uncertainty. For example, the uncertainty could be described by probabilistic factors, by semiquantitative factors, or entirely qualitative factors, such as high, medium, or low. Whatever factors are chosen, it is important to include the uncertainty with any risk assessment. In addition, it is important to explain any assumptions which were used in the analysis, the scientific uncertainties, and their strengths and weaknesses.

Risk characterization is carried out by scientific and technical experts, but it is not limited to them. Risk assessors and risk managers are again actively involved in the process, as during problem formulation. This is because issues might have arisen which necessitate a reiteration of problem formulation and a repeat of the problem analysis.

Risk Communication

A final responsibility for everyone involved in managing risk is risk communication. This is an ongoing process at the local level and usually involves a government agency, represented by risk managers, industry and other stakeholders, and the public at large.

The objective of risk communication is to maximize the transparency of every activity related to the risk through interaction with the broadest range of interested parties (Figure 2). This objective includes risk identification, analysis, assessment, implementation of the decision, and subsequent monitoring. It is important that the communication process is begun as soon as possible, preferably with an announcement of the project itself.

Risk communication is carried out in a variety of ways. Productive communication is invariably conducted at public hearings when, in theory, everyone listens carefully to each other without any prejudgment of the issue. But this is not always the case, and it is important for the risk managers representing government agencies at such hearings to maintain public trust by their independence and impartiality. Good communication is also achieved by regularly circulating published materials.

Some aspects of risk assessment are scientific and very technical, and therefore it is important that the data and all methods of collection, any models and assumptions that have been applied, and any conclusions drawn are reviewed by peers.

Monitoring for Subsequent Risk

Decisions can be made by the risk manager based on the historical and current information gathered by the team of risk assessors and stakeholders. If the potential risk is assessed as being unlikely, or small, then the risk manager can authorize the project to go ahead. However, it is important that the baseline does not change in such a way that the risk can in fact occur at a later time, and therefore the risk manager usually qualifies any decision with the requirement for the continual monitoring of certain site parameters. The task of carrying out the monitoring program may be the responsibility of the regulatory agency, the owners or managers of the project in question, or both.

It is important that any monitoring program is designed around the measurement of:

- standards identified by national legislation and regulation, and
- those parameters relevant to the indication of any increasing risk to the biological end points that have been identified.

Fundamental also to every monitoring program is an exact specification of the methodology. This, for the most part, should have been established during the baseline survey. In other words, reference stations and site stations will be located and fixed along transects on the seabed or at set surface or mid-water distances from identifiable points (such as the perimeter of a facility), and all based on the predominant direction of the current. In addition, the frequency and methods of sampling will be specified, and the methods of analysis will be identified together, where necessary, with laboratory instrumentation.

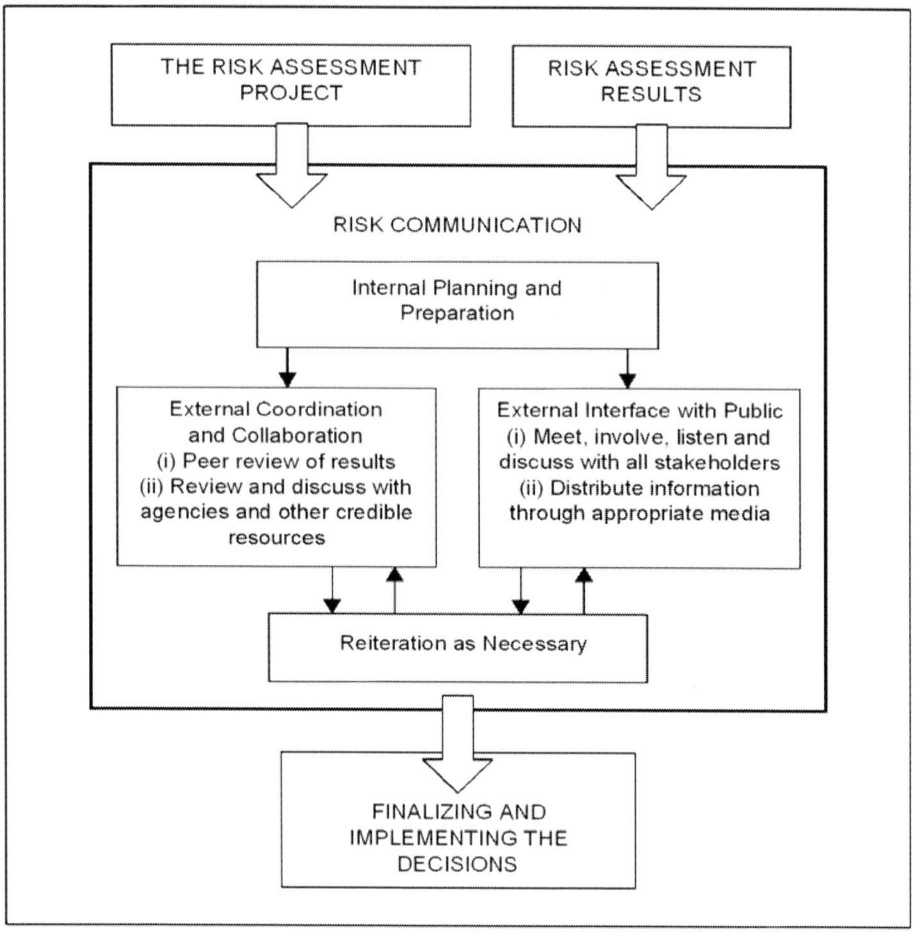

Figure 2. The process of risk communication for the project and the results.

GLOBAL APPLICATION OF THE FRAMEWORK

Physical Demands of Marine Fish Aquaculture

For the foreseeable future, intensive marine fish aquaculture will be limited to waters of the continental shelf, which is often defined as lying above the 200-m contour. However, for the practical reasons of engineering cost, operational management, and profitability, marine fish aquaculture takes place reasonably close to shore, provided that water quality conditions are suitable.

Selection of a location depends on the proposed fish farming system and practice. Again, because of the investment cost, only intensive fish production is economically feasible, and the options are floating net-pen complexes and buoyant individual cages designed to remain at the surface or to be submerged as required. Net-pen complexes are therefore usually located in coastal estuaries, sounds, and lagoons that have rapid marine water exchange, have some shelter, and provide anchorages that are less than 40 m deep. Individual buoyant cages can be located in less-sheltered waters, and submersible cages can be deployed in deeper water to avoid storms. However, submersible cages have limitations. Although wave energy attenuates with depth, the scale of each unit is limited by potential fatigue of the materials, the capacity of the automated feeders, and the need for regular surveillance and service operations by scuba divers. Scuba divers can operate safely down to a depth of 30 m, but operate most economically around 10–15 m, and working in pairs. Currently, submersible cages are being operated at depths of less than 100 m, but this may still be up to 30 km offshore. Net-pen complexes are anchored by many separate cables, depending on their formation and size. Additional lines may anchor predator nets. Individual buoyant cages are anchored by four discrete lines which maintain tension all around continuously. Single-point anchor systems have also been used, but at some time the line will become slack, which puts a burden on the cage/line interface. The preferred substrate for the anchors themselves is sand or mud. Anchors can be bolted into rocky substrates, but the practice is costly.

Buoyant cages are designed to operate in currents up to 90 cm/sec, or about 1.74 knots. This is above what is desirable for the fish, which, when confined in strong currents, expend too much energy maintaining their position in the cage instead of growth.

Environmental Demands of Marine Fish Aquaculture

Successful marine fish aquaculture depends on a synergism between the aquaculture site and the farmer. The environmental qualities or parameters of the site must be conducive to the life history and physiology of the species of fish in culture, and the operator must provide an appropriate living space for the fish, meet all their nutritional requirements, and maintain their health. Site selection for an aquaculture facility is therefore a critical task. It is made difficult because the range of marine ecosystems in which it may be located is diverse, and the suitability of their physical and chemical properties depend significantly on the species and culture practice to be implemented. For example, there are different site demands for submersible cages containing cobia 3–5 km from the coast of Puerto Rico, pens for growing-out tuna in coastal waters within 2 km of the shoreline of Australia, and enclosures for rearing sea bream in shallow marine embayments in the Mediterranean.

The hydrodynamics, nutrient levels, types of pollution, and other environmental parameters found in these locations are all very different. Consequently, there will be differences in the biological end points and their attributes resulting from aquaculture operations that characterize the potential risks to the environment. For example, the risk of eutrophication and change in species diversity in the benthic environment in the poorly flushed lagoons of the Mediterranean is higher than the offshore waters of either Puerto Rico or Australia where there are greater depths and high water exchange rates. Because of all these differences, each ecological risk assessment has to be tailored to an individual location, and an individual species and aquaculture practice. However, the categories of potential ecological risks and their fundamental methods of assessment are common, and it is only their relative importance that will vary.

A Matrix Approach to Guide the Application of Risk Assessments

In selecting a suitable site for marine fish culture, the ideal requirement is a pollution-free environment in the epipelagic zone with good water quality parameters. Primarily this means year-round high ambient levels of oxygen combined with salinities and temperatures that are between the middle and upper end of the ranges tolerated by the respective farm species, and maintained by a modest current and average tidal rise and fall. Unfortunately the ideal cannot always be found, and the parameters are so diverse that most sites are selected for reasons somewhere between ideal water quality parameters and operational cost and convenience. As marine fish aquaculture is still in its infancy in most countries, and the locations where it is practiced at the present time are few, for the purpose of these guidelines it is proposed to classify the typical marine aquaculture environment into categories of biogeographical regions or zones and categories of marine epipelagic ecosystem. The definitions of the zones and categories are as follows: are:

The two biogeographical zones suitable for marine aquaculture (as illustrated in Figure 3)

- Temperate waters (10–18°C). Typically cold waters with intrusions of some warmer waters from the subtropics. Temperate waters can be rich in nutrients and highly productive (waters off Australia being an exception), and consequently characterized by low light intensity levels. Temperate waters often support substantial fisheries, together with their dependent populations of birds and marine mammals.
- Tropical waters (>18°C). Typically warm waters with intrusions of some colder waters from the subtropics. Tropical waters are biologically very rich but nutrient poor and characterized by high light levels. Tropical waters often support migratory populations.

The three epipelagic ecosystems are:

1. Offshore waters. Typically 3 km or more from the coast, or up to 100 m in depth, and suitable for submersible cages.
2. Coastal waters. Typically less than 3 km from the coast, or up to 30 m in depth, suitable for submersible cages and floating cages, with strong tidal interchange.
3. Inshore water bodies. Typically semienclosed but large coastal sounds, lagoons, and estuaries, relatively shallow in depth, suitable for floating cages and fixed enclosures, with good tidal flushing.

The 10 categories of risk can then be evaluated in broad terms against each of the 6 generalized marine ecosystems in the form of a matrix (Table 4). The objective is to indicate probable differences in priority relative to each type of ecosystem, and to assist risk managers and risk assessors with their problem formulation. However, the information presented in the matrix does not rule out the uniqueness of some ecosystems, and this most always be considered.

Table 4. Matrix to guide the application of risk assessments in the waters of different biogeographic zones

Category of observed or perceived risk	Epipelagic ecosystem in temperate waters (10–18°C)			Epipelagic ecosystem in tropical waters (>18°C)		
	Inshore	Coastal	Offshore	Inshore	Coastal	Offshore
1. Increased organic loading	*****	**	*	*****	***	*
2. Increased inorganic loading	*****	**	*	*****	***	*
3. Residual heavy metals	*	*	*	**	*	*
4. Transmission of disease organisms	***	**	**	***	**	**
5. Residual therapeutants	**	*	*	**	*	*
6. Biological interactions of escapes with wild populations	**	**	*	**	**	*
7. Physical interactions with marine wildlife	**	**	*	**	**	*
8. Physical impact on marine habitat	**	*	*	**	*	*
9. Using wild juveniles for grow-out	**	**	*	***	***	**
10. Harvesting industrial fisheries for fish feed	**	**	***	***	***	***

Key: Potential for ecological change without management action.
***** Significantly high.
**** High.
*** Medium.
** Low.
* Little or none.

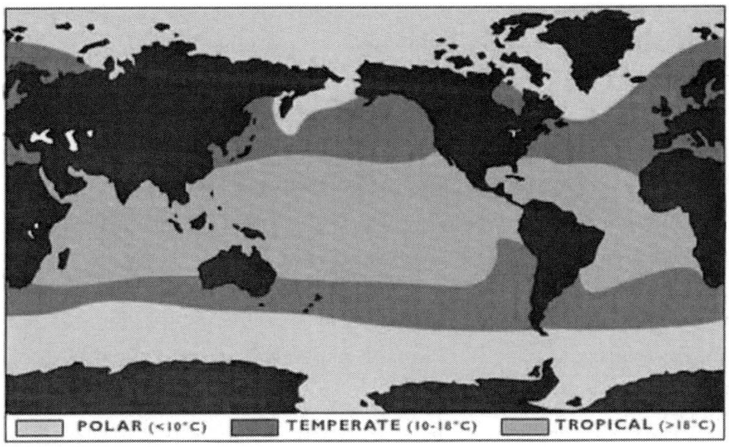

Figure 3. Broad biogeographical zones for marine aquaculture (courtesy of the Gulf of Maine Research Institute).

Glossary

Risk Assessment Terms

Adverse ecological effects. Changes that are considered undesirable because they alter valued structural or functional characteristics of ecosystems or their components. An evaluation of adversity may consider the type, intensity, and scale of the effect as well as the potential for recovery.

Assessment end point. An explicit expression of the environmental value that is to be protected, operationally defined by an ecological entity and its attributes. For example, marine turtles are valued ecological entities, and the survival of individual migrating turtles is an important attribute.

Attribute. A quality or characteristic of an ecological entity. An attribute is one component of an assessment end point.

Characterization of ecological effects. A portion of the analysis phase of ecological risk assessment that evaluates the ability of stressor(s) to cause adverse effects under a particular set of circumstances.

Characterization of exposure. A portion of the analysis phase of ecological risk assessment that evaluates the interaction of the stressor with one or more ecological entities. Exposure can be expressed as co-occurrence or contact, depending on the stressor and ecological component involved.

Community. An assemblage of populations of different species within a specified location in space and time.

Conceptual model. In problem formulation, a visual representation and written description of predicted relationships between ecological entities and the stressors to which they may be exposed.

Ecological entity. A general term that may refer to a species, a group of species, an ecosystem function or characteristic, or a specific habitat. An ecological entity is one component of an assessment end point.

Ecological risk assessment. The process that evaluates the likelihood that adverse ecological effects may occur or are occurring as a result of exposure to one or more stressors.

Ecosystem. The biotic community and abiotic environment within a specified location in space and time.

Exposure. The contact or co-occurrence of a stressor with a receptor.

LC50. A statistically or graphically estimated concentration that is expected to be lethal to 50% of a group of organisms under specified conditions.

Measure of effect. A change in an attribute of an assessment end point or its surrogate in response to a stressor to which it is exposed.

Measure of exposure. A measure of stressor existence and movement in the environment and its contact or co-existence with the assessment end point.

Population. An aggregate of individuals of a species within a specified location in space and time.

Receptor. The ecological entity exposed to the stressor.

Recovery. The rate and extent of return of a population or community to some aspect(s) of its previous condition.

Risk characterization. A phase of ecological risk assessment that integrates the exposure and stressor-response profiles to evaluate the likelihood of adverse ecological effects associated with exposure to a stressor.

Source. An entity or action that releases to the environment or imposes on the environment a chemical, physical, or biological stressor or stressors.

Stressor. Any physical, chemical, or biological entity that can induce an adverse response.

(Source of risk assessment terms: U.S. EPA, 1992, Guidelines for Ecological Risk Assessment)

Marine Terms

Benthos. Collectively all those animals and plants living on or in sediments at the bottom of the sea. Benthic animals are usually described by their position in the sediment relative to the surface, and their size, i.e.:

1. **Infauna.** Fauna living within (burrowing in) the sediments,
2. **Epifauna.** Fauna living at or on the sediment surface. They can be sessile or slow moving, and may spend some time in the water column.

Bioremediation. Biological recovery.

Demersal. Living on or near the bottom of the sea.

Epipelagic. Pertaining to the community of suspended organisms inhabiting an aquatic environment between the surface and a depth of 200 m.

Halocline. Well-defined vertical salinity gradient in the water column.

Nekton. Collectively the macroscopic animals suspended in the sea, moving about independently of currents (includes fishes and whales).

Pelagic. Of or pertaining to the open waters of the sea (beyond 20 m depth).

Porewater. The water retained in the pores between the grains of the sediment.

Pycnocline. Well-defined vertical density gradient in the water column.

Seston. Collectively all living and dead suspended microscopic animals and particulate matter in the sea.

APPENDICES

[Editor's note: Appendices A through J are templates that outline the approach for conducting a risk assessment for each of the 10 areas of marine fish aquaculture perceived by the public and public administrators to be of most concern. Figures A-1 through J-1 are flowcharts for the conceptual models. The templates are not presented here in any order of priority.]

APPENDIX A. INCREASED ORGANIC LOADING

Risk Hypothesis

Particulate organic waste produced by the high concentration of biomass at any marine fish farm is deposited directly on the substrate beneath the farm, or is distributed downstream in suspension or solution. The organic matter is mostly made up of particulate fecal matter and uneaten feed, but may also include the accumulation of marine fouling organisms falling or purposely removed from the units, together with decomposed mortalities and possibly some harvest wastes. The perceived risk is that the accumulation of organic matter may reach a point where the alterations in sediment chemistry become noxious to a particularly valuable macrobenthic fauna or to exclude an unacceptable portion of the benthic community.

Background Experience

Physico-chemical and Biological Effects of Organic Enrichment

Organic wastes from aquaculture are formed mainly from feces and uneaten feed and are primarily made up of carbon compounds. These wastes break down quickly, increasing the biological oxygen demand (BOD), which can in turn reduce the oxygen content of the sediment. Naturally occurring bacteria then strip oxygen from sulfate, leaving sulfide, a common form of sulfur found in low concentrations in marine environments. Some carbon and sulfide enrichment of benthic sediments should be expected in nearly all forms of intensive aquatic animal production, but the response by the benthic invertebrates differs widely. When enrichment becomes too great, sensitive species may be excluded and opportunistic species proliferate, thus changing benthic communities.

Organic enrichment associated with labile aquaculture waste creates BOD in sediments that can exceed the assimilative capacity of the local environment, especially in a less-dispersive environment. In these instances, the added organic material, that is, biodeposits measured as total volatile solids (TVS), or organic matter content is an additional source of food for macrobenthos, leading to increased community abundance, and increased macrobenthic biomass. Increasing BOD creates physicochemical changes in sediments characterized by increased TVS and free sulfides, and decreased redox potential. These physico-chemical changes can force biological changes on specific taxa, which can alter macrobenthic communities. Benthic faunas sensitive to enrichment, such as brittlestars (Ophiuroidea), may be extinguished from the area below a net-pen farm, and replaced by organic-tolerant opportunistic annelids and capitellids.

Spatial Extent of Effects

There is equivocal evidence that the effects of volatile organic material may extend beyond the perimeter of a farm complex, especially when fish stocking densities are high. These effects, which can be measured during point-in-time surveys, are generally referred to as near- field effects, and their extent and degree of benthic impact are determined primarily by the degree of flushing in and around a facility, and to a lesser extent with the biomass of the animals being cultured.

Appropriate siting of intensive aquaculture facilities is critical to the management of any potential benthic effects. Computer models, such as DEPOMOD, are available to assist in predicting the extent and degree of organic deposition at proposed facilities. The spatial extent of measurable effects has only been documented from the area under intensive systems to about 205 m down current from their perimeter. Significant adverse effects, when they occur, have generally been restricted to distances less than 60 m from the perimeter of intensive cultures.

As the number of aquaculture facilities in a coastal ecosystem increases, the potential for far-field effects increases. These effects are the result of all of the cumulative organic discharges from both upland and water-dependent activities. While near-field effects generally occur and remediate quickly, far-field effects are more subtle and require long-term monitoring programs or computer modeling for detection. However, because far-field effects can be widespread and may be long-lasting, a precautionary approach to their management is warranted, increasing the need for careful risk analysis as aquaculture industries expand and mature—particularly in small or restricted bodies of coastal water.

Temporal Aspects of Benthic Effects

In terms of risk, prolonged case studies reveal that the physico-chemical and biological changes in the benthic sediments associated with aquaculture sites are reversible. Chemical recovery, as characterized by a return to preproduction concentrations of TVS, reduced free sulfide concentrations, and increased redox potential to positive values, normally occurs within periods of less than a year for return to pristine conditions at well-flushed sites, but longer at those which are poorly flushed. Biological recovery occurs concurrently with chemical recovery, but frequently takes an additional invertebrate recruiting season to develop fully.

In summary, the effects of organic enrichment in sediments of well-chosen sites appear to be local and reversible, provided that appropriate management practices, such as stocking density and fallowing regimes, are followed. However, as the number of facilities and the overall biomass of cultured organisms increase, there is need for risk analyses assessing far-field, cumulative effects associated with all forms of organic discharge to coastal ecosystems.

Building the Conceptual Model

The sources of organic wastes from marine fish aquaculture operations are the fecal material of the farm fish, any uneaten fish feed passing through the enclosure, dead cultured fish that might be left to decompose, and biofouling debris that might fall naturally from the accumulation on the structures or during cleaning. This waste is dispersed by currents.

During its fall to the sea bed, or immediately following disturbance of the sea bed, a small part of the solid organic waste will become suspended as particulate matter or be dissolved in solution. The majority of the solid waste, however, will fall directly and accumulate in the underlying sediments.

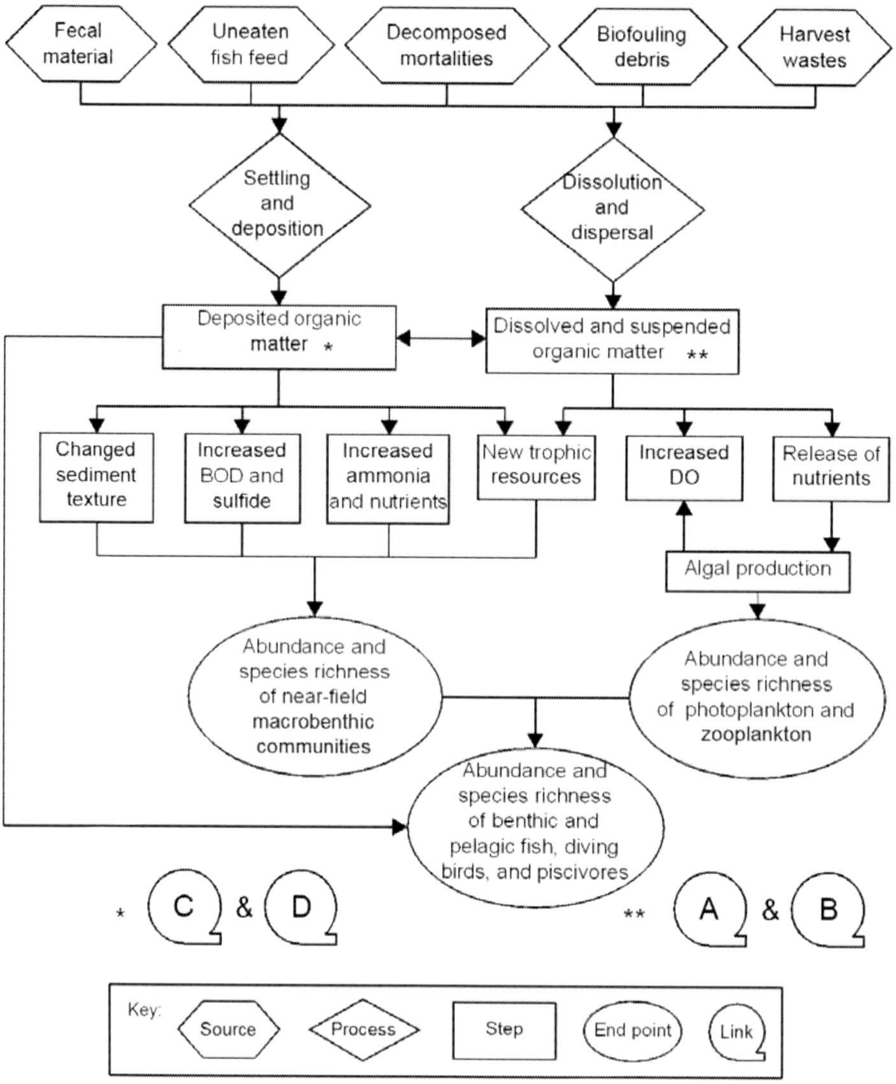

Figure A-1. A conceptual model for increased organic loading.

The organic matter deposited on the sea bed has a number of effects. It may change the texture of the sediment physically by covering and infiltrating the original sediment. It may also change the chemistry of the sediment by introducing metals, such as copper and zinc (see Appendix C), which may be toxic, and by increasing the BOD, ultimately inducing anaerobic conditions. These effects in turn lead to changes in the microbial communities with nitrates reduced to ammonia and sulfates to hydrogen sulfide. Finally, an excess of biodeposits can cause dramatic changes to the sediments biologically by attracting different species and eliminating others.

The first end point with its attributes of all these effects is the species richness and abundance of the benthic microinvertebrates, including both infauna and epifauna. As these fauna are at the base of aquatic food webs, subsequent end points and their attributes are the species richness of benthic fish and pelagic fish, together with diving birds and other piscivores. These populations can also be affected directly by the added food resource provided by feces and uneaten feed.

For the most part, the small volume of solid organic waste that becomes suspended particulate matter or is dissolved has effects which are described in more detail in Appendix B. The first effects include the release of nutrients into the water column, particularly nitrogen and phosphorus; oxygen levels may be affected, and the changes may attract different trophic resources. Secondly, the availability of more nutrients increases the production of phytoplankton, but these effects are much less than those caused by the direct release of ammonia and other nutrients into the water column from fish metabolism. However, sediments act as a reservoir of potential nutrients and can be a source of efflux long after the fish pens have been removed.

Analysis and Characterization

Although marine fish farms are preferably sited where a current maintains a high rate of water exchange, the evidence is that nearly all particulate organic waste is deposited within a few tens of meters from the perimeter of the containment structure. Therefore the effects are localized and accumulative.

The biological end points and their attributes for protection are:

1) the abundance and species richness of the near-field macrobenthic communities,
2) the abundance and species richness of phytoplankton and zooplankton, and
3) the abundance and species richness of benthic and pelagic fish, diving birds, and piscivores.

Macrobenthic Communities

Near-field effects associated with particulate organic waste have been studied in detail for over 20 years, and the effects are well understood. Consequently there is a rich body of literature describing the physicochemical and biological response to varying degrees of sediment enrichment.

The affected community is the macrobenthos—including infauna and epifauna. Exceeding the assimilative capacity of the benthos leads to physico-chemical changes in the sediment that include increased BOD and decreased redox potential. Biological changes include a shift to anaerobic microbial communities (e.g., *Desulfovibrio* spp. and ultimately *Beggiatoa* spp. mats), resulting in increased concentrations of free sulfide and ammonia, which are toxic to individual taxa at varying concentrations. It has been shown that an increase in free sulfides ($S^{=}$, HS, and H_2S) from background concentrations of less than 100–300 μM to 1,000 μM results in the exclusion of half the reference number of taxa in annelid-dominated communities. In mollusk-dominated communities, increases from background to 450 μM have resulted in a similar decrease in species richness of 50%. The extent of these changes depends on fish production levels, water currents, sediment grain size, water depth, and other factors.

Computer models, such as DEPOMOD, are readily available to predict the deposition of organic matter around aquaculture facilities, given that appropriate information is available describing bathymetry and hydrodynamics of the site and the necessary inputs regarding the physiology of the cultured fish species. Measurable effects, based on the evidence from coastal salmon aquaculture, typically extend to distances of 150–200 m from the perimeter of the culture containment system, at water depths of 30–100 m, average current speeds of 3–10 cm/sec, and maximum current speeds of 25–75 cm/sec.

Performance standards prescribed by regulatory authorities are usually defined to establish the limits of the allowable effects. The output from a model such as DEPOMOD can then be used to assess the probability that the characteristics of a proposed aquaculture facility (such as species, size at harvest, total production, type of feed, feed conversion ratio, etc.) will be able to comply with the stated performance standards.

Phytoplankton and Zooplankton

The effects of increased organic loading on phytoplankton and zooplankton populations in the water column are generally much less than those attributable to direct nutrient release into the water column, and should be considered as part of increased inorganic loading (see Appendix B). While some specific effects can be noted, such as an increase in benthic copepods and isopods that feed directly on the deposited organic matter, these are usually of limited concern for the whole ecosystem. Some phytoplankton species migrate vertically through the water column and these may be able to access benthic nutrients not readily available to species higher in the water column, thus conferring an ecological advantage.

Benthic and Pelagic Fish

Studies with baited traps indicate that scavenger fish (usually benthic species) can be attracted to the food accumulating under net-pens. This can lead to increased populations of specific species and altered food-web dynamics. It would be unusual for these effects to lead to system-wide changes, although clearly the number of fish pens and amounts of uneaten feed are key factors here.

Biological Opinion

Organic wastes deposited on the marine substrate can have definite effects on sediment chemistry and macrobenthic invertebrates, and therefore sediment organic enrichment is the primary local effect of concern for most marine fish aquaculture sites. The principal constituents of wastes from fish farm enclosures are feces and uneaten feed. About 12% by weight of ingested feed is ejected as feces, as modern diets are about 87–88% digestible, and 3–5% of feed can be lost to the environment. As feed costs are the highest operating cost, feeding statistics are rigorously followed and monitored in detail by calculating conversion rates, observing with underwater telemetry, and using new technologies of feed delivery to minimize wastage.

Other possible constituents of organic wastes and biodeposits are minor by comparison. Few fish carcasses remain to disintegrate and pass down to the sediment. Mortalities, which are typically less than 10% annually, are usually removed manually every day and disposed of appropriately. Similarly, modern best management practices require all nets to be removed and cleaned of fouling organisms onshore, and therefore there is no longer a significant residue being placed on the benthos, and harvesting wastes and waters are typically retained on the harvest vessel or at an onshore facility.

There are local effects associated with high-density aggregations of marine fish. But risks can be minimized through proper site selection, which can be enhanced by adequate preproduction risk assessments ensuring that effects on particularly valuable resources are avoided and that environmental risks to the cultured species are acceptable. Near-field hazards can be effectively managed using performance standards that producers are required to meet. Inexpensive physicochemical surrogates have been frequently substituted for more expensive and time consuming biological evaluations and standards. The costs of near-field risk assessments and monitoring are normally defrayed by the producer.

In summary, benthic effects created by intensive aquaculture are local and reversible. The effects can be minimized by careful site selection and site operation, and observing best management practices, with particular attention to:

- proper siting in areas where the waste does not accumulate deeply,
- avoiding sensitive and valuable habitats,
- maintaining appropriate stocking densities,
- removing carcasses daily,
- collecting all harvest wastes,
- cleaning nets of biofouling organisms regularly on vessels or on land, and transferring debris to approved landfills,
- monitoring feed regimens to reduce any loss, and
- rotation of farm sites (fallowing) to enable bioremediation.

Benthic effects are monitored and managed through the establishment of performance standards defining allowable effects. Chemical changes are best assessed inexpensively by monitoring the TVS, Eh, and free sulfides. Biological changes can be monitored by benthic faunal assessment using either manual sorting and identification, or much cheaper video techniques. Compliance monitoring is often required as part of a permitting process or waste management programs. Ongoing studies and monitoring results have demonstrated that these management programs are effective tools for managing environmental costs associated with this form of producing food.

Further Information

- Aquaculture (2002) 219:355–377.
- Aquaculture (2003) 226:165–180.
- Aquaculture (2005) 243:159–174.
- Aquafin CRC. 2004. Final Report Project 4.1. Fisheries Research and Development Corporation, Canberra, Australia.
- Canadian Technical Report of Fisheries and Aquatic Sciences 2286.
- Ecological Modelling 7 (1994) 5:509–615.
- Environmental Science and Pollution Research International (2003) 10:287–295.
- Fisheries Research (2003) 62:255–293.

- ICES. 2003. Working Group on Environmental Interactions of Mariculture. ICES CM 2003/F:04, Ref. ACME.
- ICES. 2005. Working Group on Benthos Ecology. ICES CM 2005/E:07, Ref. ACME.ACE.B.
- Journal of Applied Ichthyology (1994) 10:225–241.
- Journal of Applied Ichthyology (2000) 16:137–229.
- Journal of Applied Ichthyology (2001) 17:181–193.
- Scottish Executive Locational Guidelines for Fish Farming. 2002. Scottish Fisheries Research Report No. 63. Fisheries Research Services, Marine Laboratory, Aberdeen, Scotland.

APPENDIX B. INCREASED INORGANIC LOADING

Risk Hypothesis

Marine fish farms present a high concentration of biomass compared with natural settings. High concentrations of nutrients are contributed to the water column from fish metabolism, and to a lesser extent from fish feed, feces, and biofouling organisms. Mostly these are untreated dissolved nitrogen and phosphorus in inorganic and organic forms. The perceived risk is that these soluble compounds may cause eutrophication of the coastal zone with concomitant reductions in dissolved oxygen levels as the additional organic matter is degraded, or production of nuisance blooms of toxic algae and reductions in light penetration resulting in the loss of submerged aquatic vegetation.

Background Experience

In most marine environments, nitrogen is the limiting nutrient. However, upwelling brings enormous amounts of nitrogen and other nutrients to the surface along the western shores of many continents. The result is that these areas are generally not nutrient limited. In the high latitudes of temperate waters, on the other hand, primary production is generally light limited, although there may be protected bays in some areas where eutrophication may be a hazard.

Fish excrete ammonia and ammonium waste primarily across gill epithelia but also in concentrated urea, which in oxygenated waters is converted into nitrite and nitrate nitrogen. Nitrogen and phosphorus are also dissolved by the water column from uneaten feed and feces during and after descent through the water column.

Ammonia and hydrogen sulfide can be released from underlying sediments covered with feces and uneaten feed as sulfate and nitrate-reducing bacteria degrade organic carbon. However, studies reveal that there is usually little soluble hydrogen sulfide gas in the water column. About 98% of the sediment gases are methane and carbon dioxide.

Dissolved oxygen levels decrease with fish respiration and the oxidation of sediment waste. Monitoring over long periods in coastal waters of the northeast Pacific Ocean reveals a maximum oxygen reduction of 2 mg/L in water passing through net-pens where large biomasses of fish were being fed, and in most cases the reduction was less than 0.5 mg/L. Theoretically, oxygen may be depressed in water overlying enriched sediments, but this effect has not been well documented.

Harmful algae vary in their response to nutrient loading, depending on the species, habitat, timing, and loading rate. Harmful algae known to kill marine fish and shellfish, or pass on risks to human health, typically originate elsewhere and not near a farm site. Some species smother the gills of aquatic organisms, while others are digested and concentrated to levels which may be toxic when consumed. The source of cells or cysts may just as well be from a shallow bay inshore, or from coastal waters well offshore, so an understanding of the phytoplankton dynamics of the entire region is important.

The causes of nuisance algal blooms are difficult to establish. The dynamics of phytoplankton nutrient uptake in temperate waters result in cell-division times of 1–2 days during which phytoplankton will be transported many kilometers from the location where the nutrients were released. Conditions that promote phytoplankton blooms include high nutrient loadings, water column stability associated with freshwater inputs (stratification), low winds, and periods of neap tides. In higher latitude temperate climates such conditions are most frequently found in sheltered bays receiving large freshwater inputs.

Building the Conceptual Model

The sources of dissolved nutrients from marine fish aquaculture operations are the physiological and respiratory functions of the farm fish themselves, and in trace elements and micronutrients (vitamins) in any uneaten feed. Additional nutrients may be contributed from breakdown of organic matter in the sediments (see Appendix A, and below).

Nutrients can pass into solution in several ways. Most nitrogenous waste is passed across the gills of fish. Excreted ammonia and urea, together with other nitrogenous and phosphorus compounds, are excreted directly into the water column. Additional small inputs may be contributed as feces and uneaten feed pellets fall to the seabed, but the majority fall and accumulate in the sediment.

Sediment bacteria are responsible for the continuous and sometimes elaborate cycling of nitrogen between ammonia, nitrite, nitrate, and dinitrogen gas, and convert sediment phosphorus into soluble inorganic phosphorus. The majority of these inorganic nutrients, together with particulate organic nitrogen and soluble organic nitrogen, are all available for assimilation by phytoplankton and other primary producers, such as macroalgae.

Most of the inorganic nitrogen and phosphorus in the sediment is ultimately remineralized following particulate adsorption or advection, and thus can be taken out of the cycle temporarily. These nutrients add to cumulative effects in the far field. Their release rates are not well characterized, but their contribution is small in relation to the dissolved nutrients released directly into the water column during fish respiration.

Analysis and Characterization

Marine fish farms are typically located at sites where a current maintains a high rate of water exchange. Therefore nutrients going into solution as a result of farming activities are quickly diluted and directed by the current away from the site. Consequently, with the possible exception of some immediate effects on local sessile flora and fauna, the effects will be more distant and cumulative, possibly culminating with impacts on marine fish and shellfish following eutrophication or oxygen supersaturation.

The biological end points and their attributes for protection are:

1) the abundance and species richness of phytoplankton species and associated grazers in the receiving waters,
2) the abundance and species richness of local epifauna and sessile communities (macroalgae, epiphytes, biofouling organisms, and filter feeders), and
3) the abundance of shellfish and fish.

Phytoplankton Species and Associated Grazers

Concentrations of dissolved phosphorus and inorganic nitrogen tend to be greatest in proximity to their sources. However, because of the temporal and spatial dynamics of nutrient uptake by suspended particulate matter in flowing water, there is little potential for measuring enhanced phytoplankton production around aquaculture facilities in open aquatic ecosystems— regardless of the amount of dissolved inorganic nutrients added to the water.

Local Epifauna and Sessile Communities

Some degree of observable effect is likely in the local epifauna and sessile communities subjected to continuous exposure. This is often demonstrated qualitatively by the increased biomass of biofouling organisms, together with changes in the species composition of attached nitrophylic green algae (*Ulva* spp. and *Enteromorpha* spp.), and sessile or localized filter feeders.

Localized effects are also more noticeable in shallow waters, where the substrate is still in the euphotic zone. Here, dissolved nutrients from aquaculture sources can increase, decrease, or extirpate communities of benthic microalgae (microphytobenthos, mainly diatoms), subsequently impacting macrobenthic grazers.

The dilution of metabolic wastes is a function of currents. Therefore a box model approach can be used to determine their potential effects. This is supported by inputs determined by mass balance with the known-effects concentrations for the species under culture, or known nutrient releases from the species as a function of size and feeding regime. Predicted concentrations will be highest on the down-current perimeter of the containment system during periods of minimal current speeds, and these can be compared against water quality criteria or published-effects concentrations. Typically, these are based on the most sensitive species, which predominantly will be the species in culture as they are usually at greatest risk.

In far-field receiving waters, effects associated with numerous point and nonpoint sources are generally experienced by all living resources in that particular hydrologic unit, but the sources cannot be identified or distinguished specifically. In confined receiving waters near to concentrated aquaculture activities, the cumulative effects from multiple sources could lead to changes detectable by long-term monitoring programs. On the other hand, far-field effects from aquaculture activities in very open bodies of water become increasingly unlikely.

The potential for far-field effects is best assessed through computer models and long-term monitoring by a coordinated consortium of contributors to the cumulative effects.

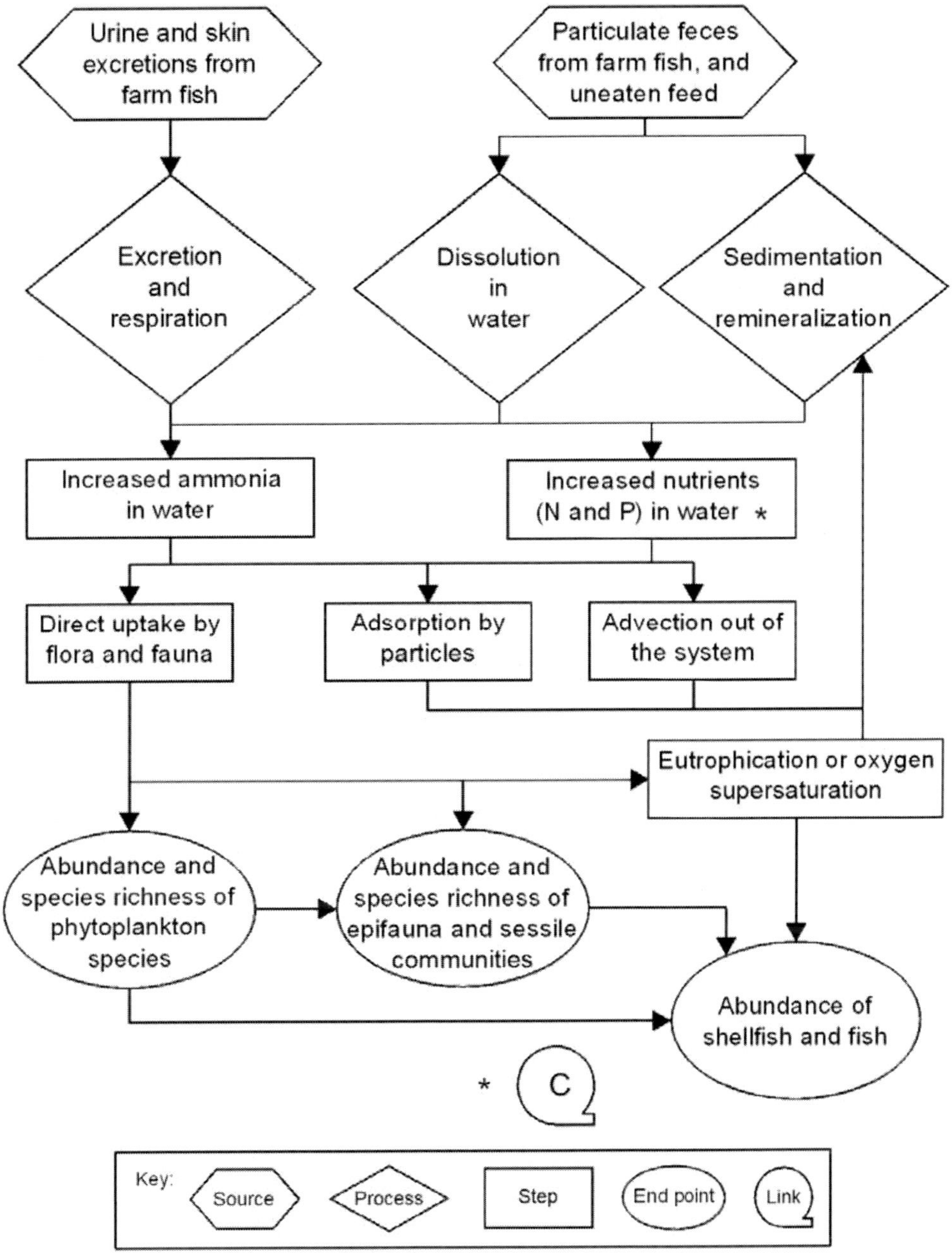

Figure B-1. A conceptual model for increased inorganic loading.

Shellfish and Fish

Because conditions of eutrophication or oxygen supersaturation generally occur on some ecosystem scale, such as in a bay or a fjord, all aquatic plants and animals resident in the system may be affected by the induced changes. In an oligotrophic system, for example, the added nutrients may result in increased productivity. However, in a nutrient sensitive environment where the assimilative capacity is already challenged, then any eutrophication can cause widespread loss of productivity at a number of trophic levels, including fish and shellfish resources and their predators.

Biological Opinion

Water flow rates through buoyant fish cages and net-pens is massive, therefore the flow and large volume means that measurable effects in the water column are hard to detect at more than a modest distance downstream. Monitoring dissolved nitrogen around net-pen farms in Europe, Canada, and the United States has revealed only small increases in levels around the perimeter of farms, and were undetectable about 30 m downstream, and monitoring dissolved oxygen has been discontinued in some areas, other than by the producers for their own use during hot weather.

The majority of coastal waters are naturally replete with nitrogen and phosphorus, but there are times (usually in spring or summer) when nutrient supply, usually but not always nitrogen, can limit the growth of marine phytoplankton populations; other factors include light availability, sinking below the compensation depth, or grazing by predators. Marine waters vary in their sensitivity to nitrogen addition. For example, well-mixed, fast-flushing habitats that are light limited are much less sensitive than vertically stratified, poorly flushed habitats, often inshore and depleted of nitrogen at the surface. Such highly sensitive areas are often characterized by prolific blooms of toxic or noxious algal species, and they are avoided by farmers siting fish enclosures as they are well known to be unsuitable.

Marine waters over the coastal shelf vary between being well-mixed and highly stratified, and any one condition may last from days to months. Periods of nutrient sensitivity do occur but invariably they are shorter in northern latitudes. Marine waters in southerly latitudes are more susceptible to algal bloom stimulation if an adequately large external source of nutrient is applied. However, the extent of any problem depends on a number of external factors, such as the nutrient ratio balance, far-field circulation patterns, the occurrence of harmful algae or cyst beds, and perhaps the proximity to coral reef habitats. On the other hand, phytoplankton are a primary base of marine food webs, so increased production in open offshore zones is sometimes beneficial. As there are currently few marine fish farms sited in semitropical coastal waters, they cannot be counted as a source of nutrients.

Dissolved nutrients may result in increased local production of attached macroalgae, but eutrophication associated with aquaculture is not a local effect. It is an ecosystem effect that can change restricted coastal areas, such as bays or estuaries. This is because it takes 1–2 days for phytoplankton to take on nutrients and divide, and 6–7 cell divisions (6–14 days) for a background phytoplankton density of 20,000 cell/L to reach a nuisance bloom density of greater than 1 million cells/L. By this time, even with small net current vector speeds of 1–2 cm/sec, nutrients released from an aquaculture facility might have traveled 10–24 km from their point of introduction.

Consequently, depending on the scale of operations, it is unlikely that a single aquaculture operation will release sufficient nutrients to affect phytoplankton production adversely in open coastal waters. However, as the number of fish farms increases, more attention should be paid to nutrient balances and the potential for increasing effects on an ecosystem basis. These system-wide effects can be cumulative, and their assessment and management requires an understanding of all of the nutrient sources to a water body, including upland and water dependent contributors. Local monitoring using point in time surveys is not recommended. Monitoring and management on a system-wide basis are required to maintain the health of coastal ecosystems. Nutrient inventories, together with mass-balance models, and system-wide, long-term monitoring programs initiated early in the development of an area for aquaculture, are the recommended strategies for management.

Further Information

- ICES. 2003. Working Group on Environmental Interactions of Mariculture. ICES CM 2003/F:04, Ref. ACME.
- Journal of Applied Ichthyology (1994) 10:225–241.
- Journal of Applied Ichthyology (2000) 16:137–229.
- Journal of Applied Ichthyology (2001) 17:181–193.
- OSPAR Commission. 2003. Integrated Report 2003, Eutrophication, ISBN 1-904426-25-5.
- Scottish Executive Locational Guidelines for Fish Farming. 2002. Scottish Fisheries Research Report No. 63. Fisheries Research Services, Marine Laboratory, Aberdeen, Scotland.

Appendix C. Residual Heavy Metals

Risk Hypothesis

Compounds of the heavy metals copper and zinc are a necessary part of marine fish farming in some form. The perceived risk to the marine environment is that they may accumulate in the sediments to toxic concentrations in a biologically available form.

Background Experience

Copper and zinc are essential micro-minerals for the nutrition of fish, and they are obtained from the natural diet or absorbed through the gills and skin. In fish husbandry they are administered as trace elements in the diet. Therefore it is highly probable that there will always be a mineral supplement in fish feeds that is compounded in an appropriate chemical form.

Trace elements in the feed after digestion are evacuated in fecal material, and they are also available in uneaten feed. Consequently, they are present in the particulate organic waste deposited on sediments near the farm. There are several pathways by which these trace elements enter the water column.

Copper, copper sulfate, or copper oxide, are the primary ingredients of many government-approved marine antifouling paints used to reduce the fouling of aquaculture structures, such as net-pens. Benefits of reduced fouling include increased water flow through the pens; decreased drag, which can compromise the moorage or structural integrity of the system; and decreased oxygen loading to the benthos around the farm.

Building the Conceptual Model

From the background experience, copper and zinc are the two heavy metals used in some form in marine aquaculture. At low to moderate concentrations, dissolved ions of these metals become toxic in the water column in sediment porewater. The primary sources of zinc from marine fish farms are the fish fecal matter and uneaten feed. Uneaten feed and fecal matter are minor sources of copper. The major source is leaching and flaking of antifouling paints that might have been used to protect equipment or applied to nets.

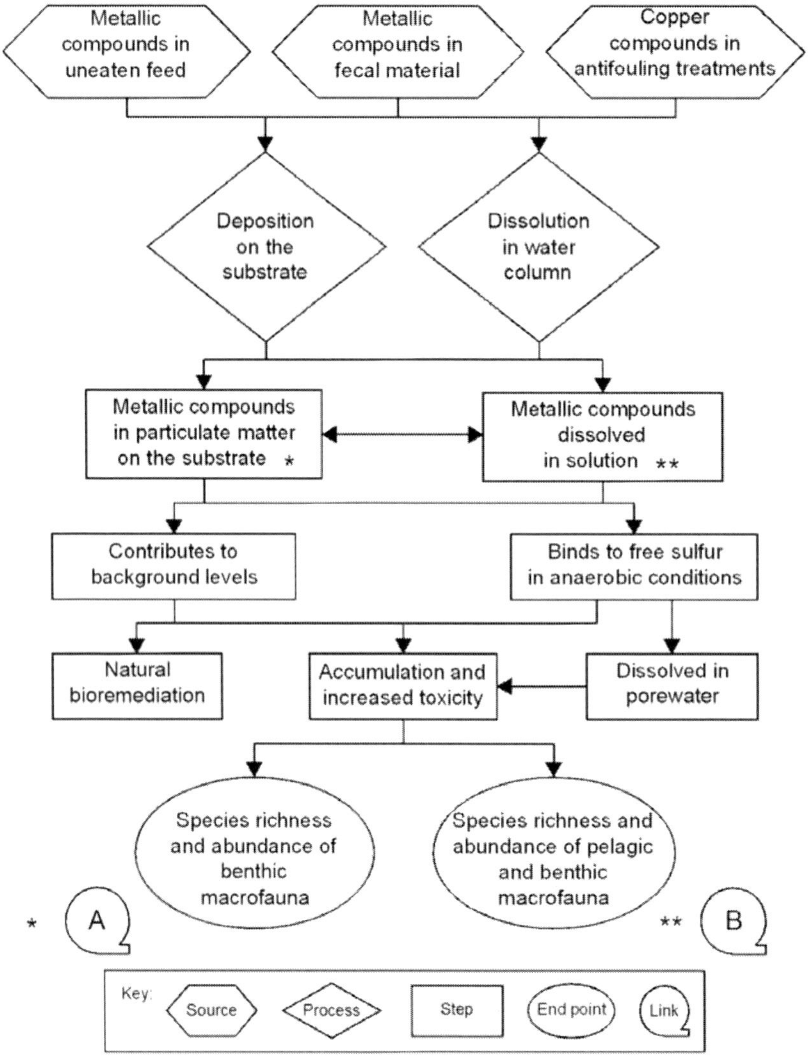

Figure C-1. A conceptual model for residual heavy metals.

There are a number of processes and intermediate steps between these sources and the biological or ecological end points that are of concern. Both particulate fecal matter and uneaten feed containing zinc settle onto the sediment. Under aerobic conditions, zinc is dissolved in the water column and contributes to the background level of the ecosystem. Even slow currents dilute these inputs, and toxicity associated with dissolved copper or zinc associated with marine aquaculture is not expected. Under anaerobic conditions, zinc is bound by sulfides in the sediment, significantly reducing its biological availability. During chemical remediation, the sulfide is oxidized to sulfate and the zinc enters that water as a dilute free ion.

Copper, similarly, from its three sources, may be dissolved in the water column contributing to the background level. In the sediment it is also bound, subsequently undergoing chemical remediation.

In areas of very poor flushing, or in the absence of appropriate best management practices, there is the remote chance that levels of copper and zinc may reach toxic concentrations. As this might have an effect on the marine ecosystem, some guidelines would be required for its protection.

Analysis and Characterization

The biological end points and their attributes for protection are:

1) the species richness and abundance of the seston and nekton, and
2) the species richness and abundance of the infauna and epifauna.

Seston and Nekton

Proximal fauna in the seston, nekton, and epifauna can be exposed to increased background levels of ionic copper and zinc in the water column. The exposure and its effect are continuous. Levels of copper and zinc in the water column are determined by standard sampling procedures and laboratory analysis using a graphite furnace atomic absorption spectrophotometer with a GTA 100 graphite system.

Many national or their provincial jurisdictions have established chronic and acute freshwater and marine water quality standards for heavy metals, including copper and zinc. The levels are usually quoted in micrograms per liter (μg/L) or parts per million (ppm).

Infauna and Epifauna

Proximal fauna in the infauna and epifauna can be exposed to increased background levels of ionic copper and zinc in the sediments following remediation. Again, both the exposure and the effect are continuous. Metals are bioconcentrated by many organisms, leading to increased concentrations in comparison with those found in the water. Levels of copper and zinc in the water column are determined by standard sampling procedures and laboratory analysis.

Numerous sediment quality benchmarks are available in various jurisdictions. However, regulatory programs for sediments have proven more difficult to develop, and few jurisdictions have published such criteria. Available benchmarks include the apparent effects threshold, threshold effects level, and probable effects level. Levels are usually quoted in micrograms of the metal per gram (μg/g) of dry sediment.

Biological Opinion

After digestion, residual metallic compounds are evacuated in feces. But their volume is extremely small, as heavy metal ingredients of manufactured feeds are absolutely minimal and they are digested efficiently. They are also available in waste feed that passes through an enclosure uneaten. However, modern feeding practices are designed to minimize waste feed and therefore that volume is also very small. This is because feed manufacturers now use proteinated forms of zinc, such as the zinc methionine analog, which are more digestible than zinc sulfate. This has greatly reduced the required concentrations in feed and therefore any subsequent environmental loading of this metal. Contamination of marine sediments by copper-based marine antifouling treatments has been greatly reduced by best management practices for net-cleaning and retreatment. These practices now call for these operations to be carried out on protected, land-based sites, with biofouling debris deposited in an approved landfill. Residual copper and zinc in sediments beneath net-pens and cages, from whatever source, are both reduced to nontoxic levels as they combine with sulfides in organically enriched sediments. High concentrations of sulfides can reduce their bioavailability quickly, and the process is particularly rapid in anaerobic conditions. Concentrations of zinc have been shown to return to background levels during chemical remediation as both sulfides are oxidized to sulfates, and there has been no buildup under the fish enclosures.

Sequestering of copper by fish is predominantly in the liver and kidneys, and not in muscular tissue, and sequestering of zinc is greatest in the eye, followed by the kidney, bone, skin, gill, and liver, with only minute traces in the muscle or gonads. Therefore these two metals, which are not particularly toxic to mammals, pose little or no human health risks.

In summary, no adverse effects to the marine environment from the presence of copper and zinc from fish farm enclosures should be anticipated providing there is:

- appropriate selection of copper or zinc compounds as mineral supplements in manufactured feeds,
- a properly managed feeding regimen in operation to minimize wastage, and
- proper use of government-approved antifoulants, with upland net cleaning and disposal of all biofoulants in an approved landfill.

This biological opinion is probably applicable to both biogeographical zones and the epipelagic waters of all three ecosystems where fish rearing in marine enclosures might occur.

Further Information

- Fisheries Research (2003) 62:295–305.
- Journal of Applied Ichthyology (2000) 16:137–229.
- Toxicology of Aquatic Pollution. 1996. Cambridge University Press, Cambridge, United Kingdom.

APPENDIX D. TRANSMISSION OF DISEASE ORGANISMS

Risk Hypothesis

Both cultured and wild fish are susceptible to the same pathogens and the same parasites in the aquatic environment, as all are part of the naturally occurring flora and fauna. The risk hypothesis is that intensive aquaculture conditions manifests the prevalence of indigenous or exotic disease agents in an area. The exposure of wild fish to disease is therefore increased directly with the water moving freely between farm enclosures and the local ecosystem, or indirectly when farmed fish escape and intermingle with wild fish (see Appendix F).

Background Experience

For disease to occur, pathogens and parasites require a susceptible host and an aquatic environment conducive to themselves but stressful to the host. Fish undergoing stress have increased susceptibility to invasion. Their immune system is overcome, and their health is then impaired. The rate of disease transfer is dependent on the virulence of each agent and the susceptibility of the host, which can, of course, be manipulated by preventative measures such as vaccines.

Intensive systems themselves are not necessarily the cause of stress among farm fish, as shoaling is invariably part of their natural behavior, but stress can occur when there are abnormal changes in ambient environmental conditions or natural meteorological phenomena: for example, continuing algal blooms caused by prolonged high water temperatures, or "winter chill" when temperatures may be unseasonably low. Stress can also be induced by certain conditions on the farm, mostly associated with poor management practices. These may include, for example, overcrowding, failure to collect mortalities daily, bleeding fish on-site at harvest, and introducing juveniles or broodstock not certified as being disease-free.

Disease can also occur if exotic pathogens and parasites are unwittingly introduced into an ecosystem from another geographic region. The origin of these exotic agents may be, for example, the ballast water discharged from ships or any unregulated movement of live aquatic animals for commercial purposes. The majority of nations enforce strict import regulations to prevent the introduction of alien pathogens via all live animals, including marine fish. But transfers can occur; for example, the white spot syndrome (Baculovirus complex) in penaeid shrimp was introduced into South America from Asia as a result of illegal movements and improper health inspections prior to shipment.

Building the Conceptual Model

The sources of the risk of increased disease transmission are the natural background pathogens and parasites in the respective marine ecosystem, and any exotic pathogens or parasites that might be introduced unknowingly.

The systems and intermediate steps in the model are few and direct. Infected wild fish first transfer their disease agents to farm fish, or infected exotic fish transfer their exotic agents to both receptive farm fish and wild fish. The result is cross-infection and cross-infestation between farmed fish and wild fish, increasing possibly to chronic or perhaps catastrophic conditions in that local ecosystem. Effects could also be extended to other, more distant ecosystems if infected or infested farm fish escape and subsequently transfer their disease agents to wild fish in another ecosystem.

At the end of these steps there are possible conditions in the marine ecosystem that put a number of biological entities at risk. These are the end points, and they are qualified by certain attributes. The health of wild fish may be debilitated as they might carry above-normal loads of pathogens and parasites, and some wild populations may be reduced in number following any epidemic. If this should occur, then the survivors would benefit by an increased natural immunity to that particular agent.

Analysis and Characterization

The biological end points and their attributes for protection are:

1) the health of wild fish,
2) the numerical strength and diversity of wild fish populations, and
3) the natural immune resistance to disease of some fish.

Health of Wild Fish

Wild fish in the proximity of fish farms may be involved in the transfer of disease agents, and may have background levels of disease organisms (pathogens and parasites) higher than normal. It is necessary to determine the natural background levels of the local disease flora and fauna, and estimate the effects of increased loads of individual pathogens and parasites on wild fish by an individual series of laboratory dose response (exposure) studies for each agent, and developing a simple probability model. In time it may be possible to model multiple disease agents, similar to those used in agriculture and human health.

Strength and Diversity of Wild Fish

In the case of a disease epidemic, catastrophic mortalities of both wild and farmed fish might occur. In order to measure such an effect, it is necessary to have some prior estimate of the natural diversity of the local fish populations and their strengths in the area. For commercial species, this information may be available from fisheries managers.

Natural Immunity of Wild Fish

The natural immunity of some wild fish populations to disease organisms can be improved if a large number of individuals susceptible to a certain disease agent are eliminated by an epidemic. Should an incident ever occur, measuring its effects on the survivors would have to be derived through combination of the analyses and characterizations of the two biological end points and their attributes, as described above.

Biological Opinion

Because of the close proximity of fish in an enclosure, it is possible for pathogens to spread through the population and multiply sufficiently to overcome the fish immune systems, causing an outbreak of disease. The trigger is stress, which may be caused by a sudden change in ambient environmental conditions, or perhaps handling during farm operations. The same phenomenon occurs in populations of wild fish. Significant disease events have been observed when large numbers of wild fish aggregate, such as at spawning time, or when there are stressful environmental conditions, such as prolonged elevated water temperatures in summer. However, scientific studies have shown that, within a few meters of a net-pen fish complex experiencing an outbreak of disease, the level of pathogens is insufficient to cause disease in nearby healthy wild or farmed fish.

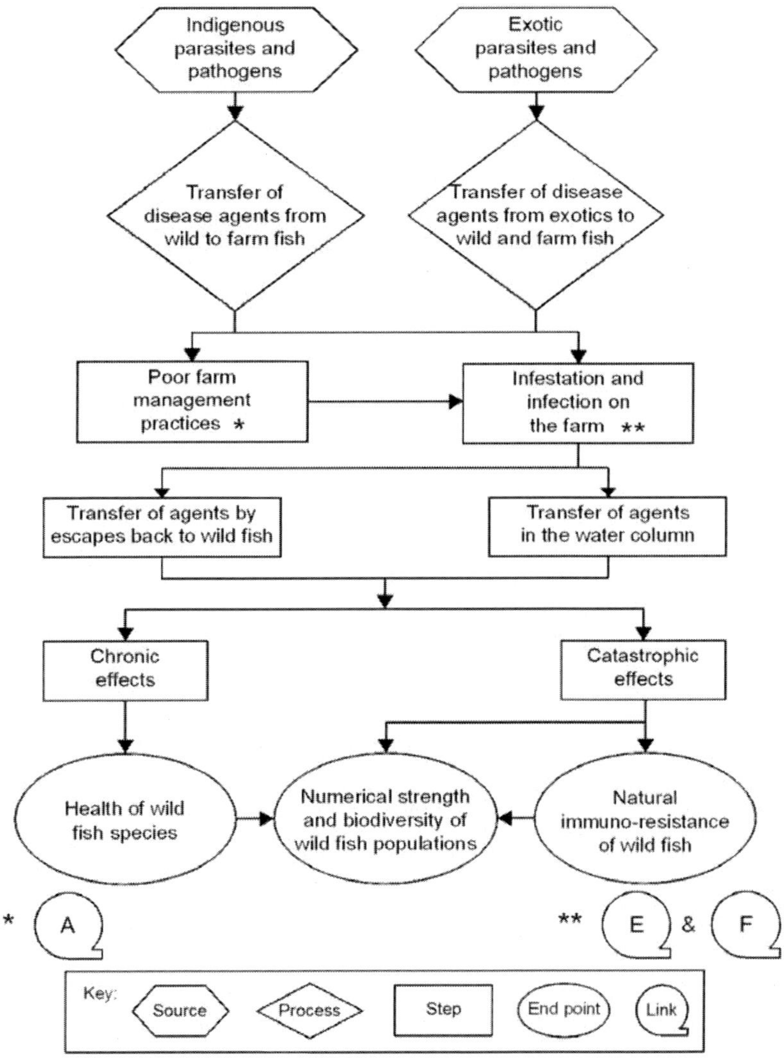

Figure D-1. A conceptual model for the transmission of disease organisms.

Experience also indicates most farmed fish that escape remain close to their enclosures waiting to be fed. Hence they become easy victims of predators. Those that escape and survive are probably not infected to begin with, but for infected fish that do survive their chances of being responsible for an outbreak in wild fish is remote. This is because they are not introducing a new pathogen to wild fish, a few escapees are unlikely to generate enough pathogens to result in an epidemic among in wild fish, and environmental factors play the larger role in triggering a disease event than the mere presence of a pathogen.

There are very many species of marine fish parasites, each with specific life histories. Very few are seriously harmful to fish or man, and their presence is only inadmissible if it affects the commercial product. One such group of parasites is the Caligidae, or copepods, commonly called sea lice, and in particular *Caligis* spp. and *Lepeophtheirus* spp. These are ectoparasites and the adults are readily visible on the skin of fish, causing disfiguration.

Some sea lice are host-specific, while others have numerous common hosts. Sea lice larvae are released into the water from eggs incubated by adult lice resident on fish hosts. The larvae drift on tides and currents as they develop until they reach the infective copepodid stage. Then the parasites must find new fish hosts or die. The time it takes for sea lice larvae to become copepodids is dependent on water temperature, and their survival is greatly reduced when the salinity is less than 30 parts per thousand (ppt).

There is no evidence that the presence of sea lice in the locations of fish farms is so great that it has caused significant disease or economic loss. Therapeutic compounds approved by authorities for use on fish farms for the control of sea lice are administered either in medicated feeds or by immersion in solution (see Appendix E).

There is an absence of scientific information to demonstrate the risk of cross-infection of sea lice between farmed fish and wild fish. It is known that juvenile fish introduced to grow-out facilities from land-based hatcheries are free of disease agents and become infected from the wild fish reservoir, but the potential for sea lice on farmed fish to increase the background infections on wild fish is uncertain. The emerging picture of sea lice associated with cultured and wild fish in the northeast Pacific suggests that, contrary to some circumstantial reports, there is no basis for expecting an increase in wild fish infections in the immediate vicinity of any source of lice larvae, including those hatched from lice at fish farms. New infections will occur only after the lice have drifted on currents some distance from their source. The distance and larval survival depends on temperature and current speeds. There is also some recent evidence that small estuarine fish, such as common sticklebacks (gasterosteids), act as an important reservoir for sea lice, and these quickly multiply when the salinity of the water increases.

Disease and its transfer are best prevented by good management practices. These include, for example, removing all mortalities daily as these are a potential source of disease, avoiding slaughter products such as blood and viscera returning into the ecosystem, and preventing fish escaping at any time.

In summary, there is low risk for the transfer of disease agents between wild fish and farm fish with subsequent ecological effects, providing:

- veterinarians and pathologists are successful in treating most common fish disease conditions, and can eliminate them,
- the health of farm stocks are maintained by appropriate monitoring and response by management,

- certified disease-free stocks are used by farms at all times,
- escapes are rigorously prevented, but if they occur there is a recovery plan, and
- all regulations concerning the movement of exotic species are strongly enforced.

This biological opinion is probably applicable equally to both biogeographic zones and the epipelagic waters of all three ecosystems where fish rearing in marine enclosures might occur.

Further Information

- Fish Medicine. 1993. W. B. Saunders Company.
- Health Maintenance and Principal Microbial Diseases of Cultured Fish. 1999. Iowa State University Press, Ames, Iowa.
- ICES. 2005. Working Group on the Introduction and Transfer of Marine Organisms. ICES CM 2005.
- Journal of Aquatic Animal Health (1998) 10:160–171.
- Journal of Applied Ichthyology (2000) 16:137–229.
- Nippon Suisan Gakkishi (2004) 70:889–895.
- Reviews in Fisheries Science (2005) 13:177–204.

Appendix E. Residual Therapeutants

Risk Hypothesis

Veterinarians prescribe and use a suite of medicines, drugs, disinfectants, and parasiticides to treat health problems of captive marine fauna, as well as chemical agents for disinfection. These therapeutants, as they are called collectively, are administered either by inoculation (which includes both injection and vaccination), orally in the feed, or externally with immersion. The perception is that residues of these therapeutants, however administered, will be taken up by the benthic infauna and epifauna to their detriment, and bioconcentrate up the food chain reducing the resistance to disease of demersal and pelagic fish.

Background Experience

Both cultured and wild fish are susceptible to the same pathogens and the same parasites in the aquatic environment as they are all part of the naturally occurring fauna, but it is a hypothesis that intensive aquaculture conditions increase their prevalence within the farm. Therefore the risk of transmission of pathogens and parasites between wild fish and cultured fish is possibly increased as water moves freely between farm enclosures and the open environment, or when farmed fish escape and intermingle with wild fish.

For disease or infestations to occur, pathogens and parasites require not only a conducive aquatic environment for themselves but also one that is stressful to the host. Fish undergoing stress have increased susceptibility to invasion. Their immune system is overcome, and their health is then impaired. Intensive systems themselves are not the cause of stress among farm fish, as shoaling is invariably part of their natural behavior, but stress can occur when there are abnormal changes in ambient environmental conditions: for example, continuing algal blooms caused by prolonged high water temperatures, or "winter chill" when temperatures may be unseasonably low.

Fish can be vaccinated to prevent disease, and most farm fish are treated successfully to prevent many disease conditions, such as bacterial infections. Therapeutic compounds may be used by fish farms for the control of some diseases and external parasites such as sea lice. These compounds may also be used to prevent the introduction of alien pathogens via live aquatic animals or their live eggs from one area to another for release, propagation, or relaying. One example is the introduction of an ectoparasite (*Gyrodactylus salaris*), which was the result of moving live smolts from Sweden to Norway without adequate treatment and health inspections prior to shipment.

Building the Conceptual Model

Therapeutic compounds emanating from aquaculture activities in the marine environment are of ecological concern. The sources of these compounds from farms are the metabolic products (feces, urine, and skin secretions) of the fish after internal or oral treatments, and the direct release into the environment of immersion waters containing a therapeutant for treatment of a disease or parasitic infestation, or uneaten medicated feed.

The processes and intermediate steps between the sources and the biological end points are few. Compounds used in solution or released in solution contribute directly to the background level of the ecosystem. From there, some will be taken up by planktonic organisms, and some may be adsorbed into the sediment. Those in particulate matter fall directly on the sediment.

Following degradation in microbial and microfaunal processes, some therapeutic compounds may still retain their integrity sufficiently to be toxic to nontarget organisms, or to reduce the immunity of some organisms to disease. There might also be bioaccumulation of some compounds up the food chain.

Analysis and Characterization

The biological end points and their attributes for protection are:

1) the species richness and abundance of the seston and nekton,
2) the species richness and abundance of the infauna and epifauna, and
3) the immune-resistance to disease of some demersal and pelagic fish.

These entities are therefore the focus and structure of the analyses and subsequent risk characterization.

Seston and Nekton

Proximal fauna in the seston and nekton can be exposed to therapeutic compounds released by the emptying of the immersion water following treatment. The effect is sudden and very localized, and of very short duration. They may also be affected by residual compounds in the urine or skin secretions, as these excretions are soluble or easily resuspended.

The effect is best determined with a sessile filter-feeding bivalve marine indicator organism, such as a mussel, suspended independently close to the facility. The EcoTox Database for marine indicator organisms will be the reference point to estimate threshold and toxicity of the respective therapeutic compound, but it is necessary to consider the mode of action of each compound to establish appropriate exposure duration. Some are broken down to inert compounds in minutes or hours and may only be acutely toxic immediately on release. A 96- hour LC50 is appropriate for all therapeutic compounds released from the immersion container, considering the immediate dispersal rate and dilution, and indicates the effects at absolute maximum concentration.

With regard to the reasonable worst-case scenario, when many farms in an embayment might be treating an epidemic and releasing immersion waters containing the same or synergistic compounds all about the same time, then reference points will have to be distributed around the embayment and monitored.

Infauna and Epifauna

Proximal infauna and epifauna close to the point of release of immersion treatment waters, or metabolic by-products in solution still containing a residual therapeutant, might be affected. However, they are more likely to be exposed to any therapeutic compound contained in the solid particulate excretory products (feces) of fish metabolism. Exposure will occur mostly within the footprint of farm but possibly dispersed further depending on the qualities of the feces, but the effects can be measured by the same procedures as in the "Seston and Nekton" subsection, above, using an appropriate benthic indicator organism.

In measuring the effect it is important to consider when treated feeds are being used, and their duration of use, as this will affect the rate of accumulation and total concentration in the sediment, as well as the breakdown and mode of action of the respective compound. The effect can be determined by measuring the concentration of the residual compound in the sediment directly, and comparing it with the maximum residual limit (MRL) or permitted residual concentration.

Demersal and Pelagic Fish

The footprint of a farm facility is rich in organic and inorganic matter, and rich in microfaunal and macrofaunal communities. It becomes a feeding ground for both demersal and pelagic fauna, and therefore residual therapeutants may bioconcentrate in the fauna at the higher trophic levels, possibly lowering their immuno-resistance to disease.

For nontarget organisms, such as large epifauna and benthic fish, it is necessary to monitor accumulation continuously by direct analysis for the therapeutants being used. The levels may be influenced by changes in farm production, the age of the fish at the time, and the need (or not) for prophylactic treatment. Monitoring should be performed in both the near field and the far field as crustaceans and fish are free to roam.

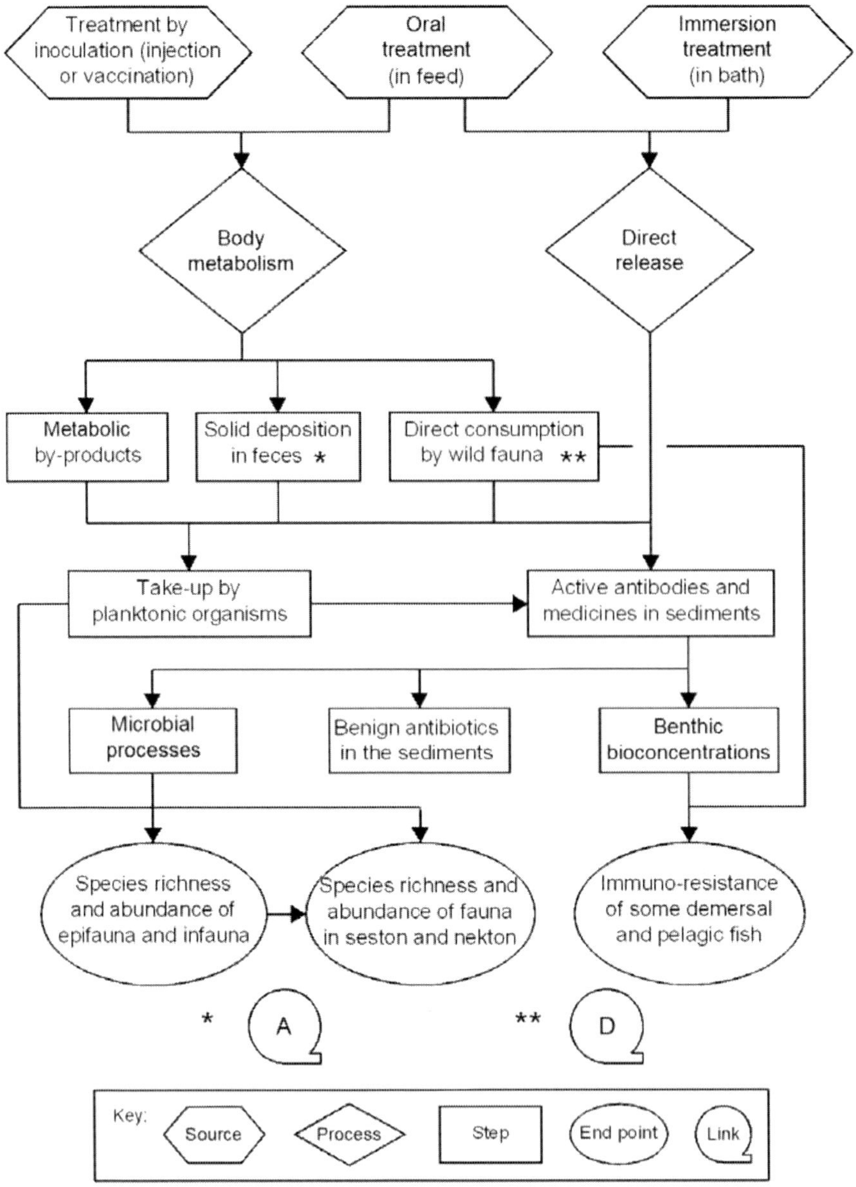

Figure E-1. A conceptual model for residual therapeutants.

Biological Opinion

Therapeutic compounds are not used on a regular basis, but only when pathogenic diseases or parasitic infestations occur, or if, from experience, they might be anticipated if conditions are conducive. Manufacturers also compound and sell medicated feeds in anticipation of certain disease outbreaks, but they too are used only when needed.

Almost every country with an active agriculture sector authorizes the use of therapeutants (veterinary medicines, pharmaceuticals, and vaccines) for cultured fish and shellfish. This they achieve by specific regulations for the manufacturers regarding their quality and efficacy, and specific regulations for their use by veterinarians and farmers. Initially their objective was to protect human health and safety, as these are animals raised for human consumption and the continuous presence of residual drugs may lead to human drug resistance. But protection of the environment is now equally important as therapeutants may continue to affect wild aquatic fauna. For example, parasiticides authorized for the control of host-specific parasitic arthropods, or sea lice, which graze on the skin of many fish, are often broad-spectrum biocides with potential to affect many phyla adversely.

Consequently, the ecological risk from the effects of any therapeutants used in marine fish aquaculture is minimized firstly by using regulated products and in the course of treatment prescribed by the veterinarians. Unfortunately, the respective quality and efficacy of some drugs, and the use of certain medicines, are not standardized around the world. Chloramphenicol, for example, has been banned by many countries from use in all food animals but it is still being used by shrimp producers in some countries.

Secondly, the risk is minimized by application of the most current and efficient treatment. In countries with valuable commercial aquaculture sectors, for example, vaccines have been developed to replace the traditional use of antibiotics. However, this is not the situation worldwide and antibiotics are still used in many hatcheries, although quantities are small, and in countries where vaccines are not readily available.

The possible misuse of therapeutants is being accommodated to some degree by the FAO Code of Conduct for Responsible Aquaculture, and is being enforced indirectly by the food safety and importation requirements of the large seafood markets. Guidelines for their use are becoming an important inclusion in the best management practices recommended for raising individual species. However, best management practices have as yet not been prepared for every farming practice, and they are in the end only nonbinding, honorary agreements, which need not necessarily be observed by every producer, and particularly one facing a disease crisis.

The best solution is disease prevention, rather than cure, and thus reducing the need for therapeutants altogether. Prevention is helped by good site selection in the first place, followed by good husbandry practices. These include, for example, the use of certified disease-free stocks, maintaining sensible stock densities for the age of the population, proper storage and timely use of feed, and effective daily monitoring for mortalities and signs of stress.

In summary, few adverse effects to the marine environment from the presence of therapeutants should be anticipated providing there is:

- only appropriate use of authorized therapeutants,
- a properly managed farm operation preventing disease, and
- the use of certified disease-free stocks.

This biological opinion is probably equally applicable to both biogeographic zones and the epipelagic waters of all three ecosystems where fish rearing in marine enclosures might occur.

Further Information

- Fish Medicine. 1993. W. B. Saunders Company, Philadelphia, Pennsylvania.
- Health Maintenance and Principal Microbial Diseases of Cultured Fish. 1999. Iowa State University Press, Ames, Iowa.
- Journal of Applied Ichthyology (2000) 16:137–229.
- Journal of Applied Ichthyology (2001) 17:173–180.
- Journal of Aquatic Animal Health (2004) 16:208–219.

APPENDIX F. BIOLOGICAL INTERACTION OF ESCAPES WITH WILD POPULATIONS

Risk Hypothesis

Escaped farmed fish, or their gametes liberated from a farm, may pose a risk to wild populations when they interact biologically. Potentially deleterious genetic impacts are perceived to be:

- interbreeding and
- competition for mates or nesting sites.

Potential ecological risks from escaped farm fish are perceived to be:

- competition for habitat and forage,
- increased predation (if piscivores),
- the introduction of exotic pathogens and parasites, and
- amplification of endemic pathogens, some of which may be antibiotic resistant.

All these possible risks are believed to pose a greater threat to natural populations (conspecifics of the escapees) than to other fish populations at large.

Background Experience

The practices of both freshwater and marine fish culture for stock enhancement or ranching have benefited from years of effort to improve the cultured stocks. In addition to the results of traditional genetic techniques used by hatchery managers, such as trait selection, inbreeding, and out-breeding, there are also the genetic influences of simply surviving in the wild. On the other hand, commercial fish culture is a relatively new field and the present generations of farmed species are still closely allied to the original wild parents. Fish populations bred in captivity have already been subjected to similar stock-improvement practices which, however small, have probably begun to change their genetic makeup. Consequently, when cultured fish are released intentionally or escape from farm enclosures into the ecosystem, they carry with them a genetic profile that can have a deleterious effect should they interact again with natural populations.

There are a number of ways for biological interactions to occur in an aquatic ecosystem where aquaculture activities are practiced. Firstly, farmed fish can escape directly from net-pens and other enclosures due to human error, damage from a catastrophic natural event such as a severe storm, or following damage to the structure by a predatory marine mammal. Secondly, some species of finfish and shellfish that spawn freely in captivity and produce pelagic eggs may release fertilized gametes into the surrounding environment. Thirdly, domestically cultured fish and shellfish raised in hatcheries can be released intentionally on a large scale in annual stock enhancement or sea-ranching programs, leaving them to migrate freely and interact with wild populations.

There is evidence that farmed fish are capable of breeding with their conspecific natural populations in the wild. Therefore escapees may present a genetic threat to a locally adapted natural population through intraspecific hybridization, resulting in a reduction in overall reproductive fitness and recruitment to the wild population. Some interspecific hybridization might also occur should farmed fish escape into an ecosystem where there are very closely related species. The use of reproductively sterile farm fish has been proposed as one means of preventing genetic interactions with wild populations, and consequently reducing their ecological impacts, but this practice is still a matter of priority research.

The introduction of exotic pathogens by the transfer and escape of farmed fish is an issue of lessening concern. This is because most countries have adopted the international protocols regarding the movement of terrestrial and aquatic species for almost any reason, and they have stringent regulations in place regarding the importation of exportation of fish or their eggs specifically to minimize the risk of transferring exotic diseases. Such precautions, however, have not always been effective. Wild fish are the reservoirs of a wide variety of common pathogens, and when certified disease-free fish or shellfish are introduced into an area for the first time they are infected by these dormant pathogens and cause the same diseases endemic to these fish in their native habitat.

Outbreaks of disease can occur at fish hatcheries, and transfer of infected fish may facilitate disease transfer between stocks. However, as the occurrence of endemic pathogens in wild fish is common, it is difficult to determine the extent that pathogen transfer occurs. Similarly, it is difficult to determine the extent to which amplification of endemic diseases occur. It has been suggested that populations of sea lice (such as *Caligis* spp. and *Lepeophtheirus* spp.) are transferred and amplified between farmed salmon and their wild populations, but no scientific evidence has been found (see Appendix D).

Building the Conceptual Model

Escapes may occur with varying frequency and intensity. Therefore, the two sources of biological interactions from the escape of cultured fish or their gametes from aquaculture facilities are catastrophic releases, or periodic natural events such as storms, and chronic releases. Their impact, however, is modified by a number of things, amongst which importantly are the numbers and the genetic characteristics of both the escapees and their resident indigenous wild populations.

Catastrophic releases are unique as they are rare and not planned, and they could involve a large number of escapees. Invariably they can be avoided or controlled if appropriate guidelines are followed for risk management (disaster prevention) and the subsequent recovery of inadvertently released animals. Although it may be impossible to anticipate the occurrence of a 100-year climatic event, a range of possible disasters can be avoided with the selection of a site concomitant with the engineering technology, and away from shipping and navigation lanes and fishing grounds, for example. The effects of a catastrophic release may also be reduced by having a plan and the appropriate equipment for retaining or recapturing escapees.

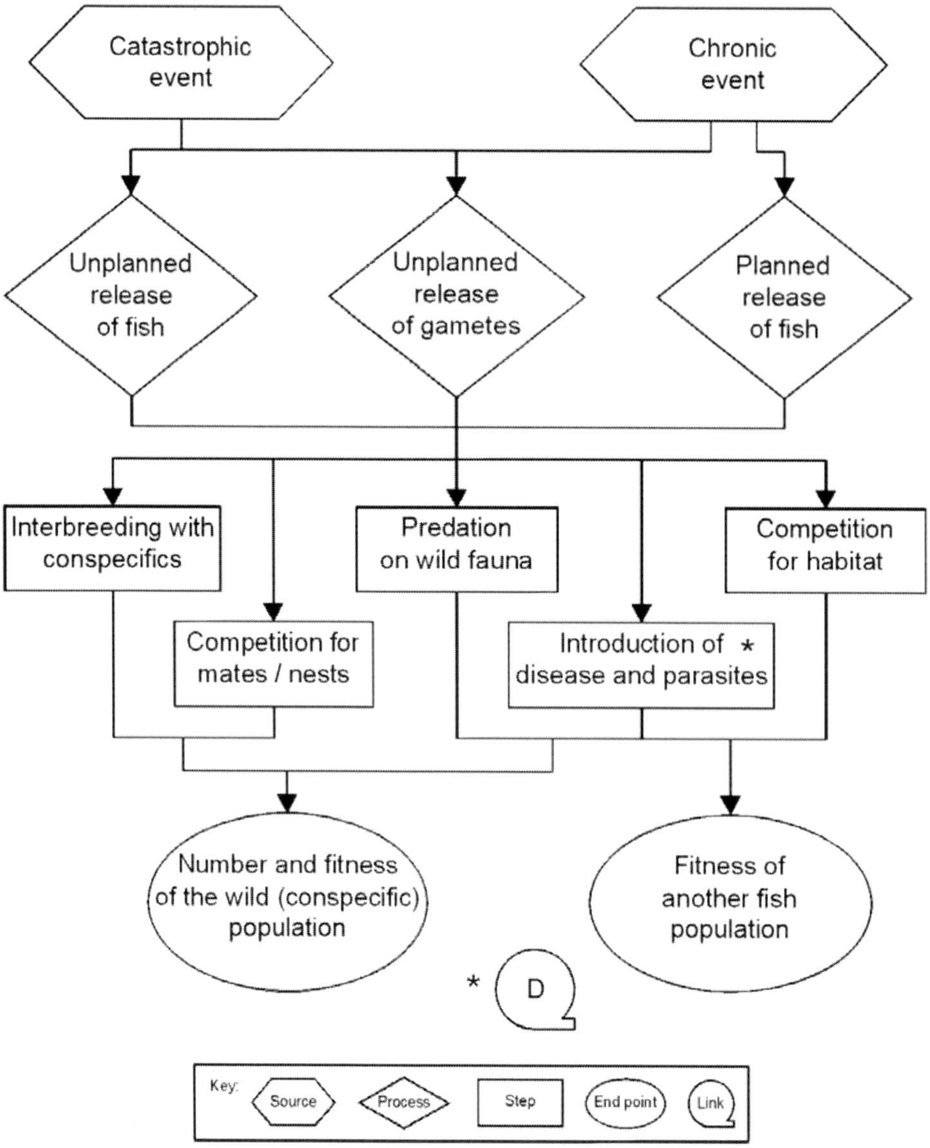

Figure F-1. A conceptual model for biological interaction of escapes with wild populations.

Chronic releases may be planned or unplanned and may involve large numbers or small numbers of escapees. Planned releases include stock enhancement and ranching programs by fisheries managers; unplanned releases include the loss of a few fish through a hole in a net made by a predator, or the release of fertilized gametes from a captive stock as a consequence of uncontrolled breeding.

Chronic releases, even due to predator attacks, are therefore often seasonal, but their potential effects for detrimental genetic and ecological interactions may be accumulative. On the other hand, the effects of planned releases of cultured fish are often minimized simply because they are target fisheries for commerce or recreation, and this reduces their potential to interact with the natural population.

Regardless of the manner of escape, escapees may affect the natural population in a number of ways. The most important and direct consequence is interbreeding, followed by the indirect consequence of competition for mates and nesting sites. The effects of interbreeding are a reduction of genetic variance between the two populations, and out-breeding depression. Some other indirect consequences in the short-term may be through competition with all species for forage and habitat space, by predation on endemic fish populations, and the introduction of bacterial or viral pathogens or parasites. The effects of these processes can be a reduction in the genetic integrity of a community or an ecosystem, and they may of course be positive or negative to both. In brief, the outcome can be a reduction in the numerical or genetic strength (fitness) of the wild population, and possibly a reduction in fitness in other fish populations.

Analysis and Characterization

The biological end points and their attributes for protection are:

1) the numerical or genetic strength (fitness) of the wild (conspecific) population, and
2) the fitness of another fish population.

Modern methodologies for measuring the size and genetic parameters of fish populations are all now carried out at the molecular level by analyzing markers, such as mitochondrial DNA and microsatellites. Consequently the techniques are sophisticated and require laboratories well- equipped with costly instrumentation. Protein electrophoresis continues to be a reliable method to detect genetic variation by identifying differences in protein allele frequencies between stocks. More recently, however, protein electrophoresis has been complemented by studies of the genome and the genetic information that can be carried and detected in a small piece of material, such as tissue from liver or muscle, for DNA identification.

Fitness of the Wild Population

Genetically effective population size (or N_e) is the most important factor to sustain a high level of genetic variation within a fish population. This is because in the actual total population (N), only a proportion (the N_e) will pass on their genetic profile to the next generation. If the total population is reduced for some reason, such as the suggested competition with cultured fish, then its original genetic profile may drift further and further away from the original. By measuring this drift, then the genetically effective population size can be calculated and conclusions drawn from the results.

However, calculating the genetically effective population size is not particularly simple. A difficult starting point is having a uniform population, so that selected fish are representative of that population with the same genetic diversity and any local adaptations. For marine fish this is made easier by the fact that few species have been subjected to the same practices of hatchery propagation, restocking, and enhancement as have freshwater fish and anadromous fish, and therefore have little or no introgression.

N_e can be estimated directly by sampling a population at two or more points in time, and separated by a specified number of generations, and it is possible to estimate N_e by the changes in allele frequencies in the interval between sampling. The usefulness of this temporal method has been increased significantly by a technique to extract genetic information from stored samples, which are usually otoliths and scales, where they exist. The polymerase chain reaction (PCR) technique can target a DNA molecule in small and old samples and amplify its genetic information. Unfortunately, fisheries biologists archived material more from freshwater and anadromous fishes than marine fishes, and therefore comparative material might be difficult to obtain.

Fitness is a measure of breeding success or survival. Relative lifetime fitness (%) is therefore the breeding success or survival of one generation to the next. However, the simplicity of this calculation is masked by several possible variables associated with any planned or unplanned releases, such as the number and timing of the release, and the suitability of the receiving ecosystem.

Annual demographic data about the population in question is also important, such as the year-class strength of successive generations. Here, there is potentially more information available for marine species than freshwater species, as demographic data has been required for some time by fisheries managers. It is also important to know when a population has substructures, as these can influence allele frequency changes and misdirect any conclusions.

Fitness in Another Population

The same procedures will be used to determine any reduced fitness in another fish population.

Biological Opinion

Escaped farm fish are not in the economic interest of producers, and there continues to be improvements in the design and operations of marine fish farms to prevent escapes occurring altogether. As many regulators now require notification of escapes, existing records show that the incidence and numbers of escapees continue to decline. However escapes can and do occur, and the escapees may interact biologically with the wild population by changing their genetic integrity or profile, introducing new or unusual genotypes, or by eroding their reproductive fitness, particularly if they are originally from nonlocal stock or selected by the breeders for certain farm traits.

Fortunately the statistical chance of these interactions occurring is affected by a number of factors, the most important of which is opportunity. Escapees are rarely sexually mature, as they are harvested by the commercial growers before nutritional energy is directed to the development of gonads. The few that might be selected as future broodstock at harvest time would be moved elsewhere—usually to a land-based hatchery. Therefore, at the time of escape, escapees are not necessarily mature enough to breed. Secondly, the escapees might not last long enough to mature in the wild and interbreed. There is considerable evidence for a variety of species that the majority of escapees, being raised in captivity on a daily routine of artificial diets, invariably remain in the vicinity of the site to be recovered or fall easy victims of predators. Thirdly, the timing of the escape might not be coincidental with the natural breeding season of the wild population. Catastrophic events may be large but they are also very rare, and chronic events may be continual but usually involve very few fish. Consequently the timing of an escape, the numbers of escapees, and the size of the wild population are all variables which play a role in defining the opportunity for biological interaction.

This is not the same for a planned release of cultured fish from a hatchery, or an unplanned release of fertile gametes from captive adults on a farm. Such events involve the release of a large number of juveniles or gametes that could mature and breed, or a few mature breeders in a restocking program in the hope that they will breed. The opportunities for biological interactions from planned releases of juveniles or broodstock, or unplanned releases of fertile gametes, are obviously considerable, and may be magnified further by the degree to which they have been selected to enhance certain traits.

The potential genetic effects of biological interactions of planned and unplanned releases may also be modified by the population structure of the wild population. For populations with a high degree of local adaptation, among which genetic variability is partitioned at the population level or on a geographical basis, then the natural population structure is particularly at risk from interbreeding with escaped conspecifics. This applies to Atlantic salmon (*Salmo salar*) and Pacific salmon (*Oncorhynchus* spp.), which are highly structured, and some Mediterranean species, such as sea bass (*Sparus auratus*).

Because of the apparent continuum of the marine environment, it has been thought for some time that most populations of marine fish species are not structured, and therefore their capacity to exert genetic effects is greatly reduced. Species such as the sea bream in the Mediterranean, for example, appear to lack structure at the population level, and gene flow across the range of such species appears extensive. Although farmed sea bream outnumber wild fish, the presence of an undifferentiated stock reduces the potential for adverse interactions. However, the increasing interest in the genetics of marine fish species for fisheries management, and increasing skills in DNA analysis, now suggest subpopulations of some marine species might in fact have remained localized for sufficient time to have developed small genetic differentiation that now are detectable. This adds to the genetic implications for releases and escapees mixing with a subpopulation of conspecifics, although, as noted above, escapees tend to remain close to the culture site, therefore selection of broodstock within the vicinity of the site would be an appropriate practice to reduce this possibility.

There is evidence that fish reared in captivity can lose any natural undiminished capacity to capture prey, and when released or escape they do not compete for forage too well. Escaped fish when recaptured invariably have empty stomachs.

In summary, ecological risks from the biological interactions of unplanned releases with wild populations can be greatly reduced, as they cannot be deleted altogether, by good management practices, such as:

- careful choice of the site;
- constant vigilance of all structures, moorings, and anchorages;
- regularly cleaning nets and predator nets;
- maintaining all navigational requirements (lights and foghorns);
- conducting any transfers with great care; and
- having a plan for escape recovery.

Genetic risks from the biological interactions of unplanned and planned releases with wild conspecific populations can be reduced by:

- selecting broodstock from within the ecosystem of the site;
- selecting marine species for farming, which have little or no substructure; and
- raising sterile animals.

Further Information

- ICES. 2002. Working Group on the Application of Genetics in Fisheries and Mariculture. ICES CM 2002/F:03, Ref. ACFM, ACME.
- ICES. 2003. Working Group on the Application of Genetics in Fisheries and Mariculture. ICES CM 2003/F:01, Ref. ACE, 1.
- ICES. 2003. Working Group on Environmental Interactions of Mariculture. ICES CM 2003/F:04, Ref. ACME.
- ICES. 2004. Working Group on the Application of Genetics in Fisheries and Mariculture. ICES CM 2004/F:04, Ref. ACFM, ACME 1.
- Journal of Applied Ichthyology (2001) 17:153–162.

Appendix G. Physical Interaction with Marine Wildlife

Risk Hypothesis

Marine aquaculture activities in nearshore and offshore environments are perceived to pose a potential risk to wildlife because the facilities incorporate structures that use nets, ropes, or twine, all of which may be opportunities for their entrapment and entanglement.

Some structures also use surface or underwater lights. These are perceived to pose a risk to the juveniles of migratory fish species, as they might be attracted to the lights and delay their migration, and unnecessarily attract forage fish and predators.

The intensity and frequency of acoustic harassment devices are perceived to be unnecessarily aversive to marine mammals, other than target species, but these are rarely used and are not considered a potential source of risk.

Background Experience

Much wildlife is attracted by marine aquaculture surface structures used for the culture of fish and shellfish. Around structures close to the shoreline, marine mammals and birds are common visitors. Pinnipeds, such as fur seals, true seals, and sea lions, may come to find food, haul-out and rest, and even give birth, while mustelids, such as sea otters and river otters, and predatory and scavenging birds, are more focused visitors coming only for food. For all these visitors, the attraction is not just the chance opportunity of the fish and shellfish being farmed, together with their feed, but the many fish that aggregate in numbers to feed on smaller aquatic animals and plants that colonize around and on the structures. Together, they all create a fertile marine habitat.

Further offshore, the more common wildlife visitors focused on food are predatory elasmobranchs, such as sharks. Then there are visitors by chance, such as whales, porpoises, and migratory marine turtles looking to shelter and rest. Commercial acoustic deterrents have not proved to be effective against any of these visitors, as both animals and birds quickly get used to them, and for the most part they are no longer used or are prohibited by law.

The danger to marine mammals and turtles for entanglement are lines or ropes that are small in diameter, slack in the water, and possibly floating near the surface. Drowning after entrapment in or around a netted structure is a possible cause of loss, together with starvation following entanglement or consumption of discarded or lost debris from a farm, such as a piece of rope or plastic.

Much of the larger marine wildlife visiting aquaculture facilities is protected by international and national conventions. Most countries have regulations requiring that the death of any marine mammal around fish farms is reported, and the information is available on public record.

Building a Conceptual Model

There are a number of physical interfaces of aquaculture facilities and activities on marine wildlife. The specific sources are any floating or submerged structures themselves, any nets, ropes, anchor lines, and anchors associated with any structures, any garbage carelessly lost by a farm, and artificial lighting.

The processes and intermediate steps between the sources and the biological end points are quite direct. In most cases, a large cross section of marine wildlife is attracted to floating or submerged structures. Some colonize, attracting others to take up residence in turn. Marine structures are primarily habitats where food can be found, and a place for rest and shelter. In other cases, wildlife may simply find them to be an obstruction in a migratory pathway or in a breeding area. Thirdly, wildlife not necessarily close to a farm may try to eat or play with garbage that has been lost from a farm. The final result of all these steps is possible entanglement or starvation.

Analysis and Characterization

The biological end points and their attributes for protection are:

1) the lives and safety of marine wildlife, particularly marine mammals, turtles, and sea birds, and
2) the natural habits and target habitats of migratory marine wildlife.

Lives and Safety of Marine Wildlife

The marine wildlife most at risk for their lives and safety are marine mammals. This is because they are more likely to remain around the site for some time, and the effect of their presence around the farm site exposes them to the risk of entanglement in nets and loose ropes. To much less extent there is a slight risk to passing cetaceans, turtles, and sea birds.

Marine mammals are capable of colonizing floating structures in substantial numbers, as marine fish farms are obvious targets not only for the opportunity to haul-out to rest, escape their own predators, and even to give birth, but also because they are potential sources of food. Initially the attraction might have been the farmed fish, but when these prove to be too well protected, marine mammals are prepared to eat any marine fish and shellfish that colonize the new habitat and add to the miniature ecosystem.

Marine turtles, on the other hand, are temporary visitors, attracted by any floating object as a place to shelter and rest. Cetaceans have no reason to visit, and for the most part any interaction is accidental. However, there may be the risk of entanglement in any net or loose rope. Both turtles and cetaceans have been known to become entangled in human jetsam, such as lengths of rope and fishing lines, pieces of fishing nets, and swallow indigestible pieces of plastic. Marine aquaculture sites can be a source of similar debris. Most sea birds around marine aquaculture structures are opportunistic scavengers availing themselves of food. The food may be mortalities from the farm, farm fish taken during any transfer or harvest maneuver, or any edible organisms washed off nets and equipment during cleaning. They also consume pelleted fish feed, which may be available at times. But some sea birds are attracted to feed on marine organisms living in the new marine habitat that is created. Here the food is small fish and crustaceans that live in and around the nets, and mollusks that settle on the nets. It is rare for sea birds to become entangled, but it is possible. For the most part, the effect of these interactions on birds is mostly beneficial, but in the long term there might be some effect on their natural behavior.

Effective management can reduce most of these risks, but the effects can only be measured *in situ* by continuous monitoring and recording actual incidents. This process is usually mandatory under the terms of the permit that is issued. Many studies on the population dynamics of several marine mammals, cetaceans, and turtles are already taking place, as are some studies on specific sea birds.

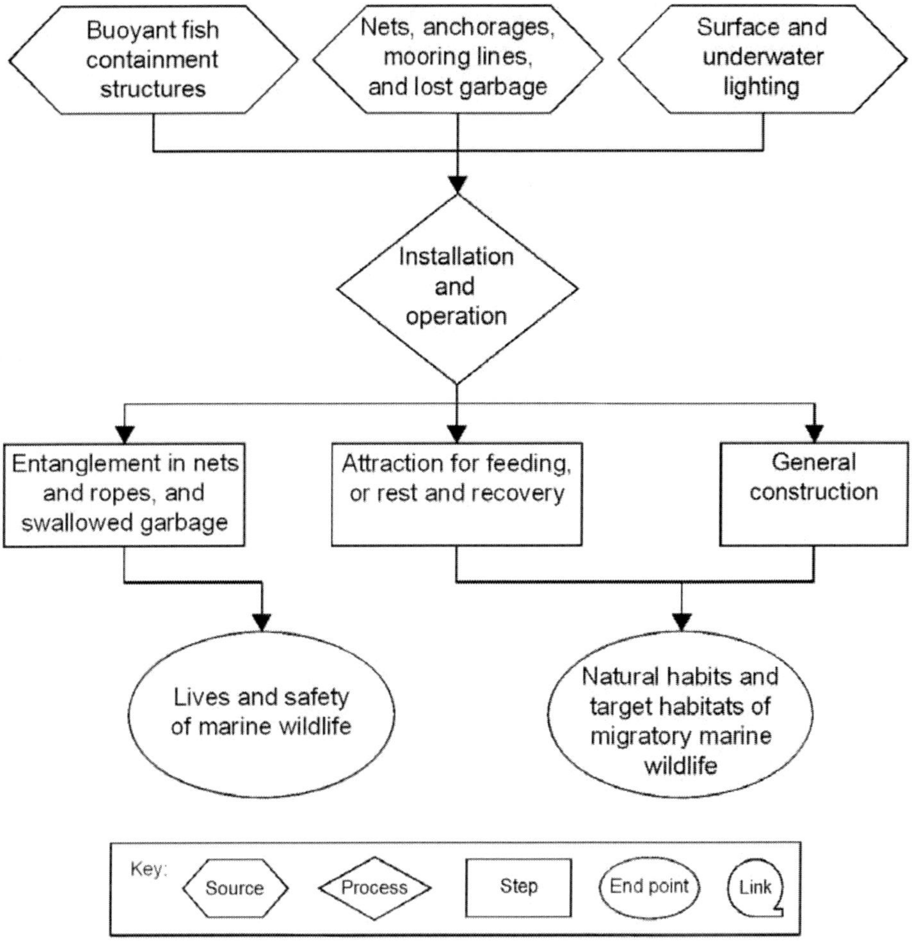

Figure G-1. A conceptual model for physical interaction with marine wildlife.

Habitats of Migratory Marine Wildlife

The route of exposure of any physical interaction between marine aquaculture facilities and migratory wildlife, such as cetaceans traveling in pods or individual marine turtles, is simply the existence of aquaculture structures in their migratory pathway, or in their target habitats such as feeding grounds or breeding grounds. Their presence may disrupt the natural patterns of migrants or perhaps make them abandon an area. Individuals might also become entangled in a loose rope or net.

These effects can only be measured by monitoring the population dynamics of the migratory species and their behavior over time, and recording any actual occurrence of entanglement and drowning. Many studies of migratory species already carried out can serve as the baseline for future studies on interactions with aquaculture facilities. Most licenses issued for operating aquaculture facilities require any accidents to marine mammals and turtles to be reported.

For migratory fish, interaction is caused by surface lighting and underwater lights used at certain times of the year to extend the photoperiod for enhancing the growth of farm fish.

Young out-migrants move in shoals close to the shoreline at night and are known to be attracted by lights. If they are attracted by the underwater lights of an aquaculture facility, the effect is to delay their migration or possibly to be drawn out into deeper water where they are more susceptible to predators, including perhaps the farm fish in the cages. Forage fish, such as shoals of clupeids, may also be attracted to the underwater lights where they could be consumed by farmed fish or their natural predators. The effect is measurable by a program to analyze the stomach contents of farmed fish throughout the day and during the most appropriate times of year, together with *in situ* research studies during the migration period.

Biological Opinion

The migratory pathways and breeding grounds of cetaceans are often within a few kilometers of shorelines. For the most part, these pathways and breeding grounds are now well known by researchers, and many of the locations have become centers of tourism. It is unlikely that marine fish farmers will choose such a location and, if they should, then it is unlikely that a permit would be provided.

The possibility of a marine mammal being entrapped and drowned in the facilities of a coastal aquaculture site is now extremely low. Modern farm complexes are no longer potential haul-out sites as the surrounding walkways are well fenced, the anti-predator nets are now much stronger, and all net-pen walls are kept rigid and taut. Floating rafts for the production of mussels (*Mytilus* spp.) and oysters (*Ostrea* spp. and *Crassostrea* spp.) are still accessible haul- out sites, but there are no predator nets to entrap them, and the vertical ropes are too thickly covered and heavy to flex and become a trap. Similarly, marine mammals present little or no risk to aquaculture facility operators. The greatest risk to producers of shellfish is fecal contamination by marine mammals, and not direct predation of the stock.

The possibility for a marine turtle drowning within or around a marine aquaculture facility is minimal providing nets and ropes are kept taught.

Modern marine aquaculture facilities do not negatively affect birds. The design of the facilities, including overhead nets to protect the farm crop from birds, prevents almost all adverse physical interactions. On the contrary, modern floating structures provide a habitat that supports many species of birds.

The greatest risk to any marine macrofauna, including birds, is probably garbage from a marine aquaculture site. Waste materials, such as a piece of rope, twine, plastic foam, plastic pipe, feed bags, or similar debris that ends up in the environment as jetsam, is more likely to entangle marine creatures or block their digestive system if eaten by mistake. The way to prevent this is by the enforcement of best management practices by the industry to deal with all potential hazardous waste.

The risk of underwater lighting from net-pen farms directly impacting the out-migration of juvenile fish or attracting forage fish and predators significantly is very low for several reasons. The intensity of underwater lighting is weak to begin with, and is attenuated to 1% within 10–20 m from the perimeter of the net-pen; net-pen complexes are in waters up to 40 m deep, and only larger juveniles would have the opportunity to interact with a complex; and when out-migrations are taking place (usually in the spring and early summer) the use of the lights is fewer than 8 hours, followed by a more than 16-hour interval of natural light, which compels the fish to move on. Finally, there is no quantitative evidence that predators concentrate above normal in areas where there is surface and underwater lighting from structures, or structures themselves. Typically, predators are territorial and only gather if the availability of prey is above average.

In summary, few negative interactions with wildlife from the location of marine aquaculture should be anticipated providing:

- all anchor lines and nets are kept taut at all times,
- predator nets are installed, or the sides of the cages are predator-resistant,
- all garbage and potential garbage is safely stowed until correct disposal, and
- the use of underwater lighting is limited to its necessity.

This biological opinion on wildlife interactions with aquaculture structures, with the exception of underwater lighting, is applicable to both biogeographical zones where fish rearing in marine enclosures might occur, and mostly concerns the epipelagic waters of coastal and offshore ecosystems. The effects of underwater lighting are more important in tropical waters where diurnal variation of light is about equal and attenuation of light is less as there is less particulate matter in suspension.

Further Information

- Aquaculture (2000) 183:307–323.
- Aquaculture Research (2003) 34:777–783.
- Aquatic Conservation of Marine and Freshwater Ecosystems (1993) 3:149–158.
- Marine Debris: Sources, Impacts, and Solutions. 1996. Springer-Verlag, New York.
- Ocean and Coastal Management (1996) 31-1:41–70.
- Wildlife Biology (2002) 8:193–199.

Appendix H. Physical Impact on Marine Habitat

Risk Hypothesis

Structures for offshore marine fish aquaculture (which specifically excludes those in the sublittoral coastal zone) are engineered and built to withstand almost all the extreme forces of the open-water, marine environment. The perceived risk is that the anchors required to hold such structures will physically destroy the sea bed habitat, and the large enclosures themselves will have an effect on current circulation and light penetration.

Background Experience

Excluding structures in the sublittoral zone, of which there are many, offshore marine fish aquaculture is currently confined to the continental shelf, which is often defined as lying above the 200-m contour. Because of the capital cost of working in marine waters, offshore fish are farmed intensively, and structures have evolved for specific species and for specific types of site. The options are floating net-pen complexes, and buoyant individual cages designed to remain at the surface or to be submerged as required.

Net-pen complexes are usually located in sheltered coastal estuaries, sounds, and lagoons that have rapid marine water exchange, and the anchorages are less than 30 m deep. A single circular or rectangular net pen may be 150–200 m^2 in surface area and 7–10 m deep. A long rectangular complex of net-pens may be up to 4,800 m^2 in surface area, with predator nets 10 m deep.

Individual buoyant cages are designed for less-sheltered waters, and submersible conical-shaped cages can be deployed in deeper water to avoid storms. Currently, submersible cages from 3,000–22,000 m^3 in volume are being operated at depths of less than 100 m, and up to 30 km offshore. Net-pen complexes are moored by many discrete taut legs or lines, depending on their formation and size. Additional lines may moor predator nets. Individual buoyant cages are moored by four discrete lines, which maintain tension all around continuously. Catenary mooring systems are still being tested.

Heavy-drag embedment anchors are used because of their high holding power. Of the three classifications of ocean bottoms (mud or silt, sand, and rock or marl), the preferred substrate is sand for its consistency and high holding power. Mud or silt is also good but can vary in consistency, and rock or marl is poor. Anchors can be bolted into rock and lava to give additional strength to the dead weight, but the practice is very costly.

Fish containment structures are designed to operate in currents up to 90 cm/sec, or about 1.74 knots. Sites with stronger currents are not desirable, as most fish must then expend too much energy maintaining their position in the cage instead of growing.

All marine aquaculture structures are regulated by international maritime laws with regard to navigational lighting.

Building the Conceptual Model

The two sources of physical impact on the offshore marine habitat are the large heavy anchors and the structures containing the fish. There is a common system for each, namely their installation and operation, and a few intermediate steps. These include the possibility of dragging anchors during heavy seas, a reduction in light penetration, and the reduction in water-mass exchange. The biological end points and their attributes include, potentially, the reduced abundance of benthic infauna and epifauna, the reduced abundance of sea grasses (*Zostera* spp. and *Phyllospadix* spp.), and the biological diversity of a confined coastal ecosystem.

Analysis and Characterization

The biological end points and their attributes for protection are:

1) the biological diversity of a confined coastal ecosystem;
2) areas of ecological importance, such as sea grasses, kelp beds, coral reef, etc.; and
3) the abundance of benthic infauna and epifauna.

Biological Diversity

Changes and reduction in water-mass exchange can indirectly affect the biological characteristics of confined ecosystems. Indicators of potential change include, for example, reduction in oxygen levels and decreased turbidity, which in turn reduces available nutrients. Changes in water quality in a confined ecosystem close to an aquaculture site can be determined by a standard water quality monitoring program, with spatial and temporal replication. If changes prove to be evident, then this can be followed by a biological monitoring program with selected indicator species.

Areas of Ecological Importance

Reduction in areas of ecological importance, in particular beds of sea grass, could be caused by fish enclosures in epipelagic waters either by reducing light penetration, or by dragging anchor lines.

Spatial and temporal Secchi disc transparency readings at regular intervals are the simplest method for comparing the extinction coefficients of light in and around an aquaculture site to determine its effect on light attenuation. Mapping the location and areas of large sea grass beds can be undertaken in a number of sophisticated ways (satellite imagery or underwater video), but for localized effects the traditional quadrant surveys by divers is appropriate. Divers might be used also to determine any chronic effects of the anchors destroying sea grasses, but in most cases the anchors will be too deep and video monitoring would be required.

Abundance of Benthic Fauna

Changes in the abundance of benthic infauna and epifauna may be the effect of dragging anchors. In coastal waters down to 40 m, divers can take standard core samples, which can be quantitatively and qualitatively analyzed in the laboratory. The results can be compared with reference stations located around the site. For deeper waters, grab samples must be used.

Biological Opinion

Floating offshore structures dampen waves by acting as surface breakwaters. However there is no evidence, as measured by levels of oxygen, that a long rectangular complex of net-pens held just below the surface has any affect on the water-mass exchange from one side to the other. Any possible impact on the water-mass exchange would occur if there were many complexes located in a sheltered water body that already had poor circulation and low rate of exchange. If this was the case then sites would not have been located there in the first place.

Buoyant cages are moored individually and are conical in shape, and therefore must provide even less obstruction to water-mass exchange. As these cages are also intended for deep coastal waters, the issue of affecting the water-mass does not arise.

For engineers, beds of kelp or coral reef suggest difficult and costly anchorage systems, and it is probable that a permit to locate an aquaculture site above a sensitive coral reef would never be requested or granted. Consequently, the most important areas of ecological importance for consideration are the many varieties of sea grasses that are rooted in sand. Sea grasses are distributed throughout both temperate and tropical waters, and all support a rich diversity of epifauna and infauna. Their growth is greatly dependent on the availability of light, and consequently the majority of sea grasses are to be found in subtidal habitats and shallow intertidal habitats down to a depth of 10 m or less. However, some species can be found at depths of 50–60 m. Any effects on these sensitive habitats from a reduction in ambient light penetration by the mass of an individual aquaculture structure or structural complex are more likely to occur in temperate waters than tropical waters. Temperate waters already have higher light extinction coefficients and shallower euphotic zones.

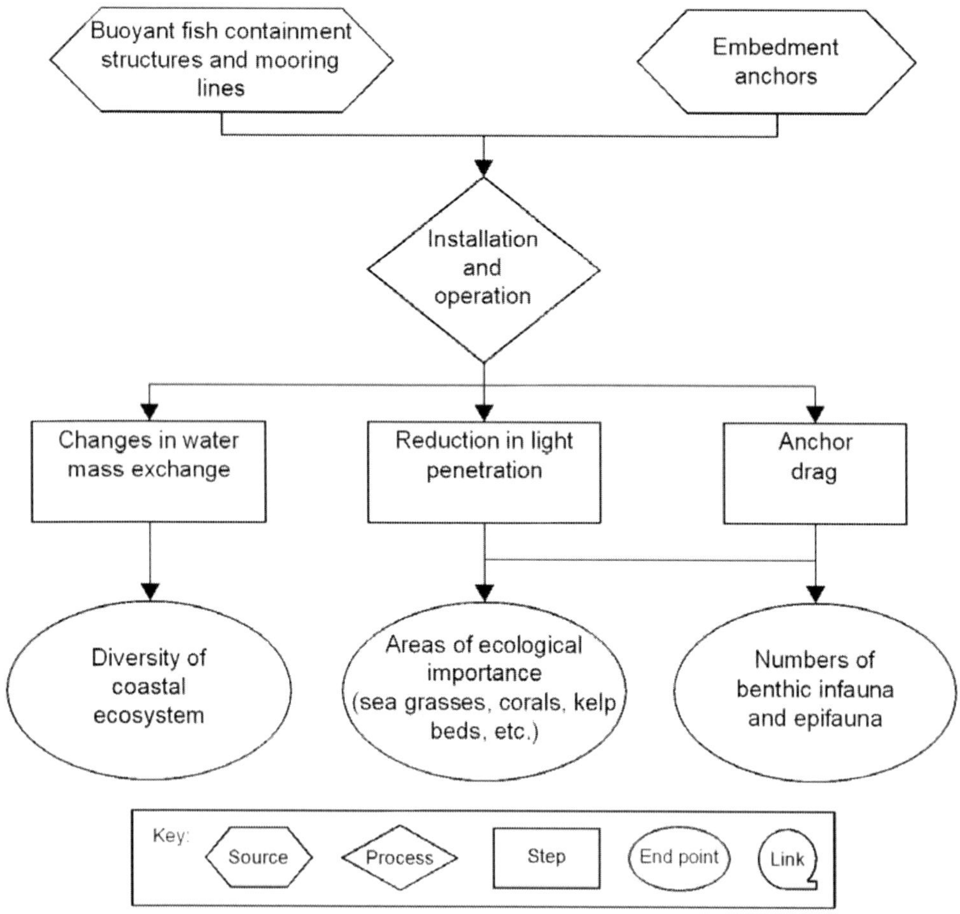

Figure H-1. A conceptual model for the physical impact on marine habitat.

In summary, the risk to marine habitat by the deployment of aquaculture structures is typically little and temporary, or none at all. Provided that any structure and its moorings are properly designed by marine engineers for the site selected, and each drag anchor is designed or purchased based on the detailed analysis of the substrate, then it is extremely unlikely that the structure will move once installed. Only during installation, as the drag anchor is being positioned by the anchor chain, may some disturbance take place of beds of sea grasses together with their epifauna and infauna, but it will be minimal and temporary. Here again, the risk can be managed by the use of the right vessel and equipment during deployment, under the supervision of professionals.

Risks to the ecosystem caused by aquaculture structures altering the water-mass exchange of an area are low at best. They can be managed, if necessary, by limiting the sizes of complexes to prevent any breakwater action, and re-siting them regularly to avoid creating a permanent footprint.

The risk of ambient light penetration shading the benthic flora and fauna of the substrate in the footprint of any complex is also minimal. Any effect is more of an issue in temperate waters.

Further Information

- Charting the Future of Ocean Farming. 1998. HAWAU-W-97-002:77-144. University of Hawaii, Honolulu.
- Engineering for Offshore Fish. 1990. Thomas Telford, Great Britain.
- Fundamentals of Aquacultural Engineering. 1995. Chapman & Hall, New York.
- Open Ocean Aquaculture. 2003. The World Aquaculture Society, Baton Rouge, Louisiana.

APPENDIX I. USING WILD JUVENILES FOR GROW-OUT

Risk Hypothesis

For centuries, large numbers of juvenile fish shoaling in coastal waters have been harvested by fishermen for bait and by farmers for stocking in fish ponds. The perceived risk is that the collection of these recruits to a wild population of marine fish can cause demographic changes in both the wild population and in other populations in that ecosystem.

Background Experience

Growing-out juveniles of wild freshwater fish populations in captivity for subsistence is part of the culture of many human societies. With the abundance of juveniles of catadromous fish, such as mullets (*Mugil* spp.) and milkfish (*C. chanos*), which congregate in brackish water lagoons to feed, the practice of out-growing provided large volumes of fish which could be preserved and stored, or marketed to others in the coastal communities. For the modern fish farming industry, harvesting wild larvae and juveniles and growing them out on a farm is only an interim measure while the technology is developed for breeding and artificial propagation of juveniles in hatcheries. Today, for example, there is now little or no collection of penaeid marine shrimp larvae as all the important species can be propagated successfully, and no further collection of some marine flatfish, such as plaice (*P. platessa*) and sole (*S. solea*), as the practice is not economical.

Dependable and economic techniques for propagating a number of valuable species, such as tunas, amberjacks and horse mackerels, mullets, and eels, among other species, have yet to developed. Therefore grow-out for market still depends on the collection of juveniles and subadults from the wild populations, which are either schooling in coastal areas to feed, or migrating to feeding grounds or breeding grounds. In most cases the numbers being harvested are being regulated by fisheries managers.

Building the Conceptual Model

The source of risk to a wild population of fish is the harvest of larvae, juveniles in their first year, or subadults migrating to feeding grounds or breeding grounds. The population may be a target species or a nontarget species.

The process of removing large numbers of juveniles from a population through one of several forms of fishing may result in consequences that offset the loss. These effects may be either positive or negative for a number of populations in the ecosystem. The direct and short-term effect of harvesting large numbers of larvae or juveniles from the wild population is that the potential number of adults will be greatly reduced. On the other hand, reducing the numbers of larvae and juveniles might reduce the competition for food and habitat and therefore improve the quality of those adults that survive. In the long-term, over several generations, the effect of harvesting larvae and juveniles may be a progressive reduction in annual recruitment to the fishery and a decline in total population strength. This could be accompanied by some compensatory behavioral changes, such as an increase in the fecundity of breeding adults.

Analysis and Characterization

The biological end points and their attributes for protection are:

1) annual recruitment to the wild target population, and
2) the adults of nontarget species taken as bycatch.

Annual Recruitment

Estimating any changes in the population dynamics of a fishery to assess the consequences of fishing pressure is a relatively routine procedure for fisheries managers. The accuracy of their assessment, however, increases with the amount and type of information available.

Consequently, monitoring the dynamics of a population of fish to assess the effects of removing larvae or juveniles in large numbers is a long-term process. For shallow coastal habitats, systematic sampling of populations with seine-nets or push-nets at locations spaced uniformly along the shoreline for morphometric data is adequate, but for offshore waters, the sampling program and gear depends greatly on the behavior of the target species. For example, many pelagic fish, such as tuna, have diurnal patterns of distribution with depth and schooling concentration. Concomitant with the sampling of the recruits, it is also necessary to have accurate or estimated information on the relative abundance, age, and size composition of the population, the age at maturity, and mortality rates, as well as relevant information about the habitat.

Armed with this information, fisheries managers can assess the strength of the target population and the structure of its community. They can then decide on a maximum quota of larvae and juveniles that can be harvested annually without jeopardizing the sustainability of the population.

Bycatch

The same techniques for the collection of information are required to assess the impact of harvesting of nontarget species.

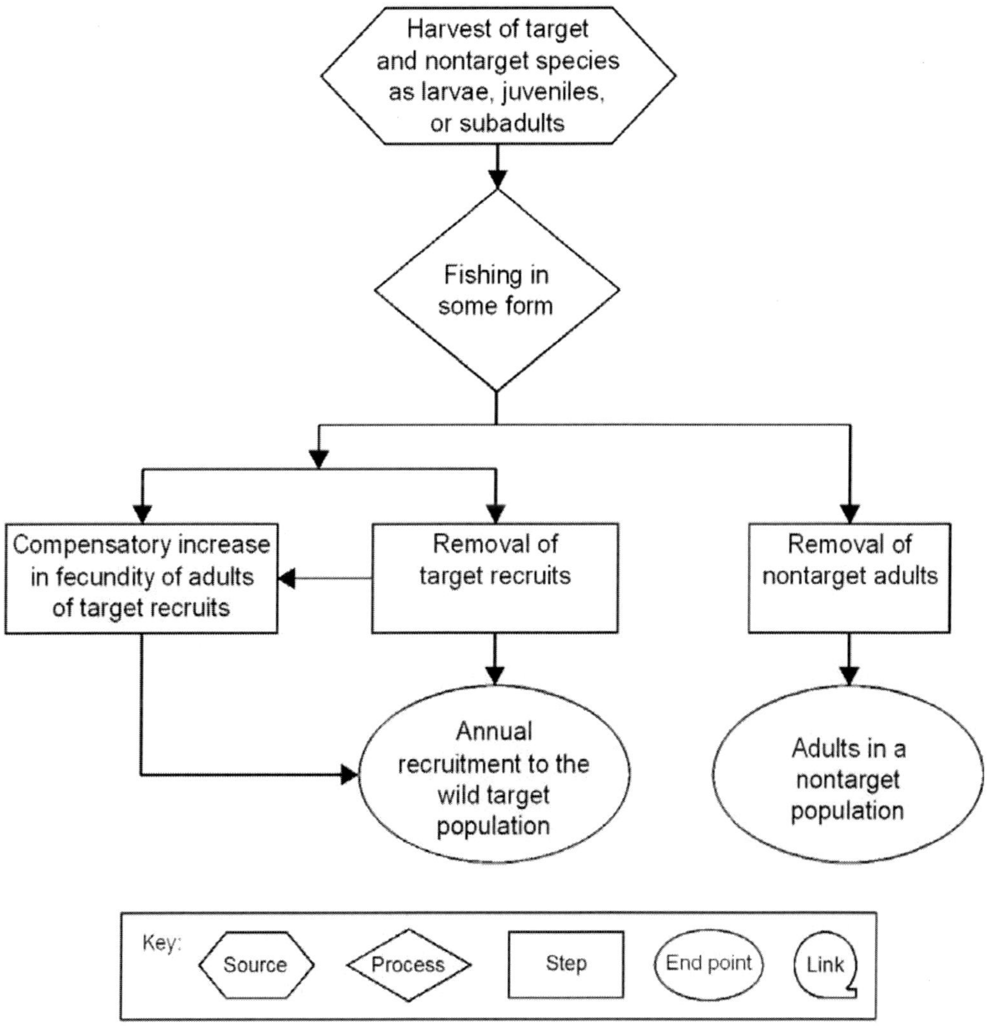

Figure I-1. A conceptual model for using wild juveniles for grow-out.

Biological Opinion

Juvenile marine fish that school in vast numbers have been harvested from their nursery grounds or on seasonal migrations for stocking and farming in coastal ponds for over two millennia. The most prolific harvests were of young mullets in the Nile Delta and lagoons of the Mediterranean Sea, followed later by some sea breams, sea bass, and eels. Egypt, for example, still collects over 125 million juvenile marine fish annually, of which over 95% are young mullets. There has also been a centuries-long tradition of harvesting many millions of juvenile milkfish in the coastal waters of the Philippines and Indonesia, although in recent years the numbers have declined as the traditional ponds have been converted to marine shrimp production. Within the last 50 years, and with the emergence of the modern aquaculture industry, the practices of harvesting juveniles has been extended to include penaeid marine shrimps in Central and South America, carangids (such as yellowtail and horse mackerel) in Japan, scombroids (such as bluefin tuna) in Australia, and anguillids (eels) and epinephalids (groupers) in Asia.

Harvesting juvenile fish has also been practiced in attempts to improve the commercial fisheries. For example, over a century ago juvenile flatfish were collected from nurseries around the coasts of Europe and transplanted to open shallow waters (sandbanks) in the middle of the North Sea to increase the resources.

The natural mortality of all fauna in the aquatic milieu decreases with age, and therefore the effect of harvesting is numerically less for larvae and juveniles as their chances for survival are low, and only a smaller percentage of the effective breeding population is being removed. This is particularly true for species of fish that school in vast numbers, and these are the species that in fact are harvested and grown out for commercial purposes. However, this is not necessarily a sound reason for fisheries managers to make a judgment because other factors must be considered, not the least of which is coastal development with its continuous reduction in the habitats that are the nurseries for many of these schools.

In summary, the harvesting of larvae, juveniles, and subadults of wild populations for growing out in marine aquaculture facilities presents a low risk of their decrease in the number of adults in both target and nontarget populations. It is possible to harvest a proportion of the young stages of a wild population without detrimental impact. The evidence is in the millions of mullets and milkfish that have been harvested annually for centuries and stocked in coastal fish ponds. In addition, the evidence is that there is little bycatch of nontarget species because most schools are very homogenous. Mullets may be an exception as the schooling populations may contain two or three species, but all of which can be targeted for harvest and grow-out. Consequently, it is the responsibility of fisheries managers to set and justify the limits for the harvest of target species or avoid any unnecessary impact on nontarget species. At the same time it is necessary for aquaculturists to find alternative sources through artificial breeding and propagation.

Further Information

- Aquaculture Economics and Management 7 (2003) 137–153.
- Canadian Journal of Aquatic Sciences (2004) 61:307–322.
- Capture-based Aquaculture. 2004. FAO, Rome, Italy.

- Interactions between Aquaculture and Capture Fisheries. 2005. General Fisheries Commission for the Mediterranean, GFCM/XXIX/2005/Dma.6. FAO, Rome, Italy.

APPENDIX J. HARVESTING INDUSTRIAL FISHERIES FOR AQUACULTURE FEEDS

Risk Hypothesis

There is a perceived risk that the projected global growth of intensive aquaculture, and its dependence on fish meal and fish oils in compounded feeds, will put excessive pressure on the commercial harvest of the industrial fisheries. As these fisheries are an important trophic link in the oceanic food chain, increased fishing effort will impact the sustainability of the resources as well as all carnivorous fauna at the higher trophic levels, such as predatory fishes, marine mammals, and sea birds.

Background Experience

The species most used for reduction to fish meal and oils are the small shoaling pelagic fish harvested from surface waters feeding at the lowest trophic level above or near to nutrient- rich oceanic upwellings. Predominantly these are the clupeid and clupeid-like industrial fishes (such as anchovies, capelin, menhaden, pilchards, sardines, and sardinellas), and sand lances (ammodytids). Their populations are very volatile, as their dynamics and location are dependent on ocean productivity, which in turn depends on the seasonal movement of some deep ocean currents. Consequently, depending on a variety of such biotic and abiotic factors, a population may appear to be at or close to its maximum sustainable exploitation rate.

These large but unstable fisheries are the principal intermediary between oceanic primary production and all the higher levels in the food chain through all the marine ecosystems. As they have a very high content of oils, they are not well-suited to processing for human consumption, and consequently they are reduced into fish meal and fish oils, predominantly for animal and poultry feeds.

The principal constituents of manufactured feeds for farmed carnivorous fish species have been fish meal and fish oils, at levels of about 25% and 30%, respectively. These two ingredients supply essential amino acids and fatty acids required by the fish for normal growth. More recently, small quantities (3–5% and 1–3%, respectively) have been included in feeds for omnivorous and herbivorous fish. In terms of use, manufactured fish feeds account for 30–35% of the fish meal and 50–59% of the fish oils produced annually. Almost all the rest is used in manufactured feeds for terrestrial farm animals and poultry.

Carnivorous fish convert these manufactured feeds to edible flesh with maximum efficiency. Farm salmon, for example, convert approximately one kilogram of feed into one kilogram of fish. For poultry the conversion is 3–5:1, and for swine it is 8:1. For this and other reasons, the farming of carnivorous marine fish is projected to increase substantially as the demand for fish increases beyond the natural resources.

Building the Conceptual Model

The source of the risk is the increased fishing pressure on the shoaling pelagic fisheries low on the marine food chain.

The process is the increasing demand by feed manufacturers for these species to reduce to fish meal and fish oils and compound into poultry and animal feeds, including fish feeds, and in the increasing demand for (oil-rich) seafood by consumers for health reasons. The short-term effect of removing a proportion of this important trophic level in the marine food chain is the reduction in numbers of carnivorous species feeding at the next level; and the long-term effect is the reduction in numbers of the higher fish predators, marine mammals, and marine birds.

Analysis and Characterization

The biological end points and their attributes for protection are:

1) a reduction in the population and breeding potential of the target species, and
2) a reduction in the population strengths of all carnivorous marine species higher up the food chain.

Reduction in Population and Breeding Potential

Because of the economic and social importance of the pelagic industrial fisheries, their population dynamics are now routinely monitored and assessed by fisheries managers and scientists worldwide. Armed with information collected over 50 years or more, fisheries managers predict each year the strength of the target populations. They then decide on a maximum quota that can be harvested without affecting the sustainability of the population.

Reduction in Population Strengths

The same techniques for assessing population strengths have also been applied for many years to species of carnivorous marine fish, marine mammals, and marine birds.

Biological Opinion

Experimental evidence indicates fishing activities can directly or indirectly affect species dependent on them. In the Shetland area of Scotland, for example, a sharp decline in breeding success of the Kittiwake (*Rissa* sp.), a seabird that generally feeds its young almost exclusively on sand eels, has paralleled an intensification of the industrial sand-eel fishery; Stellar sea lion (*Eumetopias jubatus*) populations in the Aleutian Islands have declined by an estimated 68% since the 1970s, possibly due to changes in the availability of preferred prey species that are fished commercially, and the decline of some bluefishes, key predators in western Atlantic fisheries, may be due to the competition with commercial fishermen for squid (*Loligo* spp.), butterfish (*Peprilus* spp.), and Atlantic menhaden (*Brevoortia tyrannus*).

Industrial fisheries for reduction into fish meal and fish oils are an important part of the global harvest, and contribute some 28–33 million mt annually. It is believed that a number of their individual resources have reached their maximum sustainable exploitation rates. As it is well known that these fisheries are important food sources for oceanic fauna at the higher trophic levels, they have been managed specifically by the six major nations involved and there is an international organization of fish meal and fish oil producers that has consultative status with FAO, the European Union, World Bank, and Codex Alimentarius.

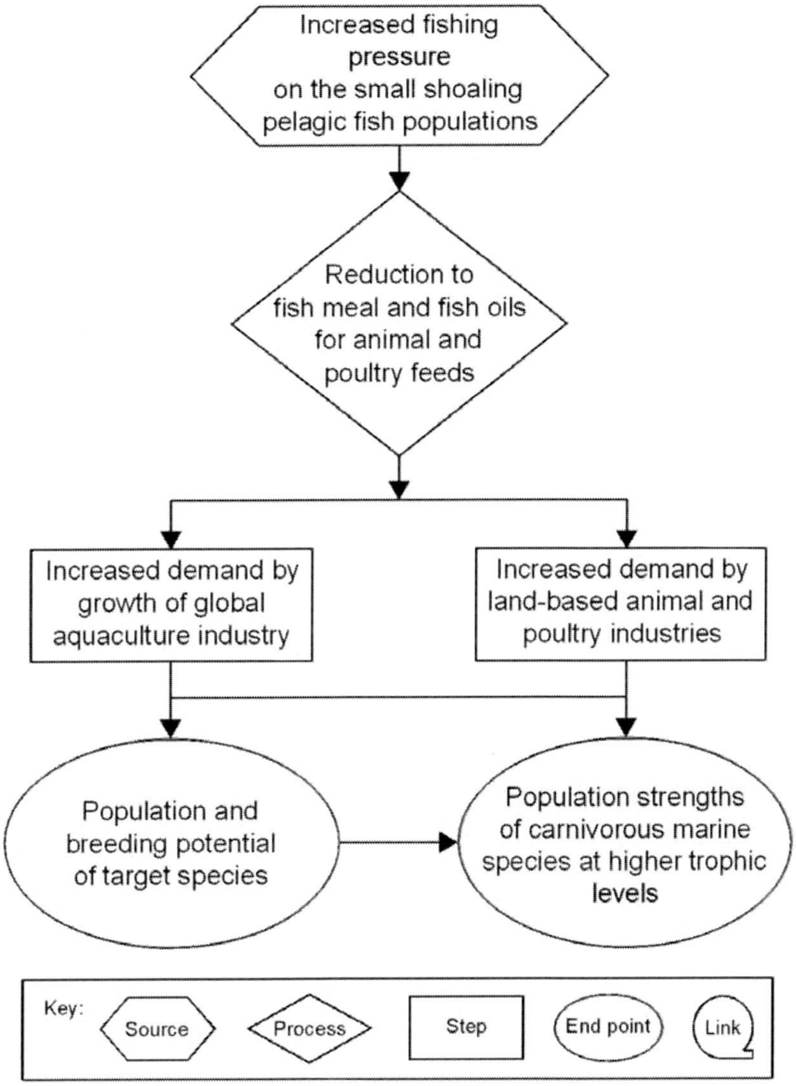

Figure J-1. A conceptual model for harvesting industrial fisheries for aquaculture feeds.

There has been intensive research and development by major feed companies and academic institutions to reduce pressure on stocks of pelagic fish traditionally used for the manufacture of fish meal and fish oils to find alternatives sources, such as fish processing waste (trimmings) and use of bycatch not currently landed for economic or regulatory reasons. In 1994, for example, annual discards at sea totaled 17.9–39.5 million mt, compared with 33 million mt of feed-grade fish harvested for reduction. More recent estimates of 20 million mt is equivalent to 25% of the reported annual harvest by marine capture fisheries. More importantly, there has been research to substitute grains and oilseed meals for fish meal as sources of protein and energy, and the development of cost-effective feeds that maximize growth rate, improve the conversion efficiency of feed to flesh, and reduce or eliminate feed wastage.

A distinct advantage of substituting a large part of fish meal and fish oils with vegetable oils in fish feeds is that it will remove most of the toxicants, such as dioxins, which are found in these pelagic fisheries. These then bioconcentrate up the food chain and can be detected in carnivorous fish, marine mammals, and sea birds, as well as farmed fish.

In summary, the projected growth of marine fish aquaculture presents little risk to the small pelagic industrial fisheries unless fisheries managers agree to target new resources in response to demand and rising prices of fish meals and fish oils. Moreover, even partial substitution of fish meal and fish oils in aquatic-animal feeds with suitable alternatives may not result in any significant reduction in the global harvest of industrial fisheries. It would reduce their costs, however, on the world market, which would be reflected in the reduced annual incomes of those communities dependent on industrial fisheries, and a possible increase in fishing effort to compensate.

Assessing and managing the risk is the responsibility of fisheries managers in the countries concerned to set and justify the limits for the harvest of target species and to avoid any unnecessary impact on nontarget species. At the same time, the feed technologists and manufacturers must be encouraged to continue to their research on substituting these ingredients as the demand for animal meats, including fish, continues to grow with the increase in world population.

Further Information

- Advances in Marine Biology (1998) 34:201–352.
- Aquaculture International (2005) 13:3–17.
- Aquatic Research, Culture, and Development (2004) 1:3–14.
- Biodiversity and Conservation (1992) 1:98–111.
- Canadian Journal of Fisheries and Aquatic Sciences (1997) 54:1342–1348.
- Fisheries Bulletin (1999) 97:776–785.
- Fisheries Bulletin (2000) 98:250–263.
- Fishmeal Information Network. 2005. FIN, Peterborough, United Kingdom.
- Marine Resource Economics (1997) 12:67–73.
- Nature (2000) 405:215–239.
- Royal Society for the Protection of Birds. 2004. 159-UK/R/01/C. Poseidon Aquatic Resources Management, Lymington, England.
- State of World Fisheries and Aquaculture. 1998 et seq. FAO, Rome, Italy.
- World Aquaculture (2000) 31:20–22.

APPENDIX K. WORKSHOP PARTICIPANTS

Kenneth M. Brooks
Aquatic Environment Sciences
644 Old Eaglemount Road
Port Townsend, WA 98368, USA
brooks@olympus.net

Peter R. Burbridge
University of Newcastle (Rtd.)
The Orchard, House of Ross
Comrie, Perthshire PH6 2JS, Scotland
p.r.burbridge@ncl.ac.uk
Stefano Cataudella

Stefano Cataudella
University of Rome Tor Vergata via Passo Lombardi 430
00100 Rome, Italy
cataudel@uniroma2.it

Brett R. Dumbauld
USDA/ARS
Hatfield Marine Science Center
2030 SE Marine Science Drive
Newport, OR 97365, USA
brett.dumbauld@oregonstate.edu

William T. Fairgrieve
NOAA Fisheries Service
Manchester Research Station
PO Box 130
Manchester, WA 98353, USA
William.Fairgrieve@noaa.gov

John R.M. Forster
Forster Consulting
533 East Park
Port Angeles,
WA 98362, USA
jforster@olypen.com

Robert N. Iwamoto
NOAA Fisheries Service
Northwest Fisheries Science Center
2725 Montlake Boulevard East
Seattle, WA 98112, USA
bob.iwamoto@noaa.gov

David F. Jackson
Marine Institute, Fisheries Research Centre
Abbotstown, Dublin 15, Ireland
dave.jackson@marine.ie

Sadasivam J. Kaushik
INRA, Quartier Ibarron
64310 Saint Pée sur Nivelle, France
kaushik@st-pee.inra.fr

Colin E. Nash (Secretariat)
NOAA Fisheries Service
Manchester Research Station
PO Box 130
Manchester, WA 98353, USA
colin.nash@noaa.gov

Michael B. Rust
NOAA Fisheries Service
Northwest Fisheries Science Center
2725 Montlake Boulevard East
Seattle, WA 98112, USA
mike.rust@noaa.gov

Philip A.D. Secretan
AUMS Ltd., Lewes Enterprise Centre
112 Malling Street, Lewes
East Sussex BN7 2RJ, England
secretan@aums.com

Karl D. Shearer
NOAA Fisheries Service
Northwest Fisheries Science Center
2725 Montlake Boulevard East
Seattle, WA 98112, USA
karl.shearer@noaa.gov

Glenn W. Suter II (Facilitator)
USEPA Facilities 117
26 West Martin Luther King Drive
Cincinnati, OH 45268, USA
suter.glenn@epamail.epa.gov

Ole J. Torrissen
Institute of Marine Research
PO Box 1870, Nordnes
5817 Bergen, Norway
ole.torrissen@imr.no

John K. Volkman
CSIRO Marine Research
GPO Box 1538
Hobart, Tasmania TAS 7001, Australia
John.Volkman@csiro.au

Masashi Yokota
Tokyo University of Marine Science and Technology
4-5-7 Konan Minato-ku
Tokyo 108-8477, Japan
masashi_yokotajp@yahoo.co.jp

APPENDIX L. SOURCES OF FURTHER INFORMATION

Black, K. D. (2001). Environmental Impacts of Aquaculture. Sheffield Academic Press, Sheffield, England.

Bridger, C. J. & B. A. Costa-Pierce. (2003). Open Ocean Aquaculture. The World Aquaculture Society, Baton Rouge, Louisiana.

Convention on Biological Diversity. (2004). Solutions for Sustainable Mariculture: Avoiding Adverse Effects of Mariculture on Biological Diversity. CBD Technical Series 12. Secretariat of the Convention on Biological Diversity, Montreal, Quebec, Canada.

Department of Fisheries and Oceans. (2000). An Evaluation of Knowledge and Gaps Related to Impacts of Freshwater and Marine Aquaculture on the Aquatic Environment. EVS Report 03-0064-41. Ottawa, Ontario, Canada.

Environmental Assessment Office. (1999). Salmon Aquaculture Review. Environmental Assessment Office. Victoria, British Columbia, Canada.

European Commission. (1999 et seq). European Commission Research on Sustainable Agriculture, Fisheries, and Forestry. Area 5.1.2 Sustainable Fisheries and Aquaculture. Online at http://europa.eu.int/comm/research/quality-of-life/ka5/en/list512_en.htm [accessed June 2005].

Relevant ongoing projects with their own web sites and publications:

AQCESS—Aquaculture and coastal economy and social sustainability.

FORM—Fish oil and meal replacement.
MEDVEG—Effects of nutritional release from Mediterranean fish farms on benthic vegetation in coastal ecosystems.
MERAMED—Development of monitoring guidelines and modelling tools for environmental effects from Mediterranean aquaculture.
PUFAfeed—Feed for aquatic animals that contains cultivated marine microorganisms as alternatives for fish oil.
RAFOA—Research alternatives to fish oil for aquaculture.

European Commission. 2004. Farmed Fish and Welfare. Directorate General for Fisheries, Research and Scientific Analysis Unit, Bruxelles, Belgium.

European Parliament. (2004). Food Safety: From Farm to Fork. Directorate General for Health and Consumer Protection. Online at http://europa.eu.int/comm/food/index_en.htm [accessed June 2005].

Hambray, J. & Southall, T. (2002). Environmental Risk Assessment and Communication in Coastal Aquaculture. Joint Group of Experts on the Scientific Aspects of Marine Environmental Protection, GESAMP WG 31. FAO, Rome, Italy.

Hayes, K. R. (1997). A Review of Ecological Risk Assessment Methodologies. CRIMP Technical Report No. 13. CSIRO, Hobart, Tasmania, Australia.

Holm, M., Dalen, M., Haga, J. A. R. & Hauge. A. (2003). The Environmental Status of Norwegian Aquaculture. Bellona Report No. 7. Bellona Foundation, Oslo, Norway.

Joint Group of Experts on the Scientific Aspects of Marine Environmental Protection. (1996). Monitoring the Ecological Effects of Coastal Aquaculture Wastes. GESAMP Reports and Studies No. 57. FAO, Rome, Italy.

Scottish Environmental Protection Agency. (2004). Freedom of Information (Scotland) Act 2002, Publication Scheme. Online at http://www.sepa.org.uk [accessed June 2005].
Relevant publications online:
Policies on Treatment Chemicals for Fish Farming.
Policies on Regulation and Expansion of Caged Fish Farming on Salmon in Scotland.
Aquaculture Guidance, Policy, and Background Information.

Scottish Executive Locational Guidelines for Fish Farming. (2002). Scottish Fisheries Research Report No. 63. Fisheries Research Services, Marine Laboratory, Aberdeen, Scotland.

University of Hawaii. (1998). Charting the Future of Ocean Farming. HAWAU-W-97-002. University of Hawaii Sea Grant Program, Honolulu.

INDEX

A

abatement, 60, 133
abiotic parameter, 66
abuse, 123
access, 66, 180
accounting, 28, 31, 62
acetic acid, 91, 123
acid, 145
adaptation(s), 204, 205
administrators, vii, 157, 175
adsorption, 183
adults, 194, 205, 216, 218
advancement, 104
adverse effects, 13, 93, 155, 164, 165, 174, 177, 190, 199
advocacy, 161
aerosols, 85
age, vii, 1, 12, 77, 130, 142, 197, 199, 217, 218
agencies, 7, 108, 169
aggregation, 5, 71
agriculture, 147, 158, 192, 198
Alaska, 100
algae, 12, 46, 47, 50, 59, 80, 85, 98, 100, 104, 182, 183, 186
allele, 203, 204
alters, 54
amino, 219
amino acid(s), 219
ammonia, 9, 33, 101, 178, 179, 182, 183
ammonium, 10, 84, 118, 141, 182
amplitude, 138
anaerobic bacteria, 33
analytical framework, viii, 158, 161
anchorage, 213
ANOVA, 167
anoxia, 40, 94
antibiotic, 82, 83, 84, 85, 116, 200
antibiotic resistance, 83
ARS, 223
arsenic, 93
Artemia, 84, 137
arthropods, 199
Asia, 20, 24, 25, 30, 43, 52, 66, 67, 83, 86, 88, 150, 191, 218
Asian countries, 66
assessment, vii, viii, 13, 16, 17, 18, 21, 27, 35, 39, 44, 57, 68, 85, 106, 121, 122, 126, 130, 132, 134, 136, 147, 157, 158, 159, 161, 162, 163, 166, 169, 172, 174, 181, 186, 216
assets, 114, 123, 139, 141
assimilation, 2, 3, 5, 9, 15, 27, 28, 46, 48, 99, 100, 109, 144, 183
Australasia, 22, 25, 29, 31, 40, 52, 66, 68
authority(s), 162, 167, 168, 180, 194
avoidance, 76, 163
avoidance behavior, 76

B

background information, 164
bacteria, 33, 52, 55, 83, 85, 89, 90, 91, 117, 142, 147, 176, 182, 183
bacterial colonies, 83
bacterial infection, 86, 196
barriers, 75, 77
base, 5, 62, 95, 98, 178, 186
baths, 165
behavioral change, 88, 216
behaviors, 54
Belgium, 226
benchmarks, 97, 160, 189
beneficial effect, 45, 59, 71
benefits, 1, 6, 89, 98, 99, 102, 107, 158, 159
benign, 82
benthic invertebrates, 58, 64, 176

bioaccumulation, 88, 94, 100, 137, 152, 196
bioavailability, 90, 91, 93, 190
biodiversity, 4, 8, 32, 39, 45, 52, 54, 55, 59, 61, 62, 64, 67, 82, 83, 85, 105, 107, 108, 111
biological consequences, 47
biological interaction, vii, 158, 201, 202, 205, 206
biological processes, 33
biological samples, 96
biomass, 3, 9, 29, 30, 46, 52, 53, 58, 59, 60, 61, 62, 63, 64, 65, 68, 69, 70, 73, 81, 100, 102, 104, 105, 106, 108, 126, 134, 148, 176, 177, 182, 184
bioremediation, 31, 92, 97, 98, 101, 102, 116, 118, 130, 131, 133, 153, 181
biotic, 19, 52, 62, 65, 66, 68, 69, 122, 132, 138, 174, 219
birds, 45, 59, 78, 79, 106, 112, 165, 172, 178, 179, 207, 208, 210, 219, 220, 222
bleaching, 138
bleeding, 191
blood, 89, 151, 165, 194
body size, 92
body weight, 62
bone, 190
breakdown, 94, 183, 197
breeding, 163, 166, 200, 201, 203, 204, 205, 207, 209, 210, 215, 216, 218, 220
brevis, 153
by-products, 197

C

cables, 171
cadmium, 92, 93, 109, 120
calcium, 65
calcium carbonate, 65
carbon, 3, 8, 19, 25, 26, 27, 28, 29, 30, 31, 32, 33, 35, 37, 41, 46, 47, 53, 57, 58, 71, 80, 92, 97, 104, 105, 110, 128, 152, 153, 154, 176, 182
carbon dioxide, 33, 35, 182
Caribbean, 24, 30, 43, 66, 113, 120, 131, 141, 145
carnivores, 71, 97
case study(s), 65, 123, 137, 148, 151, 154, 155, 177
catfish, 83
CBD, 225
cell division, 186
certification, 123
challenges, 1, 102, 103, 107, 141
chemical(s), 3, 5, 6, 19, 25, 32, 35, 36, 39, 45, 54, 55, 57, 58, 62, 66, 67, 72, 82, 86, 88, 89, 90, 91, 95, 96, 97, 99, 107, 108, 109, 116, 118, 119, 126, 131, 136, 146, 148, 153, 167, 168, 171, 175, 176, 177, 187, 189, 190, 195
chemical properties, 136, 146, 171
chemotherapy, 117
Chile, 10, 11, 14, 22, 26, 27, 32, 37, 49, 51, 60, 70, 82, 83, 91, 92, 100, 116, 123, 138, 146, 148
China, 26, 48, 50, 51, 67, 70, 85, 102, 130, 141, 144, 155, 156
chitin, 88
chlorine, 89
chlorophyll, 46, 47, 48, 49, 51, 60, 166
chromium, 93, 109
circulation, 19, 37, 42, 49, 60, 96, 186, 211, 213
clarity, 18, 81
classes, 71
classification, 35, 126, 142
cleaning, 77, 107, 177, 181, 190, 206, 208
climate(s), 7, 142, 161, 183
clustering, 62
CO_2, 34
coastal communities, 215
coastal ecosystems, 1, 151, 177, 187, 226
coastal management, 155
coastal region, 111, 122
coatings, 91
collaboration, 8
colonization, 46, 96, 144
commerce, 203
commercial, 19, 31, 46, 53, 57, 58, 65, 72, 73, 75, 93, 95, 97, 98, 99, 100, 102, 103, 109, 121, 150, 158, 166, 191, 192, 194, 199, 200, 204, 218, 219, 220
communication, 169, 170
comparative analysis, 2
compensation, 186
competition, 200, 203, 216, 220
compilation, 7
complexity, 168
composition, 3, 4, 18, 19, 53, 54, 59, 63, 65, 72, 80, 111, 121, 122, 133, 184, 217
compounds, 31, 33, 50, 88, 89, 90, 97, 120, 131, 147, 152, 165, 176, 182, 183, 190, 194, 196, 197, 198
computer, 2, 53, 64, 72, 94, 106, 108, 167, 168, 177, 184
computer simulations, 2
conceptual model, 118, 163, 175, 178, 185, 188, 193, 198, 202, 209, 214, 217, 221
conditioning, 75
conflict, 5, 73
consensus, 163
conservation, 73, 74, 75, 78, 80, 117, 121, 122, 127, 148, 151
conserving, 103
constituents, 180, 219
construction, 81
consumers, 72, 220

consumption, 72, 100, 102, 158, 199, 207, 219
contaminant, 92, 130, 131
contamination, 91, 92, 93, 102, 120, 137, 150, 210
contour, 170, 211
Convention on Biological Diversity, 225
conversion rate, 180
cooperation, 168
coordination, 168
copper, 82, 90, 91, 92, 93, 94, 97, 115, 125, 148, 155, 166, 178, 187, 188, 189, 190
coral reefs, 45, 79, 135, 166
correlation(s), 6, 9, 36, 43, 63, 68, 73, 98, 104, 108
cost, 3, 39, 41, 42, 44, 53, 54, 56, 59, 91, 97, 102, 105, 107, 146, 167, 170, 171, 172, 180, 211, 222
cost effectiveness, 56
covering, 7, 8, 43, 69, 178
crabs, 84, 88, 90
critical value, 43
Croatia, 42, 138
crop(s), 3, 5, 104, 107, 210
cultivars, 100
cultivation, 100, 101, 118
cure, 199
cycles, 96, 99, 105, 123
cycling, 21, 98, 120, 152, 183
cyst, 186
cytochrome, 80

D

damages, 75
danger, 207
data collection, 166
data set, 54, 58, 69
database, 109
DDT, 94
decision makers, vii, 157
decision-making process, 161
decomposition, 3, 19, 33, 34, 36, 39, 40
deficiency, 140
degradation, 3, 25, 41, 53, 57, 59, 89, 96, 105, 160, 196
Delta, 218
demographic change, 215
demographic data, 204
denitrification, 118
Denmark, 83, 129, 130
dependent populations, 172
dependent variable, 69, 167
deposition, 3, 25, 27, 29, 36, 62, 71, 120, 167, 177, 179
deposition rate, 29
deposits, 27, 37
depression, 203
depth, 3, 6, 9, 18, 21, 29, 30, 31, 37, 43, 49, 55, 61, 66, 69, 81, 98, 104, 105, 108, 166, 171, 172, 175, 179, 186, 214, 216
derivatives, 89
detectable, 49, 63, 96, 184, 205
detection, 3, 5, 26, 69, 105, 106, 150, 155, 177
deterrence, 150
developing countries, 150
deviation, 80
dialysis, 48
diatoms, 184
diet, 28, 37, 94, 100, 187
diffusion, 33, 152
digestibility, 15
digestion, 187, 190
dioxin, 92
direct measure, 51
disaster, 147, 202
discharges, 12, 24, 126, 177
discomfort, 75
discontinuity, 35
disease organisms, vii, 158, 165, 173, 192, 193
diseases, 85, 147, 196, 198, 201
disinfection, 195
dispersion, 9, 10, 14, 39, 43, 46, 53, 58, 61, 102, 104, 117, 121
dissolved oxygen, 2, 8, 16, 17, 33, 89, 104, 149, 163, 166, 182, 186
distinctness, 67
distribution, 41, 71, 85, 111, 127, 132, 136, 141, 146, 147, 164, 167, 168, 216
diversification, 118
diversity, 4, 53, 57, 58, 59, 61, 62, 63, 64, 65, 66, 67, 68, 69, 70, 71, 73, 96, 97, 109, 127, 159, 167, 171, 192, 212, 214
DNA, 203, 204, 205
dogs, 77
DOI, 121, 123, 143, 155
dominance, 57, 59, 63, 64, 67
doppler, 166
dosage, 83
draft, 8
drawing, 69
drug resistance, 199
drugs, 85, 86, 88, 89, 195, 199
drying, 91
dumping, 88, 167

E

ecological effects, vii, 84, 85, 157, 174, 175, 194
ecological processes, 7, 55

ecological risks, vii, 157, 159, 172, 200, 205
ecological roles, 72, 106
ecology, 153
economic incentives, 1
ecotoxicological, 93
editors, 112, 113, 117, 122, 127, 128, 134, 143, 145, 147, 148, 150, 151, 152, 155
education, 2
effluent(s), 4, 5, 12, 15, 45, 46, 47, 48, 49, 50, 51, 52, 79, 80, 81, 97, 102, 107, 118, 123, 130, 131, 133, 137, 140
Egypt, 218
electrodes, 33
electrons, 33
electrophoresis, 203
emission, 141
endangered, 79
endangered species, 79
energy, 28, 100, 102, 106, 107, 171, 204, 212, 222
energy input, 107
enforcement, 210
engineering, 102, 152, 162, 168, 170, 202
England, 222, 224, 225
entrapment, 206, 207
environmental change, 150, 156
environmental conditions, 26, 54, 60, 191, 193, 196
environmental effects, 2, 5, 6, 7, 8, 9, 10, 11, 24, 25, 86, 88, 90, 91, 94, 103, 104, 107, 108, 116, 126, 144, 145, 226
environmental factors, 61, 89, 194
environmental impact, vii, 2, 6, 8, 12, 13, 17, 19, 20, 21, 22, 27, 29, 31, 41, 44, 52, 60, 63, 65, 68, 69, 72, 73, 78, 82, 85, 86, 91, 93, 95, 96, 97, 98, 105, 108, 113, 115, 116, 125, 126, 127, 128, 129, 136, 137, 140, 142, 154, 155
environmental issues, 2, 7
environmental management, 152
Environmental Protection Agency, 28, 226
environmental quality, 28, 80, 81, 102, 130, 131, 152
environmental risks, vii, 1, 88, 161, 181
environmental sustainability, 98
environmental variables, 61
enzymatic activity, 53, 97
enzyme, 88
EPA, 28, 89, 175
epidemic, 192, 194, 197
epithelia, 182
equilibrium, 107
equipment, 7, 41, 79, 82, 90, 188, 202, 208, 215
erosion, 34
eukaryotic, 53
Europe, 8, 10, 19, 20, 22, 25, 26, 28, 32, 38, 45, 46, 52, 60, 61, 70, 82, 85, 86, 88, 90, 96, 98, 115, 119, 186, 218
European Commission, 128, 129, 225, 226
European Parliament, 226
European Union, 80, 81, 91, 135, 221
evidence, 28, 29, 46, 48, 51, 59, 96, 176, 179, 180, 194, 201, 205, 210, 213, 218, 220
evolution, 88, 140
exchange rate, 50, 171
exclusion, 5, 66, 71, 75, 76, 123, 179
excretion, 102, 104
expertise, 7, 53, 108, 122, 162
exploitation, 219, 221
explosives, 75
exposure, viii, 50, 72, 79, 83, 84, 85, 100, 137, 158, 160, 163, 164, 168, 174, 175, 184, 189, 191, 192, 197, 209
extinction, 213, 214
extraction, 134, 159, 167

F

farm efficiency, 2
farm size, 41, 108
farmers, 10, 17, 73, 77, 145, 161, 163, 186, 199, 210, 215
fatty acids, 144, 219
fauna, 4, 57, 59, 60, 61, 66, 67, 80, 84, 95, 121, 123, 140, 142, 176, 178, 189, 195, 197, 199, 218, 219, 221
FDA, 82, 83, 86
feces, 9, 13, 19, 20, 25, 28, 31, 35, 84, 86, 94, 176, 179, 180, 182, 183, 190, 196, 197
federal agency, 1
fencing, 78
fertilization, 48, 94
fidelity, 77, 121, 142
field trials, 53
filtration, 101
financial, 75, 99, 127
Finland, 39, 51, 60, 83, 96
fish oil, 19, 94, 219, 220, 221, 222, 226
fisheries, vii, 6, 50, 58, 73, 78, 128, 129, 132, 135, 147, 158, 164, 165, 172, 173, 192, 203, 204, 205, 216, 217, 218, 219, 220, 221, 222
fishing, 73, 124, 137, 159, 161, 165, 166, 202, 208, 216, 219, 220, 222
fitness, 164, 201, 203, 204
flex, 210
flora, 184, 191, 192, 215
flora and fauna, 184, 191, 192, 215
fluctuations, 152

Food and Drug Administration, 82
food chain, 48, 88, 91, 93, 195, 196, 219, 220, 222
food industry, 93, 102
food production, 101, 109
food safety, 199
food security, 103, 109
food web, 4, 46, 48, 54, 58, 59, 65, 72, 83, 104, 106, 143, 178, 186
force, 150, 176
Ford, 141
formation, 65, 171, 212
fouling, 5, 65, 71, 72, 78, 82, 91, 106, 107, 109, 119, 176, 180, 187
France, 224
freshwater, 18, 78, 83, 88, 93, 100, 107, 148, 183, 189, 200, 204, 215
freshwater species, 204
fungi, 89

G

garbage, 207, 210, 211
genetic diversity, 204
genetic information, 203, 204
genetics, 205
genome, 203
gill, 50, 136, 182, 190
global scale, 132
glutathione, 124
gonads, 190, 205
governments, 85
grain size, 42, 43, 166, 179
graph, 66, 87
graphite, 189
grasss(), 63, 122, 163, 164, 212, 213, 214
grazers, 48, 143, 184
grazing, 91, 101, 186
Great Britain, 215
Greece, 19, 24, 42, 49, 51, 63, 70, 73, 112, 132, 133, 136
green alga, 47, 184
grouping, 8
growth, 6, 12, 21, 46, 47, 50, 72, 74, 75, 79, 81, 85, 90, 93, 99, 100, 102, 109, 114, 121, 146, 151, 152, 153, 156, 171, 186, 209, 214, 219, 222
growth rate, 79, 102, 151, 222
Guangdong, 50
guidance, 9, 54, 98, 103
guidelines, vii, 18, 90, 92, 94, 97, 123, 157, 159, 160, 161, 164, 172, 189, 202, 226
Gulf of Mexico, 6, 125
Gulf of Trieste, 110, 133

H

habitat quality, 55, 59
half-life, 83
harassment, 75, 78, 141, 150, 206
harmful effects, 5, 81, 107
harvesting, vii, 77, 102, 158, 181, 215, 216, 217, 218, 221
Hawaii, 11, 14, 15, 18, 43, 51, 67, 70, 73, 117, 127, 134, 135, 142, 146, 152, 215, 226
hazardous waste, 210
hazards, 181
health, 1, 2, 5, 6, 85, 87, 89, 95, 107, 116, 122, 127, 148, 149, 171, 187, 191, 192, 194, 195, 196, 220
health problems, 195
heat shock protein, 80
heavy metals, vii, 8, 91, 92, 93, 94, 107, 112, 137, 150, 157, 165, 173, 187, 188, 189
hematology, 151
historical data, 65
history, 50, 148, 149, 158, 166, 167, 171
homogeneity, 167
Hong Kong, 30, 67, 94, 123, 155
host, 4, 191, 194, 195, 199
House, 223
human, 4, 24, 65, 67, 69, 72, 81, 83, 85, 93, 94, 100, 103, 106, 158, 159, 161, 167, 183, 190, 192, 199, 201, 208, 215, 219
human health, 83, 85, 159, 161, 183, 190, 192, 199
Hunter, 128, 129, 135
husbandry, 5, 74, 85, 89, 107, 187, 199
hybridization, 201
hydrocarbons, 166
hydrogen, 34, 37, 86, 88, 89, 136, 147, 151, 155, 178, 182
hydrogen peroxide, 86, 88, 89, 136, 147, 151, 155
hydrogen sulfide, 34, 37, 178, 182
hydroid, 125
hydrological conditions, 51
hypothesis, 46, 191, 195
hypoxia, 39

I

Iceland, 148
ideal, viii, 5, 158, 172
identification, 34, 57, 60, 64, 163, 166, 169, 181, 203
image(s), 35, 39, 42, 55, 57, 63, 65, 69
image analysis, 35, 39, 42, 63, 65
imagery, 28, 42, 57, 61, 132, 213
immersion, 82, 88, 89, 91, 194, 195, 196, 197
immune response, 137

immune system, 191, 193, 196
immunity, 192, 196
IMO, 131
impact assessment, 78, 115, 138
imported products, 109
improvements, 2, 5, 18, 21, 25, 31, 43, 89, 92, 96, 101, 104, 105, 204
in vitro, 116
inadmissible, 194
inbreeding, 200
incidence, 166, 204
independence, 169
India, 67, 70, 115, 129, 144
individuals, vii, 60, 63, 73, 75, 76, 77, 157, 160, 174, 192
Indonesia, 72, 150, 218
industry(s), 1, 2, 4, 6, 7, 8, 19, 21, 24, 32, 44, 53, 58, 60, 69, 74, 75, 85, 95, 98, 101, 103, 104, 108, 109, 118, 126, 138, 139, 141, 152, 159, 168, 169, 177, 210, 215, 218
infancy, 172
infection, 165, 192, 194
inferences, 68
infestations, 126, 195, 198
ingest, 79
ingestion, 42
ingredients, 187, 190, 219, 222
initiation, 57, 153
injury, 74, 75
inoculation, 165, 195
insecticide, 88
inspections, 191, 196
institutions, 221
integration, 39, 99, 102, 103
integrity, 159, 188, 196, 203, 204
intensive aquaculture, 59, 137, 177, 181, 191, 195, 219
intensive farming, 158
interface, 27, 37, 171
intervention, 75, 105, 159, 167
intrusions, 172
invertebrates, 5, 58, 85, 100, 117, 121, 122, 124, 180
investment, 99, 171
investors, 7
ions, 188
Iowa, 112, 113, 152, 195, 200
Ireland, 82, 137, 224
iron, 33, 34, 92, 93
Islam, 9, 13, 20, 26, 130
isopods, 180
isotope, 23, 28, 35, 41, 46, 47, 64, 80, 100, 121, 122, 146
Israel, 42, 65, 70
issues, viii, 2, 6, 37, 45, 72, 111, 114, 115, 127, 158, 159, 163, 169
Italy, 11, 14, 23, 29, 32, 41, 42, 49, 51, 62, 63, 70, 77, 97, 99, 114, 115, 117, 121, 125, 131, 134, 218, 219, 222, 223, 226

J

Japan, 30, 32, 43, 90, 102, 135, 144, 150, 155, 218, 225
justification, 130
juveniles, vii, 158, 165, 173, 191, 205, 206, 210, 215, 216, 217, 218

K

k-dominance curves, 54, 57, 62, 63
kidney(s), 190
kill, 91, 183
knots, 171, 212
Korea, 66, 101, 102, 132, 142

L

laboratory studies, 88, 89
landfills, 181
landings, 73, 135
language barrier, 108
larvae, 90, 91, 100, 122, 165, 194, 215, 216, 217, 218
larynx, 125
Latin America, 73, 145
laws, 212
leaching, 117, 188
lead, 14, 25, 31, 46, 81, 93, 109, 178, 180, 184, 199
legislation, 169
legs, 212
lice, 88, 89, 113, 126, 150, 155, 194, 196, 199, 201
lifestyle decisions, 158
lifetime, 6, 108, 204
light, 4, 18, 45, 48, 49, 81, 83, 89, 108, 172, 182, 186, 210, 211, 212, 213, 214, 215
lipids, 2, 8, 19
lithium, 75, 93
liver, 90, 190, 203
local community, 159
local conditions, 96
Louisiana, 149, 215, 225
low risk, 4, 78, 89, 94, 194, 218
lying, 170, 211

M

macroalgae, 12, 102, 116, 127, 166, 183, 184, 186
macrobenthos, 57, 67, 68, 97, 110, 122, 135, 176, 179
magnitude, 4, 21, 51
majority, 16, 52, 158, 164, 167, 177, 183, 186, 191, 205, 214
Malaysia, 81, 110, 125
mammal(s), 4, 5, 45, 74, 75, 76, 77, 78, 79, 106, 122, 139, 155, 164, 165, 172, 190, 201, 206, 207, 208, 209, 210, 219, 220, 222
man, 194
manganese, 33, 92
mangrove forests, 5, 45, 79, 106, 110
mangroves, 81, 106
manure, 93, 94
mapping, 57, 132
marine ecosystems, vii, 1, 13, 45, 104, 105, 157, 159, 162, 163, 164, 171, 172, 219
marine environment, vii, 1, 4, 5, 7, 12, 15, 20, 45, 48, 52, 64, 82, 85, 88, 89, 91, 94, 95, 103, 106, 107, 108, 118, 123, 142, 147, 150, 167, 176, 182, 187, 190, 196, 199, 205, 211
marine finfish aquaculture, vii, 1, 2, 6, 7, 86, 88, 98, 103, 104, 111, 116, 117, 122, 126, 128, 137, 144, 147, 148, 149, 150, 151, 154, 155
Marine Mammal Protection Act, 7, 77, 78
marine species, 204, 205, 206, 220
marketability, 100
Maryland, 139
mass, 9, 13, 15, 21, 22, 26, 27, 184, 187, 212, 213, 214, 215
materials, 35, 77, 79, 169, 171, 210
matrix, viii, 158, 160, 172
matter, 3, 35, 36, 37, 39, 41, 42, 114, 150, 165, 175, 176, 177, 179, 184, 188, 189, 196, 197, 201, 211
measurement(s), 32, 36, 37, 38, 39, 40, 48, 67, 69, 71, 147, 164, 166, 167, 169
media, 19, 147
mercury, 93, 94, 109, 120
meta-analysis, 17, 69, 146
Metabolic, 152
metabolism, 16, 54, 179, 182, 197
metals, 5, 82, 91, 92, 93, 94, 118, 147, 166, 178, 188, 190
methodology, 27, 39, 42, 48, 57, 170
Mexico, 16, 29, 61, 70, 121, 125
microbial community(s), 52, 80, 140, 178, 179
micrograms, 189, 190
micronutrients, 165, 183
microorganisms, 31, 80, 97, 226
microsatellites, 203
migrants, 209, 210
migration, 77, 206, 210
migration routes, 77
military, 167
mineralization, 27, 40, 44, 118
miniature, 208
Minneapolis, 113
Missouri, 123
misuse, 89, 107, 199
mitochondrial DNA, 203
mixing, 24, 41, 48, 60, 205
modelling, 117, 120, 122, 146, 226
models, 2, 7, 17, 35, 39, 53, 72, 75, 94, 102, 106, 140, 144, 167, 168, 169, 177, 179, 184, 187
mollusks, 54, 64, 90, 208
MOM, 60, 125, 126
momentum, 98
morphometric, 216
mortality, 77, 88, 138, 217, 218
mortality rate, 217
muscular tissue, 190
mussels, 5, 99, 100, 102, 135, 144, 149, 210

N

natural resources, 219
negative effects, 95
nematode, 59, 62, 63, 65, 67, 88, 96
Netherlands, 138, 141
neutral, 114
New Zealand, 19, 22, 29, 32, 40, 68, 77, 90, 94, 97, 118, 119, 136, 138
Nile, 218
nitrates, 178
nitrification, 33
nitrite, 182, 183
nitrogen, 2, 8, 9, 10, 11, 12, 13, 15, 19, 20, 21, 22, 23, 24, 25, 27, 39, 40, 41, 44, 45, 46, 47, 48, 64, 69, 71, 80, 98, 99, 100, 101, 102, 104, 112, 122, 132, 134, 140, 148, 149, 152, 153, 154, 179, 182, 183, 184, 186
NOAA, 1, 2, 6, 8, 28, 47, 50, 53, 66, 74, 76, 78, 86, 101, 103, 108, 127, 131, 138, 139, 140, 146, 157, 223, 224
North America, 17, 21, 23, 26, 29, 34, 56, 58, 61, 70, 88, 92, 135, 152
Norway, 28, 32, 38, 60, 82, 83, 85, 92, 117, 125, 141, 196, 225, 226
nuisance, 164, 182, 183, 186
nutrient concentrations, 100, 112, 166
nutrient enrichment, 3, 5, 20, 21, 22, 23, 25, 31, 32, 48, 49, 51, 53, 71, 81, 107, 123
nutrient transfer, 72

nutrition, 152, 162, 187

O

obstruction, 207, 213
oceans, 13, 46, 124
OH, 225
OIE, 110
oil, 19, 94, 167, 220, 221, 226
oilseed, 222
oligotrophic conditions, 99
operations, 3, 4, 6, 8, 9, 10, 12, 16, 17, 18, 19, 23, 30, 41, 44, 58, 67, 74, 77, 79, 80, 81, 82, 84, 86, 88, 91, 92, 95, 100, 101, 104, 105, 106, 107, 108, 160, 171, 177, 183, 186, 190, 193, 204
opportunities, 2, 99, 152, 205, 206
organic matter, 3, 19, 27, 28, 33, 34, 35, 36, 37, 39, 40, 41, 42, 43, 44, 47, 52, 54, 57, 58, 63, 68, 69, 71, 99, 104, 105, 117, 120, 138, 146, 152, 153, 154, 166, 176, 178, 179, 180, 182, 183
organism, 9, 52, 59, 197
organotin compounds, 144
outreach, 2
overlap, 132
oversight, 83
ox, 182
oxidation, 33, 80, 182
oxidative stress, 80
oxygen, 2, 16, 17, 19, 25, 27, 31, 33, 35, 37, 39, 40, 42, 43, 44, 52, 70, 83, 90, 92, 97, 102, 140, 141, 166, 172, 176, 179, 182, 184, 185, 188, 213
oxygen consumption, 43, 44
oyster(s), 5, 91, 100, 102, 109, 120, 137, 146, 210

P

Pacific, 11, 13, 14, 16, 17, 27, 32, 43, 50, 67, 73, 75, 82, 85, 95, 115, 120, 122, 130, 131, 138, 139, 142, 143, 150, 182, 194, 205
paints, 187, 188
parallel, 52
parasite(s), 88, 89, 95, 165, 166, 191, 192, 194, 195, 196, 200, 203
parents, 200
participants, viii, 158, 162
pathogens, 83, 95, 98, 107, 165, 191, 192, 193, 194, 195, 196, 200, 201
pathophysiology, 136
pathways, 83, 166, 187, 210
PCBs, 92, 94, 127
PCR, 204
peer review, 8
permission, 56, 84
permit, 7, 168, 208, 210, 213
peroxide, 89
personal communication, 85
pH, 33, 37, 38, 39, 83, 89
pharmaceutical(s), 88, 125, 168, 198
pharmacokinetics, 145
Philadelphia, 200
Philippines, 25, 31, 32, 43, 79, 128, 153, 218
phosphate, 13, 40
phosphorous, 132
phosphorus, 8, 13, 14, 15, 19, 20, 21, 22, 23, 24, 25, 26, 39, 45, 46, 47, 48, 60, 80, 98, 100, 102, 104, 106, 130, 134, 136, 165, 179, 182, 183, 184, 186
photographs, 166
photosynthesis, 18, 79, 146
physical interaction, vii, 158, 209, 210
physico-chemical changes, 176, 179
physiology, 79, 153, 171, 180
phytoplankton, 4, 10, 12, 15, 18, 45, 48, 50, 84, 88, 90, 104, 106, 120, 143, 148, 164, 166, 179, 180, 183, 184, 186
pigmentation, 84
pilot study, 41, 67, 100, 109, 129
plankton, 12, 48, 60
plants, 45, 47, 84, 88, 91, 159, 175, 185, 207
polar, 7
policy, 103, 124, 140, 163
policy options, 124
pollution, 31, 60, 77, 94, 132, 133, 142, 145, 171, 172
polychlorinated biphenyl(s) (PCBs), 94, 166
polycyclic aromatic hydrocarbon, 94, 166
polymerase, 204
polymerase chain reaction, 204
ponds, 215, 218
population, 50, 59, 62, 74, 75, 76, 94, 109, 121, 164, 174, 193, 199, 201, 203, 204, 205, 208, 209, 215, 216, 217, 218, 219, 220, 222
population control, 75, 76
population size, 203, 204
population structure, 205
positive interactions, 4
potential benefits, 70, 99
poultry, 219, 220
practical knowledge, vii, 157, 163
predation, 5, 72, 74, 75, 76, 77, 106, 200, 203, 210
predators, 4, 5, 77, 94, 106, 186, 194, 205, 206, 208, 210, 220
present value, 99
preservation, 163
prevention, 199, 202
probability, 163, 180, 192

probiotics, 85, 109, 120, 121
producers, 100, 128, 168, 181, 183, 186, 199, 204, 210, 221
professionals, 215
profitability, 6, 99, 100, 107, 170
project, 40, 41, 61, 91, 96, 99, 113, 141, 169, 170
propagation, 196, 204, 215, 218
prophylactic, 197
protected areas, 166
protection, 75, 163, 164, 179, 184, 189, 192, 196, 199, 203, 208, 212, 216, 220
proteins, 80
prototype, 91
public health, 50, 159
public welfare, 124
Puerto Rico, 11, 14, 15, 16, 18, 24, 26, 30, 32, 43, 49, 51, 67, 70, 72, 73, 95, 100, 120, 141, 151, 171

Q

Queensland, 12, 15, 17, 18, 25, 31, 72, 137

R

radius, 44, 59, 69
reactants, 33
reactions, 33, 80
reactivity, 32
recognition, 76
recommendations, 2, 79, 80, 81, 89, 94
recovery, 3, 5, 36, 38, 39, 40, 41, 53, 57, 58, 60, 61, 62, 65, 67, 68, 90, 95, 96, 97, 98, 105, 107, 111, 134, 135, 136, 138, 174, 175, 177, 195, 202, 206
recovery plan, 195
recovery process(s), 58, 61
recreation, 203
recreational, 73, 159, 161, 163, 166, 167
recreational areas, 166
recruiting, 177
recycling, 91, 99, 134
regions of the world, 7
regression, 36, 69, 92, 118, 167
regression analysis, 36, 167
regression model, 36, 92, 118
regulations, 1, 75, 116, 191, 195, 199, 201, 207
regulatory agencies, 1, 2, 69, 95
regulatory oversight, 6, 108
regulatory requirements, 51, 66
relevance, viii, 118, 158, 160
reliability, 40
remediation, 3, 36, 42, 57, 58, 60, 95, 96, 97, 98, 101, 107, 115, 116, 118, 159, 189, 190
repair, 72
replication, 68, 213
reproduction, 79
reptile, 164
requirements, 7, 40, 54, 69, 78, 98, 171, 199, 206
research institutions, 7
researchers, 13, 24, 35, 36, 47, 54, 59, 60, 68, 72, 73, 93, 96, 210
residues, 84, 109, 117, 131, 135, 195
resilience, 1
resistance, 85, 110, 111, 117, 132, 137, 147, 164, 192, 195, 196, 197
resources, 1, 103, 147, 158, 159, 160, 166, 179, 181, 184, 186, 218, 219, 221, 222
respiration, 16, 17, 19, 28, 33, 52, 79, 182, 183
response, 4, 6, 32, 46, 47, 51, 52, 53, 54, 61, 62, 63, 65, 71, 73, 77, 80, 81, 82, 97, 98, 105, 108, 122, 128, 138, 140, 143, 153, 163, 164, 168, 174, 175, 176, 179, 183, 192, 194, 222
restoration, 97, 134
revenue, 99
rhizome, 81
risk assessment, viii, 139, 158, 159, 160, 161, 162, 163, 164, 167, 169, 172, 173, 174, 175, 181
risk management, 139, 202
root, 81
root zones, 81
routes, 163
routines, 77
Royal Society, 222

S

safety, 122, 199, 208
saline water, 159
salinity, 80, 122, 165, 166, 175, 194
salinity levels, 80
sanctuaries, 166
Scandinavia, 85
scavengers, 208
school, 218
schooling, 216, 218
science, vii, 1, 103, 157, 158, 163
scientific knowledge, 2, 108
scope, 2, 7, 95, 103, 104, 105, 161, 163, 168
seafood, 6, 7, 99, 109, 135, 199, 220
seasonal changes, 43
seasonal factors, 45
seasonal flu, 63
sedimentation, 3, 5, 19, 27, 32, 33, 35, 37, 39, 40, 42, 43, 60, 71, 79, 81, 95, 97, 104, 106, 151, 156
sensitivity, 154, 186
services, 81, 127

sewage, 23, 47, 80, 131
shape, 213
shear, 3, 27, 44
shellfish, vii, 36, 98, 99, 100, 115, 144, 146, 149, 157, 166, 183, 184, 186, 198, 201, 207, 208, 210
shelter, 4, 70, 72, 171, 207, 208
shoot, 81
shoreline, 46, 171, 207, 210, 216
shores, 182
showing, 37, 47, 60, 92, 95
shrimp, 59, 81, 88, 124, 191, 199, 215, 218
side effects, 89
signs, 199
simulation(s), 13, 15, 53, 64, 72, 94, 108, 146
skin, 187, 190, 194, 196, 197, 199
SO_4^{2-}, 34
social acceptance, 98, 107
software, 35
solid waste, 19, 24, 31, 116, 117, 177
solubility, 35, 83
solution, 106, 121, 155, 167, 176, 177, 183, 194, 196, 197, 199
South Africa, 73, 147
South America, 19, 20, 22, 25, 27, 32, 37, 45, 46, 59, 191, 218
South Korea, 110
Spain, 23, 26, 41, 49, 51, 62, 63, 70, 71, 83, 93, 100, 122, 130, 132, 138, 156
species richness, 60, 61, 64, 67, 72, 164, 178, 179, 184, 189, 196
specific tax, 176
stability, 183
stakeholders, 1, 6, 103, 108, 122, 148, 151, 159, 161, 169
starvation, 207
state(s), 2, 3, 6, 7, 33, 35, 40, 44, 54, 59, 71, 80, 83, 85, 92, 98, 110, 140, 141, 144
statistics, 180
sterile, 201, 206
stimulation, 10, 50, 186
stock, 104, 108, 199, 200, 201, 203, 204, 205
stomach, 210
storage, 115, 199
storms, 171, 201, 212
stratification, 166, 183
stress, 36, 72, 74, 75, 80, 85, 105, 148, 160, 191, 193, 196, 199
stressors, 79, 135, 161, 164, 174, 175
structure, 7, 54, 55, 57, 58, 60, 62, 66, 67, 71, 73, 95, 100, 132, 133, 140, 142, 153, 167, 179, 196, 201, 205, 207, 214, 217
subsistence, 158, 215
substitution, 222
substrate(s), 39, 40, 59, 73, 98, 167, 171, 176, 180, 184, 212, 214, 215
succession, 135, 142
sulfate, 33, 36, 54, 176, 182, 187, 189
sulfur, 33, 40, 176
supervision, 215
surface area, 81, 212
surface chemistry, 37
surface structure, 207
surrogates, 181
surveillance, 171
survival, 79, 85, 109, 114, 145, 174, 194, 204, 218
survivors, 192
susceptibility, 74, 148, 191, 196
sustainability, 103, 112, 152, 154, 217, 219, 220, 225
sustainable development, 130
Sweden, 196
syndrome, 191
synthesis, 113, 120

T

Taiwan, 11, 12, 14, 15, 30, 32, 48, 128, 137
tanks, 102
target, 4, 5, 75, 77, 83, 88, 89, 106, 165, 203, 204, 206, 208, 209, 216, 217, 218, 220, 222
target population(s), 216, 217, 220
Task Force, 143
taxa, 54, 61, 62, 65, 68, 95, 96, 179
techniques, 8, 28, 31, 41, 42, 43, 56, 57, 59, 60, 61, 69, 75, 76, 102, 107, 117, 181, 200, 203, 216, 217, 220
technological advancement, 31
technology(s), 3, 5, 7, 26, 35, 37, 44, 57, 59, 69, 91, 97, 100, 101, 103, 106, 107, 132, 138, 158, 180, 202, 215
temperature, 36, 43, 83, 89, 108, 166, 194
tension, 171, 212
terrestrial ecosystems, 159
territorial, 211
texture, 178
Thailand, 112, 131
therapy, 132
threats, 79, 81
tides, 31, 153, 166, 183, 194
time frame, 158
tin, 90
tissue, 23, 90, 92, 138, 203
total product, 180
tourism, 159, 210
toxic effect, 88, 90, 93
toxicity, 19, 50, 54, 84, 88, 89, 90, 91, 121, 189, 197
toxicology, 93

trace elements, 183, 187
tracks, 61, 64, 72
trade, 7, 161
training, 57, 167
traits, 204, 205
transmission, vii, 157, 165, 191, 193, 195
transparency, 169, 213
transport, 27, 33, 34, 35, 83, 84, 151
treatment, 84, 88, 89, 91, 116, 126, 136, 196, 197, 199
trial, 60
tributyltin chloride, 121
trimmings, 222
turbulence, 101
Turkey, 17, 18, 30, 49, 51, 73, 93, 109, 110, 112, 120, 155
turnover, 21
turtle, 210

U

U.S. Army Corps of Engineers, 7
U.S. Department of Commerce, 138, 139
U.S. Geological Survey, 147
UNESCO, 131
uniform, 204
United, 1, 6, 86, 98, 99, 109, 111, 115, 117, 124, 131, 133, 135, 141, 148, 149, 161, 186, 190, 222
United Kingdom (UK), 82, 99, 109, 113, 115, 190, 222
United Nations (UN), 86, 98, 117, 131, 133, 148, 161
United States, 1, 6, 111, 124, 131, 135, 141, 149, 186
urban, 166
urban areas, 166
urea, 9, 182, 183
urine, 196, 197
USA, 11, 14, 109, 117, 223, 224, 225
USDA, 146, 223
USGS, 24

V

validation, 26, 35, 42, 131
variables, 43, 55, 68, 69, 166, 204, 205
variations, 12, 60
varieties, 213
vector, 186
vegetable oil, 19, 222
vegetation, 18, 182, 226
velocity, 17, 29, 66, 156
vessels, 181
victims, 194, 205
viral pathogens, 203
viscera, 194
vitamins, 165, 183

W

Washington, 21, 83, 119, 124, 141, 145
waste disposal, 167
waste management, 31, 97, 181
waste water, 127
water chemistry, 110
water quality, 2, 3, 6, 7, 8, 9, 10, 11, 12, 13, 15, 16, 17, 47, 48, 49, 90, 95, 102, 104, 105, 112, 126, 140, 141, 149, 160, 163, 166, 171, 172, 184, 189, 213
water quality standards, 189
wealth, 167
web, 58, 109, 180, 225
web sites, 225
whales, 5, 75, 77, 175, 207
wildlife, vii, 158, 165, 173, 206, 207, 208, 209, 211
wind farm, 100, 137
Wisconsin, 147
withdrawal, 85
workers, 85
World Bank, 221
World Health Organization (WHO), viii, 110, 131, 158, 161, 162
worldwide, 9, 32, 44, 46, 86, 90, 199, 220
worms, 164

Y

yeast, 85
yield, 91

Z

zinc, 46, 82, 91, 92, 93, 94, 97, 115, 120, 155, 166, 178, 187, 188, 189, 190
zinc sulfate, 190
zooplankton, 88, 106, 155, 179, 180